Unfolding a Mountain

*A Historical Archaeology of Modern and Contemporary Cave
Use on Mount Pelion*

Unfolding a Mountain

A Historical Archaeology of Modern and Contemporary Cave Use on Mount Pelion

Edited by

*Niels H. Andreasen, Nota Pantzou,
Dimitris Papadopoulos & Andreas Darlas*

 Monographs of the Danish Institute at Athens,
Volume 19

Unfolding a Mountain
© Aarhus University Press and The Danish Institute at Athens 2017

Monographs of the Danish Institute at Athens, Volume 19

Series editor: Kristina Winther-Jacobsen
Type setting: Ryevad Grafisk
This book is typeset in Minion Pro and Warnock Pro and printed on Luxo Satin 130g.
Cover: Ryevad Grafisk
Cover illustration: Bourdovanou (MOU-7). Photo: Dimitris Papadopoulos
Printed at Narayana Press, Denmark

ISBN 978 87 7124 379 6
ISSN 1397 1433

Distributed by:
AARHUS UNIVERSITY PRESS
Finlandsgade 29
8200 Aarhus N
www.unipress.au.dk

Gazelle Book Services Ltd.
White Cross Mills, Hightown
Lancaster LA1 4XS, England
www.gazellebooks.com

ISD
70 Enterprise Drive
Bristol, CT 06010
USA
www.isdistribution.com

**PEER
REVIEWED**

/ In accordance with requirements of the Danish Ministry of Higher Education and Science, the certification
means that a PhD level peer has made a written assessment justifying this book's scientific quality.

The publication of this book was financed by:
The Danish Research Council for Culture and Communication

Contents

Preface

Greek caves were left outside the field of archaeo-logical research for a long part of its history. Rare exceptions from this longstanding disregard were those distinguished caves used as significant sanc-tuaries in Antiquity, such as Idaion Antron, or those situated within important archaeological sites, such as the Acropolis. For some decades now, however, caves have increasingly attracted the attention of new research projects, though still to a limited extent and predominantly as part of inquiries targeting prehis-tory, and more specifically the Palaeolithic and the Neolithic periods. In addition to this increasing in-terest, excavations recently carried out in some of the most impressive caves as part of works conducted for their development as tourist sites have indeed revealed important antiquities. The ubiquitous small caves and rock shelters, however, have hardly ever attracted the serious interest of systematic archaeo-logical research. Yet these caves preserve multiple and diverse kinds of past remains, and are able to provide valuable knowledge on past human use.

Caves overall have constituted a terrain of dia-chronic human activity. Their occupation is not at all limited to the remote Stone Age – though the prevailing concepts of the early twentieth century suggest otherwise, recalling the mythical character of the Caveman. Contrastingly, caves in fact show evidence of human activity through the ages, ex-tending from the start of Prehistory to as far as the most recent past; this activity is more specifically intense, independently of any time period, in those

regions where caves are abundant. The phenomenon is clearly the result of human adaptation to the natu-ral environment, and of exploitation of the available resources. Finally, the use of caves should not be attributed to any lack of skill on the part of humans to create individual structures on their own, but is conversely the outcome of their high competence to take full advantage of the available natural shelters and employ the most suitable means of sustaining a broad range of life subsistence imperatives in their context.

Cave deposits and even their rock walls (for their paintings and graffiti, among other evidence) have preserved an invaluable record of human presence from all periods of the past. Cave sites have thus functioned as protective shells which have greatly secured and maintained this sensitive corpus of cul-tural data up to our present, providing a material base for the articulation of archaeological and his-torical discourses. Conclusively, caves have a signifi-cant research potential, and for this reason should be considered as worth of scholarly investigations.

Within these promising perspectives, *The Pelion Cave Project*, on Mount Pelion in Thessaly, has ini-tiated an innovative interdisciplinary research pro-gramme, which has revealed significant new knowl-edge on the local past, and more particularly on the recent past of the region. The project arose out of the desire to develop a more detailed and interdiscipli-nary approach with a focus on the various uses and meanings of caves in late post-medieval Greece. The present publication is the outcome of this research project. The data published here were collected during three field seasons, from 2006 to 2008.

In the scope above described, this book deals with how people took advantage of caves in the late Modern period in Greece. It addresses some deceptively complex questions about cave use, such as when and why they were occupied, modified, reused and abandoned. The answers are rarely straightforward, but clues can be found in changes in rural practices on a local level, and in the relationship between cave use and wider social and historico-political processes on a national level during the last couple of centuries.

The Modern Greek nation-state was established in the first half of the 1800s. It was a century marked by emerging technologies, which were to affect the lives of all Greeks. By the twentieth century, the transformations in political, economic and social structures came in rapid succession. New industries and the urban diaspora, the arrival of the motor car, the diesel lorry and new building technologies combined to alter the face of the countryside permanently.

The emergence of modernity in the Greek countryside did not fundamentally disrupt the relationship between people and caves. Caves continued to be occupied as convenient shelters linked to a variety of activities, and to frame everyday practices. Because of the permanence of these natural features and their generally isolated yet accessible positions in the landscape, caves are today a valuable resource for archaeological knowledge, regional history and local, living heritage.

However, archaeologists mostly avoid or ignore material remains from the recent past in Greek caves, the argument being that Modern surface debris has little scientific value. The limited appeal of such remains has meant that they are mostly left untouched and intact – since caves are rarely cleaned out. In other cases, the importance and perception of a cave is often not recognizable when Modern use does not produce a material record (not sur-prisingly, cave users reduce their equipment to the most necessary), or when the material record is not preserved. The frequent lack of material evidence is corroborated in the often intangible relations between people and caves, including the role of stories in constructing meaningful places. For instance, memories of atrocities and conflict are part of the reason for the secrecy surrounding the use of caves during the World War II and the Civil War, which persist today (see Chapter 2.9).

Despite the multitude of caves known in Greece, basic information is still lacking for many of them. Satellite and GPS systems now make it possible to locate caves precisely, but considerable time and effort is needed to relocate many of them. Furthermore, many caves have not been mapped or adequately sampled from an archaeological standpoint. Even a small controlled surface collection and one or two test pits in each cave could potentially provide important data regarding such basic questions as stratigraphy, chronology and use.

By using a blend of archaeology, archival sources, graffiti studies, literary evidence, placename studies, myths, historical texts and ethnographic interviews, fieldwork has produced a set of richly varied data that is the foundation for the analyses and syntheses presented in the following pages. Some of the project caves stand out because of their prominent topographical settings or their dramatic entrances, but the majority are inconspicuous. The often modest nature of these features attracted our attention precisely because they are not obvious focal points in the landscape. In fact, many of them were so small or inaccessible that they seemed unlikely candidates for human use.

As the synthesis of these diverse approaches, *Unfolding a Mountain* also presents reflections on approaches within later historical archaeology in Greece and underlines the potential of historical archaeology and ethnoarchaeology to collaborate on the discussion of the social issues arising from the emergence of modernity in the Greek countryside.

To some extent, it offers a proposal that expands and facilitates innovative and unfamiliar ways of looking at the last few centuries of Greek rural history. Not least, we see this kind of study as a field in which archaeology and ethnology are able to work together successfully. Particularly during fieldwork, it became clear that archaeology was better at identifying economic uses while habitation and non-economic uses were more frequently identified by ethnographic queries. This has potential implications for archaeological investigations, which may fail entirely to detect evidence of habitation and sheltering. We hope our multifaceted approach disrupts some of the traditional taxonomies and succeeds in adding complexity to the simple caricature of caves as humble shelters for shepherds or as hideouts during wartime.

Finally, the premise for this monograph is that caves in Greece need to be studied more consciously and comprehensively in context. The necessity of adopting a contextual approach to the study of the human use of caves has been emphasised over the past couple of decades, but still needs reiterating today. We believe that cave studies can inspire and be of significance to wider contemporary archaeological research agendas, particularly when a contextual approach is adopted.

Unfolding a Mountain is aimed at a broad audience that includes both academics and students of archaeology, ethnology, history and landscape studies, as well as the public with an interest in the rural facets of Modern Greek history. Although the geographic focus of this book is a portion of the eastern Greek mainland, many of the themes are relevant to the wider Mediterranean region, where caves are abundant. We envision the volume as an invitation for dialogue, and encourage discussions of the implications of the Pelion case study within wider archaeological agendas. The material studies in Chapter 2 have Greek summaries to allow non-English speakers to access the information that forms the basis for our conclusions.

In addition, we would also expect this volume to stimulate increased attention to heritage preservation objectives. The range of theoretical and practical approaches achieved by the present publication stands as a strong affirmation that the plethora of Pelion caves constitute a great cultural heritage resource alongside their historic value. This is an important realisation, given that these monuments, which have survived through the ages to the present, are now severely menaced by intense building activity and uncontrolled extensive operations on the landscape, which have become more prevalent during recent years and altogether entail the unprecedented and irreversible destruction of this resource. Undertaking all possible measures for their protection is therefore an obvious imperative, which demands immediate action in order that all of what nature has respectfully preserved for tens of decades will not vanish within the short while of our present era.

Andreas Darlas
Niels Henrik Andreasen
Co-directors of the Danish/Greek survey on Pelion

Acknowledgements

I am grateful to everyone who has supported the project financially and practically. The director of the Danish Institute at Athens at the time, Dr Erik Hallager, received my initial ideas with enthusiasm and wholeheartedly supported the project. The succeeding director, Dr Rune Frederiksen, continued this support in full. The project was set up as a *collaboration* between the Danish Institute at Athens and the Ephorate of Palaeoanthropology and Speleology of Northern Greece. We are most grateful to the Greek Ministry of Culture and Sciences for granting us a fieldwork permit through the Danish Institute at Athens and to Andreas Darlas, Director of EPSA, for facilitating the project at every stage.

I am obliged to Evi Margariti for her help in getting the project off the ground in 2006 and for her assistance during its initial stages. I learned a lot about the mountain's caves and its geology from Markos Vaxevanopoulos, who first took me to some caves there in 2005. I am grateful to Markos for that first visit and his subsequent contributions to the project. Without Polyxeni Boni's patient translation from Greek into English of various articles and book chapters, it would have taken me much longer to uncover relevant information in the literature. Thank you to Theodosios Touisouzoglou for digitizing the cave plans and to Yuki Furuya Watanabe and Odysseas Metaxas for drawing the artefacts. Finally, I could not have put this book together without the efforts of my co-editors Nota Pantzou and Dimitris Papadopoulos.

The Pelion Cave Project was made possible by generous financial support from the J.F. Costopoulos Foundation (2006, 2007, 2008), the Institute of Aegean Prehistory (2006), Her Majesty Queen Margrethe II's Archaeological Fund (2007, 2008), and the Augustinus Foundation (2007, 2008). Analysis of the survey data and their subsequent transformation into this publication was made possible by a generous grant from the Danish Research Council for Culture and Communication (Humanities).

The following people participated in the fieldwork on Pelion:

2006:
Niels Henrik Andreasen, Pernille Foss, Silas Michelakas, Evi Margariti, Markos Vaxevanopoulos, Kostas Filis, Aris Bachlas, Giannis Vlastaridis

2007:
Niels Henrik Andreasen, Markos Vaxevanopoulos, Nota Pantzou, Dimitris Papadopoulos, Odysseas Metaxas, Ioannis Voskos, Kostas Filis, Giannis Vlastaridis, Dimitris Agnousiotis, Kostas Vouzaksakis

2008:
Niels Henrik Andreasen, Markos Vaxevanopoulos, Nota Pantzou, Dimitris Papadopoulos, Katerina Ragkou, Ioannis Voskos, Giorgos Papamichalakis, Michalis Kontos, Kostas Filis, Giannis Vlastaridis.

Part of the Pelion Cave Project team at *Pidima tis Grias* **rockshelter below the village of Agios Georgios, autumn 2007. Markos Vaxevanopoulos, Dimitris Papadopoulos and Nota Pantzou.**

Copenhagen, 2016
Niels H. Andreasen

Abbreviations

my	Million years
m.a.s.	Metres above sea level
EH, MH, LH	Early, Middle, Late Helladic
EBA, MBA, LBA	Early, Middle, Late Bronze Age

Abbreviations employed for survey districts; site codes consist of a district abbreviation and a number (e.g. KAR-20):

AFE:	Afetes
AGR:	Agria
ART:	Artemida
KAR:	Karla
KER:	Keramidi
MAK:	Makrinitsa
MIL:	Milies
MOU:	Mouresiou
VOL:	Volos
ZAG:	Zagora

Glossary

andarte	resistance fighter
arnolivado	enclosure or area for breeding sheep
EAM	Ethniko Apeleftherotiko Metopo – *National Liberation Front*
ELAS	Ethnikos Laikos Apeleftherotikos Stratos – *National People's Liberation Army*

galaropoti	enclosure or area for nursing sheep or sheep intensively managed for high milk production
gennolivado	enclosure or area for sheep ready to give birth
giataki	expediently constructed "bed" for pastoralists, made of branches. Alternatively, a spot in the shade suitable for sleeping, either in the open air or inside caves.
haravlo	small hole in the ground or in a rock that exudes cold air
kalivi	seasonal habitation, such as those used by shepherds
kapetan	the leader of a guerilla band
kafeneion	café
kalderimi	cobbled footpath or road
kerantzides	mule driver
konaki	makeshift shelter for sheep and shepherds
mandri	animal enclosure, fold, pen or corral, constructed of stones
melistra	clearing (local dialect)
milovotos	pasture
mitato	traditional shepherd's hut
perifheria	the area around a village which is owned and used by the villagers
pisteria	"pigeon shelter" – folk name for cave
provato	sheep
schismi	crevice

skaftiades	workers (low social class)
skoutia	fabric made of sheep or goat wool, for capes and jerseys
stavlos	stable sheepfold
stani	sheep-pen or station
strouga	a milking pen
TEA	Tagmata Ethnofylanis Amynis – National Security Battalion
trypa	hole
tyrias	cheese-making facility and storage, also specifically used for caves in the highlands where cheese is kept cool during summer
varathro	vertical cave ("abyss/chasm")
vlachopimenas	Vlach shepherd
voskos	shepherd
vrachoskepi	rockshelter
xeimadio	area in the lowlands for the wintering of flocks

Chapter 1

The Pelion Cave Project (PCP): Research background

Niels H. Andreasen, Nota Pantzou & Dimitris Papadopoulos

Greek Summary

Μοναστικό βουνό αρχικά, με την οικιστική ανάπτυξη και επέκταση των οικισμών στο Πήλιο να ξεκινά περίπου το 1550. Τα προνόμια, τα οποία παραχωρήθηκαν από τις Οθωμανικές αρχές, πυροδότησαν την οικονομική ανάπτυξη από το δεύτερο μισό του 17ου αιώνα, όταν το Πήλιο ευημερούσε ως ένα ισχυρό οικονομικό κέντρο. Στην διάρκεια του 18ου και 19ου αιώνα, εξελίχθηκε στην πιο πλούσια και πυκνοκατοικημένη ορεινή περιοχή της Ελλάδας. Η ανοικοδόμηση και εν γένει οι οικονομικές δραστηριότητες παρήκμασαν μετά το μέσον του 19ου αιώνα, κυρίως λόγω της σταδιακής ανάπτυξης του Βόλου ως αστικού και βιομηχανικού κέντρου. Ωστόσο, οι αγροτικές καλλιέργειες εντάθηκαν στην ύπαιθρο, και τα πεδινά χωριά του Δυτικού Πηλίου άρχισαν να αναπτύσσονται ταχύτατα μετά την απελευθέρωση της Θεσσαλίας από τους Οθωμανούς.

Μετακινήσεις πληθυσμών, ανησυχία και συγκρούσεις σημάδεψαν το πρώτο μισό του 20ου αιώνα. Κατά τη διάρκεια του Δεύτερου Παγκοσμίου Πολέμου στο ορεινό Πήλιο είχε οργανωθεί ισχυρή αντίσταση στην κατοχή των δυνάμεων του Άξονα, ενώ η δράση ανταρτικών ομάδων συνεχίστηκε και στην διάρκεια του Ελληνικού Εμφυλίου Πολέμου. Πολλοί κάτοικοι του Πηλίου μετανάστευσαν στον Βόλο και στην Δυτική Ευρώπη μετά τον Εμφύλιο Πόλεμο, ενώ η ταχεία ανάπτυξη και αστικοποίηση του Βόλου συνέβαλε επίσης στην φθίνουσα πορεία των χωριών του Πηλίου.

Η Ελλάδα έγινε μέλος της Ευρωπαϊκής Κοινότητας το 1981. Όσο σημαντικές ήταν οι αγροτικές επιδοτήσεις για τους γεωργούς και κτηνοτρόφους των πεδινών του Βορείου Πηλίου, άλλο τόσο σημαντικά ήταν τα έσοδα από τον τουρισμό για το υπόλοιπο Πήλιο. Τα τελευταία χρόνια το Πήλιο γνωρίζει έναν συνδυασμό διαφορετικών κερδοφόρων στρατηγικών, συμπεριλαμβανομένου του τουρισμού και της εμπορικής γεωργίας με επίκεντρο τα οπωροφόρα δένδρα.

Από αρχαιολογικής πλευράς, το Πήλιο είναι μια περιοχή στην οποία δεν έχει δοθεί ιδιαίτερη ερευνητική προσοχή. Ιδίως τα πολυάριθμα σπήλαια του βουνού παραμένουν άγνωστα στην αρχαιολογική κοινότητα. Η άρτια καταγεγραμμένη ιστορία του Πηλίου προσφέρει γόνιμο έδαφος για έρευνα των ποικίλων χρήσεων και της σημασίας των σπηλαίων από την μεταβυζαντινή περίοδο μέχρι σήμερα. Η εθνοαρχαιολογική έρευνα με επίκεντρο το Πήλιο ξεκίνησε επισήμως τον Σεπτέμβρη του 2007 από το Ινστιτούτο της Δανίας στην Αθήνα, σε συνεργασία με την Εφορεία Σπηλαιολογίας και Παλαιοανθρωπολογίας του Υπουργείου Πολιτισμού. Το ερευνητικό πρόγραμμα για τα σπήλαια στο όρος Πήλιο επικεντρώνεται στη λειτουργική, οικονομική και πνευματική χρήση των σπηλαίων κατά τα μεταβυζαντινά και νεώτερα χρόνια, και εξετάζει τη δυναμική των σπηλαίων ως αξιόπιστη πηγή αρχαιολογικής γνώσης, τοπικής ιστορίας και ζωντανής πολιτισμικής κληρονομιάς κάθε περιοχής.

Καθώς ορισμένες κτηνοτροφικές ή μη χρήσεις γίνονται πλήρως κατανοητές κατόπιν σύνδεσης με τις ευρύτερες ιστορικές και οικονομικές εξελίξεις, το ερευνητικό πρόγραμμα καταγράφει ορισμένους από τους τρόπους, μέσω των οποίων οι τοπικές, εθνικές και διεθνείς οικονομικές εξελίξεις και τεχνολογικοί μετασχηματισμοί επέδρασαν στις παραδοσιακές μεθόδους παραγωγής και στην κοινωνική δυναμική των κατά τόπους κοινοτήτων. Η τοπικής κλίμακας μελέτη σπηλαίων και βραχοσκεπών επιτρέπει να συνεκτιμηθούν η αναδιάρθρωση ή η εγκατάλειψη της γης ως αποτέλεσμα αλλαγών στην αγροτική οικονομία και της αυξανόμενης εκβιομηχάνισης. Επομένως, το ερευνητικό πρόγραμμα για τα σπήλαια του Πηλίου αποτελεί ένα χρήσιμο αντίβαρο των ερευνών σε υπαίθριους οικισμούς στην Ελλάδα.

Η εθνοαρχαιολογική προσέγγιση στοχεύει τόσο στην αποκάλυψη των υλικών συσχετισμών, οι οποίοι θα μπορούσαν να απαντήσουν αρχαιολογικά ερωτήματα, όσο και στη διερεύνηση της ιστορικής και κοινωνικά δυναμικής σχέσης μεταξύ των κατά τόπους κοινοτήτων και του τοπίου που τις περιβάλλει.

1.1 Setting the field: The spatiotemporal context

Popular images of Mount Pelion include green forests, rich dark blue seas, numerous streams with fresh cold drinking water and spectacular stone mansions. Beyond this consensus view lie differing perceptions of the mountain and contrasts depending on whether one is a tourist, village dweller, migrant day-worker, transhumant shepherd or archaeologist. The physical environment on Pelion is due not only to the mountain's particular geology, landforms, vegetation and climate, but also to ease of transport and the presence of economically valuable rocks. All these factors influenced the cultural landscape in the past and continue to do so today.

The mountain ranges of Olympus, Ossa, Mavrovouni and Pelion run in an almost continuous chain from the western shore of the Thermaic Gulf to the Aegean shores of Thessaly. Pelion, the most fertile of the four mountains, extends into a hook-like peninsula between the Pagasetic Gulf and the Aegean Sea. Seven of the mountain's summits reach heights of around 1500 m and the highest among them is Pourianos Stavros at 1624 m.

The main bulk of Pelion is in the north and here the mountain consists of karstic limestone with schist-chert formations and enclosed ophiolitic bodies (Vaxevanopoulos, this volume). Rocky outcrops in the central and south part of the mountain are mainly fertile schists with marble intercalations. Pelion has notable differences between its north and

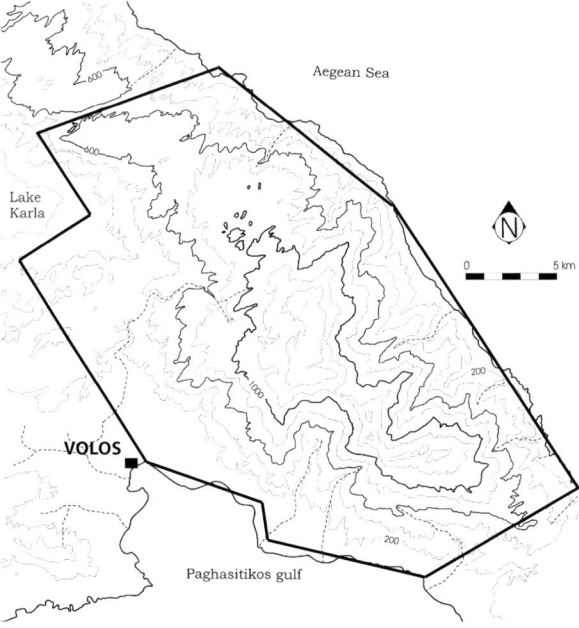

Fig. 1.1. Map of Pelion.

Fig. 1.2. Panorama of Northeast Pelion, facing toward the southeast, with the Flamouri Monastery on the opposite side of the gorge.

south in geology, topography and vegetation, and the summits that divide the peninsula further define an east–west boundary between the environments of the maritime and continental sides of the mountain. Pelion can, therefore, be divided into four areas, each of which has its own characteristics. Common to the whole region is the fact that few places are far from either sea or mountain, and it therefore offers an environmental mix capable of supporting a variety of economic strategies.

Due to its elevation and geographical location, Pelion receives a large amount of rain. Gorges transect both sides of the mountain and streams on East Pelion can become extremely active during downpours and in the spring when the snow is melting. Every year, local torrents carry millions of cubic metres of water into the Aegean Sea along with large amounts of sediment and forest debris. Small beaches have formed where the gorges empty into the Aegean. Local toponyms, such as *Kakoskali* and *Kakia Skala* ("Bad Stairs"), may suggest the potential force of some of these streams.

Volos, at the foot of Northwest Pelion, is the capital of Magnesia and the major commercial centre. It is also the only outlet towards the sea from Thessaly,

the country's largest agricultural region. While the entire upper part of the mountain remains unpopulated, a number of mountain villages are scattered on the slopes of the mountain up to 700 m.a.s. On the western side, settlements are more numerous and larger than on the eastern side, where these are mainly small fishing hamlets with some recent tourist developments. Mid-altitude villages above 500 m.a.s. are also more widespread on the western side. Villages all over the mountain range in size from small semi-abandoned hamlets to the largest village, Zagora, with about 4000 inhabitants. Settlements in East Pelion tend to be more spread out on the slopes than on West Pelion, possibly due to easier availability of water and difficulties involved in building on the steep slopes.

Small roads and an extensive web of cobblestone trails interconnect all the villages.[1] A winding asphalt road leads from Portaria above Volos to Chania at the top of the mountain, before descending on the eastern side. An alternative route extends from Volos to the southeast along the Pelian foothills where it leads through a string of settlements and olive groves situated on the narrow coastal stretch

---

1   Haratsis 2003.

along the Pagasetic Gulf. A parallel route on the mountain passes through the villages of Pinakates and Vyzitsa before both routes merge at Milies. From Milies, a road continues over the mountain to the Aegean side where it becomes circuitous as it navigates several deep ravines. It passes through all the villages as it snakes northeastwards towards Zagora. Access to the Aegean side has always been slow and at times impossible during severe weather. An improvement to the Pelion infrastructure was made in 2010 when an asphalted road was constructed across the mountain from Kissos to Chania. A new extension of the highway bypassing the centre of Volos has been tunnelled through the Goritsa Hill to Agria at the Pagasetic Gulf, with the purpose of making access to the Pelion peninsula easier.

The area north of Zagora has few settlements (Keramidi, Veneto and Pouri) and still almost no roads (Fig. 1.2). This mountainous, barely populated and inaccessible area represents half of the natural habitat that covers all 24 Pelion villages and is protected under *Natura 2000*, an ecological network of protected areas within the European Union.[2] The northwest part of the research area includes the Pelion foothills along the eastern edge of Lake Karla in the Thessaly plain. Called "Voiveis" in Antiquity, Lake Karla was referred to by a number of ancient writers.[3] With an extent of 25,000 ha and a depth of 6 m, it was one of the most extensive wetlands in Greece and the most important in Thessaly. Until recently, the lake was drained for the production of cereals, cotton and vegetables, but part of it has now been re-established.

Apart from its mythology, ancient writers mentioned Pelion for its pleasant climate and excep-

tionally rich vegetation.[4] The climate on Pelion is Mediterranean continental with a large temperature difference between seasons. Summer months are warm and humid and the average temperature reaches 26° C, but highs of over 30° C are common during July and August.[5] Spring and autumn see temperatures of 16° C–23° C, and up to 10 hours of sunshine every day. The average winter temperature is 4° C, and it can get lower than -5° C. The winter months from November to February are cold and wet, during which the monthly average rainfall can exceed 63 mm. Snowfalls are frequent and usually observed until early spring. The wind during the winter months is predominantly from the west and northwest. The rest of the year, a gentler and warm breeze blows in from the Aegean. The Aegean Sea strongly affects the local climate on the east side of the mountain. While the whole region is prone to torrential downpours, precipitation is much greater on the Aegean side. This, along with differences in wind direction and topography, creates varying conditions for vegetation and agriculture.

Pelion supports abundant vegetation with a diverse array of plant species. Much of the mountain is covered in woods consisting mainly of broadleaved deciduous trees such as beech, oak, maple, wild chestnut and a range of fruit trees. Especially the eastern part of the mountain is densely forested, and one can see plane trees, alders, poplars and willows covering the banks of many streams on this side.

Pelion has three vegetation zones. Typical Mediterranean maquis shrubland covers the low altitudes (0-600 m.a.s.). This zone includes mostly self-sown aromatic and pharmaceutical taxa, such as sage, thyme, mountain tea etc. Pelian flora includes at least 50 aromatic and pharmaceutical herb species. Above this is the para-Mediterranean broad-leaved decid-

---

2   European Commission 2009. Code GR1430001 signifies Mount Pelion and its coastal areas.

3   Herodotos: Book 7, Ch. 129; Pindar: The Pythian Odes 3, 60; Euripides: The Alcestis, str. 591; Homer: The Iliad, Book 2, 712.

---

4   For Pelion and its rich natural landscape, c.f. Homer 2.2.755; Eur. Med. 1.

5   www.hnms.gr

uous tree zone (600-1200 m.a.s.), which includes mainly oak and chestnut forests. The beech forest zone covers the areas above the para-Mediterranean zone up to the tree-limit zone (800-1600 m.a.s.).[6] Pelion also includes grasslands, phrygana and agricultural land. The cultivated species are fruit trees (oranges, lemons, apples, apricots, kiwis, pears and cherries), walnuts, almonds and vine. The lowlands on the west side of the mountain have extensive olive groves.[7] The fruit trees are not recent introductions; they were mentioned by nineteenth-century travellers.[8]

1.2 Major events and historical trends on Pelion

1200-1423: High Medieval period:[9] Forced transfer of all Thessaly from the Venetians to the Turks from 1411 to 1423. General Tuired Bey occupied Thessaly under Sultan Murat II and Thessaly, with Pelion, became a province of the Ottoman Empire. Pelion was a monastic mountain and several important monasteries were established on its slopes.[10]

1423-1668: Establishment of mountain villages: Most villages on Northeast Pelion were originally seaside villages. Inhabitants were mainly seafarers and, to a smaller extent, farmers with land extending up the mountain. Frequent pirate attacks and the arrival of the Turks led to the abandonment of seaside settlements. Instead, the inhabitants sought protection higher up on the mountain where new villages were formed near the monasteries. Further develop-

ment and expansion of most Pelian settlements took place from around 1550.[11] Systematic cultivation of olive trees was introduced around 1600, and in 1615, all land was divided into two distinct categories. *Vakoufia* was the Turkish term for fields owned by religious institutions and schools, including fields from which the profit was dedicated to these institutions. After the liberation of Greece (and the Lausanne agreement in 1922), these religious trust properties were declared exchangeable and a special service was formed to deal with this under the National Bank of Greece. *Chasia* is land where the tax was due directly to the Sultan or to state officials. The latter would pay the palace a specific amount of money and in return receive the majority of the tax revenue from their region.[12]

1668-1821: Progress and prosperity: From 1668, special privileges granted to upland villages by the Ottoman authorities as part of an economic growth package stimulated production, commerce and economic expansion. At the same time, this attracted many Greek immigrants from the lowlands and nearby islands who were eager to escape high taxation and the constant threat from raiding pirates in the coastal areas. Immigrants included city dwellers, manufacturers, merchants and seamen, and because of the composition of the labour force, Pelion prospered into a powerful economic centre showing rapid growth in productivity.[13] Along with other mountainous communities in Ottoman Greece and Anatolia, it became a hub for mobile artisans and traders.[14] Trade on the mountain was on a scale sufficient to sustain specialist carriers (muleteers) and during the 1700s and 1800s it was the wealthiest and most densely populated mountainous area

---

6   www.iama.gr/ethno/faskomilo/Fwtiadis.pdf
7   Thomas (1966, 60) mentions that olive trees grew on Pelion from 1600 onwards.
8   E.g. Magnitos 1860, 36.
9   For Prehistory and Antiquity, see Leake 1835 368-99, 426-33; Mézières 1854; Wace 1906, 143-68; Theocharis 1967b; Feuer 1992, 286-7.
10   Makris 1982, 181.

---

11   Makris 1982.
12   See also Asdrachas 2005, 14-5.
13   For instance, during 1760-1770 many Moscopolites (today the Albanian town of Voskopoje) settled on East Pelion after Moscopolis' decline (Mackridge 2009, 58).
14   Tsotsoros 1986; Asdrachas 2003, 357-67.

in Greece. Particularly Zagora steadily grew into an important commercial and manufacturing centre.[15] Cultivation of silk (30-40 tonnes per annum), tanning industries and fur and copper processing generated significant economic wealth. Wool was imported from different areas of Greece (Levadia in Boeotia delivered almost all of its annual production of wool to Zagora), mixed with local qualities, and then made into woven fabric at the Zagorian workshops.

Following the Russian–Turkish Treaty of Kuchuk Kainarji (or Küçük Kaynarca) in 1774, which ensured free navigation for Eastern Orthodox Christians in the Mediterranean under the Russian flag, Pelion's autonomy and relative independence made it possible for Greek seamen to organise a commercial shipping fleet. Silk and cloth could then be shipped out from East Pelion's port at Trikeri (Horefto area) and the products were sent to many important trading centres throughout Europe. This further added to Pelion's status as an important centre for industry and trade.

Economic and cultural progress caused a steep increase in construction activity and led to early "urbanization" on the mountain.[16] Examples are the works of architecture on Pelion (bridges, cobbled paths, monasteries, watermills, schools, etc.) and multi-storied, finely decorated private houses.

Outside the villages, olive oil production was also intensified.

However, despite economic progress, a range of problems plagued the region. The population in the marsh villages on the Thessalian plain suffered greatly from malaria and other epidemics in the second half of the eighteenth century. The death rates were so high that they had an impact on the shaping of land ownership patterns in Thessaly.[17] Malaria was such a widespread problem in the region that it could support a specialised production of mosquito nets in Portaria in the nineteenth century.[18] Epidemics were also experienced on Pelion itself, and infected individuals were in some cases isolated outside the villages (e.g. at Agios Lavrendios).

While greater security had been a motive in abandoning seaside settlements and founding villages higher on the mountain, the coastal waters around Pelion and neighbouring Mount Ossa continued to be plagued by piracy and brigands and had a reputation for being wild and lawless places. Pouqueville, who travelled in Thessaly between 1806 and 1815, described the problems: "Mount Ossa, the head-quarters of those bands of robbers and plunderers lay Thessaly under contribution".[19] And he continues: "The peasants of this country, and those of Mount Pelion, have preserved a sort of fierce courage, which leads them often to engage in the piratical adventures of the people of Trikeri, at the entrance of the gulf of Volo".[20] Brigandry intensified in Thessaly during the struggle against Ottoman rule,[21] and various bands of brigands reputedly used Pelion as a base of operations well into the twentieth century. In some mountainous regions, these "cattle rustlers and brigands who preyed upon the countryside" were only eradicated by the emergence of ELAS during

15 We prefer the term "manufacturing centre" or "village industrialization" to describe the Pelion economy during this period. Although the villages did not manufacture value-added goods or experience a wider modernization process, it can be argued that a form of early industrialization took place that led to important social and economic changes on Pelion. This, among other things, meant the re-organisation of the economy for manufacturing and the development of metallurgy production. Industry structures used for the large smelting industry processing iron ore are still visible near the Taxiarches monastery. Zagora merchants would likely have been involved in the exportation of the ore.

16 Makris 1982.

17 Skouvaras 1959, 23.
18 Magnitos 1860, 56.
19 Pouqueville 1820, 117.
20 Pouqueville 1820, 117.
21 Koliopoulos 1981.

the Second World War.[22] Tales of brigands pervade Pelion folklore and traditional songs ("brigand's songs") refer to both historically confirmed raids by brigands and to the relationship between villagers and brigands in general.[23]

1821-1881: Struggle and decline: Pelion joined the 1821 Greek revolution against Ottoman rule, but the revolution was crushed and in 1823, the Pelian villages of Agios Lavrentios, Pinakates, Vyzitsa and Mitzela were burned. In 1854, a series of uprisings were organised in Epirus and Thessaly with support from independent Greece, but Ottoman, British and French forces suppressed the revolt. A Greek revolt erupted in Thessaly and Epirus during the Russo-Turkish war of 1877-1878, but the Ottomans soon stamped out the rebellion. In the end, however, Thessaly was incorporated into the Greek kingdom in 1881.[24]

From the mid-nineteenth century, building, construction and economic activity declined on Pelion, mostly due to the gradual development of Volos as an urban and industrial centre.[25] However, at this time, foundries and smelting constructions were established in Zagora and facilities for producing silkworm cocoons at Lechonia. Around this time, 50,000 inhabitants lived in the 24 villages on Pelion, according to Mézières.[26]

Before the middle of the nineteenth century, people had started moving closer to the coast and building warehouses and shops, but the coastal settlements on western Pelion were still insignificant. Agria, for instance, had only a few buildings, such as a hostel for caravans and a toll station. The settlement then belonged to the villages of Drakeia and Agios Laurentios and functioned as a port from which these

---

22 Sarafis 1980, 312-3.

23 Liapi 2006, 235-84.

24 Greece crossed the border in January 1878 with a force of 24,000 infantry, 300 horses and 24 artillery pieces, without having first declared war on the Ottoman Empire. The Greek Army reached Domokos and then retreated (before entering the Thessalian plain) because

meanwhile the Russo-Turkish War had ended. While there were on this occasion rebel skirmishes against Turkish forces, no actual battles took place in Thessaly between the Greek and the Ottoman armies; see Kofos 1977, 339-40; Seisanis 1879.

25 Makris 1982.

26 Mézières 1854.

and other villages distributed agricultural and craft products.

In the nineteenth century, travellers noticed the small number of villages in Thessaly and the lack of agricultural activity. Nevertheless, there were exceptions. Despite the scarcity of arable land on Central Pelion, five municipalities had a density higher than 100 inhabitants per km² and were much more populated than the plains.[27] In 1881, the population density on Pelion was the highest in any district in Thessaly.

1881-1910: Growing importance of Volos and the bay area: With the annexation of Thessaly/Magnesia to independent Greece, the Muslim population started leaving the area. The growing urban centre at Volos experienced increased industrialisation and new workshops and factories appeared. The first pottery workshop opened in Volos in 1884.[28] The Pelion Diaspora and the arrival of Epirotes, Agrafiotes and islanders initiated much of the new development.[29]

Cultivation intensified in the countryside and the lowland villages on West Pelion started to grow rapidly after the annexation of Thessaly from the Ottomans. Further impetus came with the construction of a coastal road and rail network. Thessaly Railways decided in the late nineteenth century to extend their network eastwards, to connect Volos with the communities of Pelion. The new line extended from Volos to Agria (1892), reaching Ano Lechonia in 1896 and Milies in 1903. The railway was the first serious public investment in the area and would continue to be influential for many years. The new connection gave a boost to local producers of seafood, olive oil and black olives in the bay area. Local businesses were founded and flourished as the packaging and trading of olives picked up. Improvement of the infrastructure also set in at other places

on Pelion after 1881, as many packhorse and foot-path bridges were built across streams and ravines.

A large earthquake in 1885 and five months of occupation by Ottoman forces in 1897 during the Greco-Turkish War only briefly halted new developments. More serious was the deep conflict between major landowners and tenant farmers that had followed the annexation of Thessaly. Tenant farmers' claim for land redistribution and the struggle against the violation of their rights constituted an intense and continuous movement throughout the period 1881-1910.[30] While this conflict was mainly focused on the Thessalian plains, traditional land use on Pelion continued to focus on its rich forest resources. Hunting and forestry (e.g. charcoal production, wood cutting) were important elements of the local economy as were seasonal resources such as wild chestnuts and a wide range of fruit trees.

1910-1949: Conflict and settlement of refugees: The first half of the twentieth century was turbulent and marked by population movements, unrest and conflict following both local developments and events on the international scene.

The Balkan Wars (1912-13) and the First World War (1914-18) had demographic and economic consequences for Pelion and these conflicts were followed by a large influx of refugees from Ionia, Pontus, Cappadocia and Eastern Thrace following the Greek/Turkish population exchange in 1922/23. Immigration continued during the 1920s and in 1928, refugees accounted for 25% of the population in Volos and Nea Ionia. Many refugees also settled in coastal settlements along the Pagasetic Gulf (e.g. Agria and Lechonia). A solution to the landownership problem of Thessaly became imperative with the massive arrival of refugees from Asia Minor and the revolutionary Plastiras government finally settled the conflict in 1923.[31]

27 Sivignon 2009, 460.
28 Vroom, this volume.
29 After 1840, see Makris 1982.

30 Patronis 2009, 469.
31 Glegle 2009, 499.

A Greek expatriate community had been founded in Egypt around the mid-nineteenth century and it continued to grow during the first half of the twentieth century.[32] Many Peliorites had settled in Alexandria and Cairo and they contributed significantly to the financial life of Egypt. Wealthy Greek industrialists, traders and bankers established a thriving commerce between Greece and Egypt and they would later donate large amounts of money for the building of schools and hospitals.

The Italian (1941-43) and German (1943-44) military occupations of Thessaly during World War II led to atrocities in the Pelian villages of Zagora, Portaria, Milies and Drakeia. Resistance on the mountain was well organizesed and partisan activity continued during the Greek Civil War (1946-49).[33]

1949-1982: Migration to lowland urban centres: Many residents of Pelion migrated to Volos and Western Europe after the occupation and the Civil War, in order to make a living. Many villages and fields were left almost deserted. As a symbol of the demographic and economic downturn on Pelion, the Volos–Milies rail connection stopped operating in 1971, when it became too uneconomical to run. Simultaneously, rapid growth and urbanisation of neighbouring Volos contributed further to the decline of the Pelion villages, as all activities shifted to the new industrial centre. In 1911, the international cement plant "AGET Heracles" had been founded just outside Volos. This industry gradually became one of the largest cement producers in the world, employing a large number of people in the area. While mass production and mass distribution of industrialised goods increasingly took place in Volos, electricity, radio and automobiles were first introduced to Pelion in the 1950s.

Industrial progress in Volos went hand in hand with a general desire for increasing productivity and a need for local agricultural products. Lake Karla in the Pelian foothills to the west was an 180-km² wetland area (the second largest in Greece) that was completely drained in 1962 (draining was initiated in 1956), both to protect surrounding farmlands from flooding and the local population from malaria, and to increase agricultural production of cereals, cotton and vegetables. Before its drainage, it was the site of a unique fishing culture, with fishermen spending some nine months of the year in reed huts that they built on the lake. The lake fisheries were an important tradition and to some extent a significant economic activity. Kanalia, which lies between the hills and the lake, used to be dependent on the lake fishing, which was strictly managed by a company. Fish from Karla ("Kalrisia") were quite famous and reached the markets of Bulgaria, with carp as the principal species. Thousands of residents around the lake lived off it (fishers and stockbreeders), since its vegetation was rich and it supported numerous species of fish and birds.

The particular way of life that characterised the shallow lake and surrounding wetlands changed drastically after the draining. Material culture related to the wetlands, such as small sailing boats, canoes and fishing equipment, became redundant as fishermen were forced to turn to farming. Unfortunately, agriculture was never successful in the saline soils of the former lakebed and the permanent loss of wetland functions and values resulted in a broad range of environmental, social and economic problems.[34]

1982-: European subsidies and tourism: Greece entered the European Community in 1982. As important as farming subsidies were to the lowland farmers and agro-pastoralists of northern Pelion, income from tourism became equally important to the rest of the mountain.

Animal husbandry is not and never was particularly developed on most of Pelion, but there are a

---

32   Kitroeff 1983, 5-15.
33   Andreasen, this volume.

34   Gialis & Laspidou 2014, 1063.

few cattle and pig farms along the former Lake Karla and the lakebed was, until its recent re-flooding, pasture for a large number of farm animals. Goat herding in particular (with some sheep) has survived into modern times as an important segment of the economy along the lake. Goats are also raised in mountainous and less wooded terrain above Volos, east and north of Lake Karla and around Veneto on Northeast Pelion.

During recent years, Pelion has been successful in combining various cash-producing strategies, including tourism and commercial agriculture with a focus on fruit trees. While overgrown agricultural terraces above villages on West Pelion still speak of the post-war decline, the villages themselves have experienced a revival through the establishment of local enterprises and small industries. The most significant non-tourist enterprises are timber cutting, quarrying of local schist stone and plant nurseries. Located in one of Greece's premier apple-growing areas, the Agricultural Cooperative at Zagora, founded in 1916, is the main contributor to this town's recent prosperity through export of the famous *Zagora* apple. Widespread apple cultivation occurred after 1950 with the introduction of Red Delicious clones, and today annual production on the mountain is around 30-40,000 tonnes from trees cultivated at 300-800 m.a.s.[35]

Herbs, fruits, olives, homemade preserves and honey are important local products and are sold in great varieties to tourists in villages all over the mountain. In 1995, after a long interruption, the Ano Lechonia–Milies railway started operating again as a tourist attraction. The tourist industry also supports many restaurants, guesthouses and shorefront facilities on both sides of the mountain.

An ambitious reclamation project that started in 2009 to refill and restore part of the former Lake Karla was finalised in 2011. Support for the project from the villages around the lake was prompted by a desire to see their lost wetland environment fully restored, as it is expected to contribute to further development of tourism in the area.

1.3 A short history of archaeological research in caves on Pelion

From an archaeological viewpoint, Pelion is quite a poorly researched region; above its foothills, Pelion was widely regarded as having little or no potential for recovery of archaeological remains. Archaeologists seem to have devoted more attention to accessible hills and foothills near the coast with its well-known and documented sites (Sesklo, Dimini, Iolkos, Demetrias and Pagasae) than to the rough and densely wooded mountain. Early in the twentieth century, the archaeologist Alan J.B. Wace travelled on Pelion and recorded primarily Classical and Hellenistic artefacts and monuments but did not comment on caves in the region.[36] Other scholars also briefly dealt with the mountain in Antiquity, mainly through placenames mentioned by ancient writers.[37]

Particularly the mountain's cave resources have remained curiously unknown to the archaeological community. Excavations have remained small-scale and partial and to our knowledge, there has been no larger, systematic excavation in a cave anywhere on Pelion. In 1910, the archaeologist Arvanitopoulos made a brief excavation in a cave below the Plaka summit of neighbouring Mount Ossa during the first decade of the twentieth century. A number of dedications to the mountain nymphs, fourth/third-century BC pottery and fragments of terracotta figurines were recovered.[38] In 1911, the same archaeologist excavated the remains of a sanctuary probably dedicated to Zeus Akraios on the Pliassidi summit

---

35   Nanos & Dianelos 2011, 4.

36   Wace 1906.

37   Leake 1835; Mézières 1854; Bursian 1862-72.

38   Arvanitopoulos 1910, 183-4; Stählin 1924, 40.

of Pelion. The remains consisted of a peribolos, two temples and a stoa. Votive pottery and weapons were recovered and suggest a date around the fifth to the fourth century. A cave was located at the periphery of the sanctuary and it possibly served some cultic function in connection with Chiron or the deity worshipped in the sanctuary.[39]

The Ephorate of Palaeoanthropology and Speleology investigated Landovitos cave (ZAG-10-e) between Pouri and Kerasia. The excavation uncovered Roman remains, but no further information is available.

In the second half of the 1960s, archaeologist Dimitris Theocharis led a programme of archaeological explorations in several caves on West and Northwest Pelion in search of Prehistoric remains.

- At Sarakinos Cave (MAK-4?) west of Makrinitsa, Theocharis in 1964-65 found several engraved stone pendants including a hunter with bow and an ibex and dancing scenes. He also recovered earrings of elephant tusk and a hairpin of anthropomorphic shape, which he interpreted as Palaeolithic.[40]
- At Kostas Cave (MAK-17) west of Makrinitsa, members of the local speleology society recovered an engraved stone plate believed to be Palaeolithic.[41]
- Theocharis found Early Bronze Age sherds in a cave ("Cave Z") between Glaphyra and Melissiatika villages in 1968.[42]
- In "Cave A" at Vigla (KAR-8), south of the Agios Athanasios hill at Lake Karla, Theocharis reported several Palaeolithic-style cave drawings depicting mammoths and other animals including a wounded cervid, and three ivory statuettes. He

made a brief excavation in the cave in 1969 and found pieces of ivory tusks and a bone pin.[43]
- Theocharis found Paleolithic artefacts in a small cave between Agios Vlasios and Ano Lechonia.[44] A stone artefact with an engraved horse was recovered in front of the cave.

Theocharis wrote about his findings from the caves in a series of short articles in a Greek archaeological journal.[45] Prior to his publication of the evidence for a pre-Neolithic presence in Thessaly, he had been warned by colleagues who disputed the authenticity of the rock paintings and artefacts, based on the style, composition and the motifs depicted.[46] Contemporary specialists such as G. Freund and A. Leroi-Gourhan examined the findings, but could not confirm their authenticity. Instead, they found indications suggesting that both the cave paintings and the mobile artefacts were the works of a local fraudster. As a consequence of this development, Theocharis suspended his research on Pelion. Apart from the forged objects and engravings, Theocharis also reported finds of "modern debris" mixed with Early Bronze Age ceramics and lithics and numerous animal bones. There is little reason to dispute that Theocharis came across genuine prehistoric material in his test trenches. Three of the above caves were located and re-visited by the Pelion Cave Project and archaeological material was collected at Theocharis' "Cave A" (KAR-8). In 2010, the Ephorate of Palaeoanthropology and Speleology of Northern Greece conducted a test excavation at the same cave. Artefacts dating to various periods from the Neolithic to Late Antiquity were recovered, but the

---

39   Arvanitopoulos 1911, 305; Stählin 1965, 41.
40   Theocharis 1966a, 76; 1966b, 255.
41   Ioannou 1964, 217-20.
42   Theocharis 1969, 223.

43   Theocharis, 1966a, 76-82; 1967a, 297-8; 1969, 222-3.
44   Theocharis, 1966a, 76-82; 1966b, 255.
45   Theocharis 1966a, 1966b, 1968, 1969. Two of these caves (Sarakinos and Agios Athanasios / "Cave A") were relocated by The Pelion Cave Project.
46   Freund 1968, 418.

stratigraphic sequence of the cave has not yet been established.[47]

In 2009 the Pelion Project's geologist investigated an underground mining gallery of possible Roman date southwest of Xourichti (MOU-2). Corridors show two faces of exploitation, probably one of the Roman period and one earlier phase. Ancient metallurgy in Pelion is mostly unknown and recent investigations of the Xourichti mine provide new clues about ancient mining practices in the region.[48]

It becomes clear from this short overview that archaeological field surveys and excavation on Pelion have remained unrelated to the cave use on the mountain of the last 1500 years. There are two main reasons why data on cave use has remained largely anecdotal. The first is that caves are often perceived as marginal sites in the historical archaeological landscape, with most interest centred on ritual uses. In economic terms, caves are usually regarded as low-status facilities. The second reason is that many aspects of Modern and contemporary heritage are not addressed within the wider archaeological community in Greece. Publications regarding cave use in recent periods tend to be restricted to site reports in local journals and there is a deficiency of synthetic overviews.[49]

1.4 What caves can tell us: Research questions

The Pelion Cave Project arose out of a desire to develop a more detailed and interdisciplinary discussion of the various uses and meanings of caves in post-Medieval and Modern Greece (Table 1.1).

Our study focuses specifically on caves and rock shelters on Mount Pelion in Thessaly. Pelion was chosen for its rich heritage of caves, known in part owing to myths surrounding the cave-dwelling Centaurs, like Chiron. This mythological heritage is still maintained through symbolically or commercially valued use in naming and depicting local administration, restaurants, hotels and local businesses. An encouraging factor was that documentation and archival resources for the Modern and contemporary economic and cultural history of the region were abundant, so they could be cross-examined and investigated along with an archaeologically produced context of data.

Early post-Medieval	16th–18th centuries
Late post-Medieval	19th–20th centuries
Early Modern	1880s–1920s
Modern I	1930s–1940s
Modern II	1950s–1970s
Contemporary	1981-present

Table 1.1. The chronological divisions used for the post-Medieval period by the Pelion Cave Project.

As part of the project's pre-fieldwork preparation, we made a catalogue of all questions, aims and objectives that were considered potentially relevant or of interest based on our level of archaeological and historical knowledge of the region. Of course, we did not expect to obtain answers or information on all of these aspects, rather we were trying to map all areas of interest. An excerpt from the list gives an idea of our intentions and expectations:

- Function. Animal housing? Human shelter? Storage facility?
- Structures and use of space. How was limited space in a cave used and what modifications were required in the form of structures around caves? What causes people to make their various spatial

---

47 http://www.taxydromos.gr/perrisotereseidhseis/ tabid/152/articleType/articleView/articleId/35191/--. aspx
48 Vaxevanopoulos, this volume.
49 But see Faure 1964.

adaptations to caves? How visible would adaptations be in the archaeological record? How are pastoral and other activities organised in and immediately around caves?

- Chronology. Site construction sequence? Chronological range and frequency of artefacts on cave floors? When (and why) were caves modified, used, reused and abandoned?
- Landscape. Relationship to road, path? Land use in surrounding area?
- Cave ownership. Multiple ownership? Personal or family cave property rights?
- Food production and resource exploitation in and around caves. The degree of production of agrarian resources (crops, animals) in caves? Exploitation of natural resources from the area around the caves? What are the socio-economic use values of cave sites?
- Status and cultural difference. Is there anything in "cave artefacts" to suggest ethnic or social differentiation (Greek/non-Greek)? Are there any gender-specific artefacts?
- Cognitive/intangible associations. Can specific intangible associations whether in ideology, traditional customs, oral history or spiritual values be traced in cave material culture? And how are these (if at all) linked to broader transitions from traditional to industrialised society? What are the aims and purposes of different kinds of stories about caves? How do changes over time affect caves, stories, and the human audiences appreciating them?
- Regional differences. Possible continuities or qualitative differences between geographic or geological zones of the mountain (e.g. East and West Pelion).

As shown by anthropological or ethnoarchaeological studies undertaken on contemporary cave use, it is possible to extract significant information from structures, artefacts and graffiti preserved in caves and rockshelters and verify the accuracy of this data

through informant interviews.[50] We intended to find evidence of land use, reuse and restructuring, or abandonment caused by changes in agriculture and local economy. At the same time, we wanted to explore contemporary daily practices in and around caves, thus gaining insight into the ways they are being used today or have been used in the recent past.

We had good reasons for wishing to employ a multi-sited, regional approach rather than a localised study. One of our basic premises was that archaeologically visible features of pastoral activities or other cave uses are the outcome of both spatially and temporally diverse rural practices both on the local level and in their interaction with wider economic and political structures. Land use transformations caused by changes in agriculture, productive processes, demographic changes and increasing tourism have had profound effects on daily life in all Pelion mountain villages. To address these diachronic processes and their intersections would necessitate a regional scale of analysis.

We also wanted to take a closer look at relations between people and caves, including the role of stories in constructing meaningful places. Stories may be told orally by narrators or by material remains; they may be permanent or temporary. Stories may be linked, for example, to the cave's topography or geology, wildlife, cultural heritage or metaphysical creatures. Such stories can be historically accurate, purposefully invented or created entirely in the cave user's mind. Caves on Pelion occupy a central place in the way that recent historical events are remembered, and they are communally acknowledged as enduring loci for the convergence of memory and meaning concerning nineteenth- and twentieth-century resistance and liberation.

Finally, a secondary aim of the project was to collect a body of data as a basis for hypotheses and pos-

---

50   Flood 1997; Gorecki 1991; Galanidou 2000; Veth *et al.* 2005.

sible analogies concerning site use and function in the past. This would allow for a deeper archaeological insight into pastoral or other cave uses through their material relations, also contributing to a wider understanding of site formation processes. However, an examination of the range of Modern sites and examples simply provides a conceptual background for attempting to think through archaeological evidence encountered in the field. A look at cave use in the Modern period can provide a more representative and diverse picture than can be gained from archaeological investigations that concentrate on earlier periods alone. For instance, excavations rarely reflect activities such as herding, shearing, milking and cold storage of cheese. Gathering and interpretation of surface finds from cave floors and documentation of structures such as drystone walls, fences and stone pavements can demonstrate these activities.

Within this scope, the Pelion Cave Project had two overriding aims:

- To obtain detailed insight into the functional, economic and spiritual use of caves on Pelion, particularly during the late post-Medieval and Modern periods
- To address the potential of cave sites as a valuable resource for archaeological knowledge, regional history and local, living heritage

To approach our research questions in an appropriately analytical manner, we needed to structure the fieldwork so that it would take full advantage of all available diverse sources and sets of data, whether archaeological, historical or ethnographic, and to develop combined methodologies as close collaborations or real-time dialogues between archaeology and ethnography.

1.5 Ethnography and archaeology: mixing methods, combining practices

The interrelationship between archaeology and ethnography has formed a tradition of scholarship, growing into different branches and taking new directions in recent years, in what Castañeda has defined as the "ethnographic turn" in archaeology.[51] In a few words, today one encounters archaeological projects employing ethnography in an effort to draw parallels between the past and present, to decode past practices, to establish a communication channel with local communities and the public or to assess the discipline's socio-economical and ideological impact. At the same time, there are also research projects that treat archaeologists themselves as subjects of ethnographic enquiry and ethnographic fieldwork projects that interrogate archaeological practices and touch upon archaeology's disciplinary ontological foundations.[52] Within this context, both anthropologists and archaeologists are carrying out ethnographic work not only to serve archaeological research purposes but also to produce insightful accounts of the archaeological practice itself as applied in the field and communicated to local communities.

Within the contemporary Greek context and under the scope of the Pelion Cave Project, three main fields were of particular interest in shaping our own research methodology: a) ethnoarchaeological projects dealing with various aspects of traditional pre-industrial local communities such as pastoralism, herding, cultivation, habitation (Chang, Halstead, Bintliff); b) long-term or diachronic archaeological survey projects that have also applied ethnographies of contemporary Greek communities;

---

51 See Gould 1978, 1980; Watson 1979, 1995; Robin & Rothschild 2002, 167; Meskell 2007; Castañeda & Matthews 2008; Hamilakis & Anagnostopoulos 2009.

52 See Meskell 2005; Edgeworth 2006; Holtorf 2006; Hamilakis & Anagnostopoulos 2009.

and c) critical, reflexive, ethnographic accounts of archaeological disciplinary practices in heritage sites and excavation projects.

In the first category, one can draw a further distinction between two branches. The first includes scholars who have attempted to find parallels for archaeological artefact production through ethnographic documentation of traditional craft activities, such as pottery making. Another branch of ethnoarchaeological research in Greece employed ethnography as a tool with which to refine archaeological approaches to the study of pastoral economies. These investigations focused principally on the morphology of pastoral settlements and functional aspects of pastoral production. Several of these studies provided a stronger focus on structural remains of Modern pastoral communities. Chang, for instance, [53] has advanced the understanding of pastoral site morphology and her research provided much-needed social and behavioural insights into pastoral land management. Halstead has also provided valuable accounts of the pastoralist practices of rural mountain communities.[54] A recent and complementary development is the implementation of scientific techniques (e.g. geoarchaeology and phytolith analysis) at Modern pastoral sites.

In the second category lie archaeological survey projects with a diachronic approach such as the Methana, Argolid and Sphakia surveys. These projects have a wide time scope but a strictly regional focus, thus featuring a research approach that is quite similar to that applied by PCP. The Argolid Exploration Project (AEP),[55] a multidisciplinary study of the natural and human environment of the south Argolid region, had an extended time frame – from prehistory to modern times. Similar in focus is the Methana Survey Project,[56] operating in a neighbouring region in the Peloponnese Peninsula. These surveys integrated ethnography as a means to explore human interaction with the landscape through economic, social and symbolic practices. Forbes in Methana, for instance, endeavoured to "present an alternative view of a set of rural landscapes, seen not from the outside, but from within".[57] In like manner, Lucia Nixon in the context of Sphakia Survey [58] produced a study of outlying churches and icon stands from the Medieval period onwards, shedding light on an extended network of landmarks of both symbolic and practical function. In the case of the AEP, efforts approaching the communities of Koilada were also initiated and diverse outreach activities were performed,[59] reminding us that an archaeologist's work and responsibility extends beyond conventional understandings/definitions of the field.

Closely related to the ethics and politics of archaeology is the third category of archaeological ethnographies, which focus on the socio-political impact of archaeological practice and heritage discourse and stress the need to bring forward local, alternative views and values as opposed to official narratives. Recent studies include Lynn Meskell's archaeological ethnography of the Kruger National Park and the ethnography of the Kalaureia Research Project.[60]

Although maintaining an "ethnoarchaeological" survey character, the Pelion Cave Project has moved beyond the term's origins and conventional conceptualisation, defined as the investigation of archaeological problems through the study of contemporary communities,[61] and has engaged in a more complex approach that integrates various elements of all the

---

53   See Chang 1981; Chang & Koster 1986, 1994.
54   See Halstead 1998.
55   See Jameson *et al.* 1994; Runnels *et al.* 1995; Sutton 2000.

56   Mee *et al.* 1997.
57   Forbes 2007, xvii.
58   Nixon 2006.
59   See Stroulia & Sutton 2010; Kamizis *et al.* 2010.
60   Meskell 2005; Hamilakis & Anagnostopoulos 2009; Hamilakis *et al.* 2009.
61   E.g. Gould 1978; 1980; Watson 1979.

research strategies identified in the categories mentioned above. As a result, the ethnography applied in PCP acquired certain features and had a certain character:

A) Fieldwork was carried out in constant, synchronic dialogue and exchange with the archaeological survey. Ethnographic and archival resources aimed to contribute to the investigation of archaeological research questions whenever possible, since the object of study was the human use and perception of cave sites from the post-Byzantine epoch to the present. On the other hand, ethnography was constantly informed by the findings of the archaeological survey, thus integrating new questions and areas to explore.

B) Ethnographic fieldwork was at the same time multi-sited and site-specific. Although ethnography was done in different types of locations (e.g. the village and town, the local library, a cave site or rockshelter), the purpose was always to reveal perceptions of and interactions with certain sites that would be identifiable by the archaeological survey team. It also maintained a strictly regional focus throughout the project's duration.[62]

C) Ethnographic fieldwork was carried out in close collaboration and interaction with the local communities in Pelion. Pelion villagers were not mere "informants", but contributors and participants often acting as guides in the field. A conscious decision was taken at the beginning of the project that PCP should go beyond the limits of a conventional archaeological survey restricted to the study of the material evidence, and try to embrace local values and perceptions of the cave sites and the mountain landscape. This was based on the acknowledgement resonating in the comments of the Koiladas mayor's with respect to the AEP: "The relationship therefore between the

archaeologists and the local community should take place on time, should be timely, it should not take place after the fact, 'after the name day has passed', as we say in Greek".[63]

D) Finally, the ethnographic fieldwork in PCP also had a reflexive scope and impact in terms of re-approaching archaeological surveying practices and disciplinary methods for knowledge production. Having archaeological backgrounds themselves, the ethnographers took on the new trends and conceptions of ethnography's role and contribution to the archaeological discipline, such as Meskell's "Archaeological Ethnography" and Castaneda's "Ethnographic Archaeology".[64] Moreover, since they were perceived as "locals" compared to the project's international members, they were also aware of the implications of doing "anthropology at home".[65]

As a result, PCP is a project where archaeology and ethnography go hand in hand, aiming at exploring patterns and changes in the contemporary historical Pelion landscape by applying an anthropocentric perspective while at the same time taking under consideration the social implications of archaeological practice.

1.6 Applying an ethnoarchaeological approach in Pelion

The ethnoarchaeological approach adopted by PCP aimed not only to reveal material relations that could provide answers to archaeological questions, but also to explore the historical and socially dynamic

---

62 The survey and research area was well-defined right from the early stages of the project.

63 Kamizis *et al.* 2010, 425.

64 See Meskell 2005 and Castañeda & Matthews 2008. Although the term "Archaeological Ethnography" has been in use since 1977, it was only in 2005 that it attained a meaning that surpassed the conventional limits of "ethnoarchaeology".

65 See Bakalaki 1997.

relationship between local communities and their landscape. This approach entailed a certain involvement of the locals in the archaeological process as field guides, informants or discussants.

The impetus for this research strategy was the realisation that in the case of caves, a number of pastoral as well as non-pastoral uses can only be properly understood when related to historical and economic developments outside the studied region. In the wider scheme of things, it is believed that PCP provided an opportunity to document some of the ways in which regional, national and international economic developments and technological transformations affected traditional modes of production and societal dynamics in local Greek communities. In particular, by studying cave and rockshelter sites on a regional scale, we wanted to evaluate the restructuring or abandonment of land resulting from changes in the agricultural economy and increasing industrialisation, a process that reshaped all aspects of local life. As such, the Pelion Cave Project offers a useful counter-balance to case studies from open-air sites in Greece. The overall aims of the project were to be achieved by means of a survey, in which archaeology and ethnography were equal partners.

To stress and explain meticulously the close tie between ethnography and archaeology in PCP, it is essential to clarify that the boundaries of research and practice between the two teams were not strict, but rather fluid and constantly overlapping. Both teams were involved in each other's work in a manner that did not disrupt the investigation process or undermine the research goals. Therefore, on several occasions, the ethnographic team participated actively in the identification, surveying, recording of cave sites, familiarising themselves with site finds and cave locations and subsequently enhancing/readdressing their research questions, etc. At the same time, the members of the survey team also took part in interviews and discussions, in this way gaining valuable insight into local history and site use, but also becoming acquainted with informants that

would navigate them around mountain tracks and show possible cave locations. Overall, this research design forced each team to think about the fieldwork in a more comprehensive way and provided an understanding of the challenges encountered by the other team.

Some discussion of procedure is necessary at this juncture, since among our goals was an attempt to demonstrate the value of information from mixed sources of data and delineate the logistics and practical aspects of combining archaeology with ethnography in such a way.

From the outset, we had a clear impression of the inadequacies of the usual methods employed by both disciplinary approaches for reaching our objectives. Refinement of these methods had to result in something that could provide more in the way of a cultural history. Therefore, the essential requirement of the survey was not merely gathering a comprehensive body of data as a basis for a quantitative and qualitative inquiry about the function of cave use in the Modern period – our challenge also lay in deciding how to establish a relation between the ethnographic/historical and archaeological sources of information.

Archaeological survey

The selection of caves for inclusion within the survey programme depended upon knowledge of the distribution of caves at the project's start and discovery of new caves during the field survey. A list of known caves was compiled from the archaeological and speleological literature, and especially the files maintained by the Ephorate of Palaeoanthropology and Speleology of Northern Greece. These records indicated that 30-40 known caves fell within the boundaries of the survey region, but coverage was partial and its representativeness and significance were also unclear. HERON, an association of speleologists in Volos, provided additional and more accurate information on a smaller number of caves.

Small rockshelters and artificial caves of limited archaeological and speleological interest were generally not included in the records, but we wished to include these features in our survey as we had previously observed that activities taking place at such sites are similar or identical to those associated with caves.

We decided to divide the caves into four categories that we found had potential relevance to the way in which caves were used (cave, vertical cave, rockshelter, artificial cave). The geological classification used in the survey is based on speleogenetics and is therefore necessarily different from the archaeological one (see Vaxevanopoulos, this volume). Nevertheless, the two classification methods supplement rather than contradict each other.

- A *cave* was defined as a natural cavity in the bedrock with an opening large enough to permit entry by humans. The cavity should penetrate further into the bedrock than the largest dimension of its opening and it should have a permanent dark zone. The orientation of the cave in space is not definitive, and a pit (or *vertical cave*) was considered a cave if it met the minimum dimensions.
- A *rockshelter* is a natural rock overhang, a hollow under a boulder or a fluvial undercut that forms a protected shelter. Rockshelters are relatively shallow and are wider than they are deep with no cave component. Rock shelters usually do not extend to total darkness. There are exceptions since both categories can be part of the same natural feature or closely associated with it.
- The term *artificial cave* was applied to openings in the natural substrate constructed by humans, such as tunnels or mines.

It was realised before the start of the project that a systematic pedestrian survey of the heavily vegetated and often steep mountain slopes would be impossible. An alternative was to follow roads and paths by car or by foot and scout for potential cave-bearing outcrops. Surveys of several gorges, ridges, beaches and part of the Milies rail line was undertaken by foot in order to spot caves. The process of locating caves based on the Ephorate files was problematic because in most cases only a cave's association with a village was stated. However, it formed a good starting point for enquiries within each village. Targeted searches for caves described to us by informants often involved a combination of motorised transport and walking.

Our approach was to try to maximise information on as many sites as possible. In terms of recording, a handheld GPS unit was used to provide fast and accurate location of sites, apart from in a few cases where the unit was affected by the landscape, such as tree cover or mountainsides. A small field team consisting of two archaeologists and a geologist/speleologist conducted the recording of each site on a standard "site form", on which archaeological and topographical features were listed. This data was then ready to be fed into an electronic database. The advantage of this approach was that limited resources were spent on the recording of each site, making exploration of the entire mountain possible within three rather short field seasons.

In the absence of excavation, our only means of estimating use-date and type of use of a given locality was through diagnostic architectural elements or portable artefacts recovered from the surface. Cave floors and areas outside the caves were therefore systematically surveyed for any artefacts (in the widest possible sense). Visibility in and around caves was sometimes poor due to vegetation cover or layers of animal excrement. Particularly vegetation cover was a serious impediment to visibility as the litter of fallen leaves, as well as living vegetation, tended to completely obscure archaeological surface remains.

Our methodology originally included employment of a metal detector to search the top layer for metal artefacts, but this plan was quickly abandoned as we anticipated considerable difficulties in obtaining permission from the relevant cultural authorities to use a detector. While metal detectors are routinely employed at archaeological excavations in Northern Europe, a stigma still surrounds the use of detectors in Greece. While the restrictions imposed on the public are understandable in the Greek context, it is not clear why detectors are not used by professional archaeologists. We have little doubt that a systematic search of the surface sediment (5 cm or so) in our case would have revealed a wide range of additional artefacts, including datable modern coins.

Criteria for selecting a cave or rockshelter for more detailed documentation were: 1) the presence of structures; 2) the presence of artefact concentrations; 3) details of its use that could be obtained from local informants. Structures, loose parts of structures and all other cultural material on the surface were recorded on plan drawings in 1:50 or 1:100 by either one or two persons with a tape measure and metre rule. While all visible artefact categories were collected, some types of non-diagnostic detritus were, for practical reasons, documented and described only in the field. Particular consideration in the form of drawing and photography was given to artefacts that might potentially date or shed light on activities carried out within or around the cave. Sometimes, people also engraved their names, initials, drawings or dates. All graffiti/engravings were digitally photographed and the images were later processed and redrawn in CorelDraw.

Initially, we discussed whether documentation of some sites should include limited excavation. Subsurface testing can help establish the extent, depth and possible age of drywall remains and other partly buried structures, or provide evidence for whether surface scatters of ancient pottery come from an exposed cultural layer. This would form a small component of the project as the primary aim was to document relatively recent features in the caves. However, our final opinion was that trial trenches would be too time-consuming and perhaps cause difficulties in future excavations.

Ethnographic fieldwork

A team consisting of two archaeologically trained, Greek-speaking ethnographers carried out the ethnographic fieldwork with a dual purpose: 1) to have a direct, synchronised exchange of information with the archaeological survey team, and 2) to be able to contextualise ethnographic data through combined pre- and post-fieldwork historical and archival research.

Interviews were conducted with local villagers to obtain a thorough understanding of the economic and social organisation and village histories as sources for explaining cave use, and to shed light on the relationship between material culture and behaviour at each cave. A basic cave use typology was developed by the project team and tested in all structured interviews or informal conversations. This typology includes 10 types of cave uses (see Chapter 4).

The ethnographic team operated in close dialogue with the archaeological survey team. This let us benefit from a feedback scheme that allowed us to 1) acquire information from informants on observed features in the caves, and 2) submit questions to informants that were directly related to observations in the cave or unidentified finds. In addition, documented topographical variables were used to generate a set of preferences for cave site location, which could be checked against informants' explanations as to why they chose specific caves for specific purposes.

We carried out semi-structured interviews and on some occasions had informal conversations with small groups in public places. In order to facilitate the categorisation, further processing and 'compatibility' of the ethnographic material with the archaeological survey, we used a structured data sheet organised in sections (e.g. personal informant data, cave placenames and locations, cave uses and practices, local history and economy, oral tradition and personal narratives). Interviews with the villagers were conducted in Greek, summarised in English for the Danish field director. At the end of the afternoon/beginning of the evening, this information was used to plan fieldwork for the next day. This research method enabled the team to discuss findings obtained during fieldwork and to verify and correct possible misinterpretations due to language problems.

Informants were typically found in the fields during the day or in village squares in the evening. After contact was established, the ethnographic team would usually arrange an interview. On several occasions, informants were interviewed 'on the spot' while in the fields, or herding their goat/sheep, thus providing an opportunity to identify sites in the vicinity visually. Some informants volunteered to guide us to certain sites, this being an ideal means of identifying, dating and interpreting cave structures, features and artefacts. We would also return to informants to have further discussions in light of the survey findings. The majority of the informants were male, over 50 years of age and occupied in agriculture, animal husbandry or logging.

Archival research was combined with ethnographic fieldwork, not only to enhance available knowledge resources and fill in research gaps, but most importantly to set the local narratives acquired through fieldwork in a wider, historical context of the Pelion region in the Modern and contemporary periods. Archives and valuable resources were found in local libraries (e.g. Milies, Zagora), central libraries and institutions (e.g. Volos, Gennadius Library in Athens) and in personal and family collections to which we were generously granted access by Peliorites.

Although employed in several sites, from the village square to the local library to a rockshelter on the mountain, the ethnographic fieldwork maintained a situated character, aiming to unfold the perception and interaction of the locals with the mountain landscape through certain placenames, sites and landmarks. Only through such a situated approach, combined with an overview of archival resources, would it be possible to tell a "bigger" story of Pelion through its caves and rockshelters.

Chapter 2

Scientific analyses and the archaeological record

2.1 Geology of the caves on Pelion

Markos Vaxevanopoulos

Greek summary

Το όρος Πήλιο παρουσιάζεται ως ένα σύνθετο γεωλογικό οικοδόμημα. Αποτελεί την προς νότο προέκταση της οροσειράς Όλυμπος – Όσσα σχηματίζοντας ορογενές διεύθυνσης ΒΒΔ-ΝΝΑ. Μεγάλοι όγκοι μεταμορφωμένων πετρωμάτων, αποτελούμενων κυρίως από μάρμαρα και σχιστολίθους, συνθέτουν τη γεωλογική εικόνα της ευρύτερης περιοχής. Εκτεταμένα κανονικά ρήγματα διακόπτουν τη συνέχεια των πετρωμάτων, και συντελούν στην κατά βάθος διάβρωση των ανθρακικών πετρωμάτων, αποκαλύπτοντας παράλληλα στοιχεία του υπογείου καρστ.

Στο πλαίσιο του προγράμματος "Pelion Cave Project" καταγράφηκαν και μελετήθηκαν συνολικά 159 σπήλαια, εκ των όποιων τα 7 είναι υπόγειες ανθρωπογενείς κατασκευές. Οι περισσότερες μορφές που μελετήθηκαν είναι οριζόντια, καρστικά σπήλαια και κυρίως βραχοσκεπές. Επιπρόσθετα ερευνήθηκαν 8 βάραθρα, δηλαδή μορφές κατακόρυφης ανάπτυξης. Τα σπήλαια του Πηλίου διανοίγονται κυρίως σε μάρμαρα ηλικίας Παλαιοζωικού – Ιουρασικού ή μερικώς ανακρυσταλλωμένους ασβεστολίθους Ανωτέρου Κρητιδικού, αποτελώντας μέρος του υπογείου καρστ. Δημιουργούνται από τη διάλυση που προκα-

λεί η κίνηση του εμπλουτισμένου σε CO_2 νερού στη μάζα των ανθρακικών πετρωμάτων και αποκαλύπτονται από την ενεργό τεκτονική. Πέραν των καρστικών σπηλαίων, αξιοσημείωτη είναι η παρουσία των σπηλαίων που δημιουργούνται από διάβρωση κοντά στις ακτές ή στις κοίτες των ρεμάτων του Πηλίου, από την επίδραση κάποιου ισχυρού διαβρωτικού μέσου. Σε αυτά ο διαβρωτικός παράγοντας είναι κυρίως το νερό που προκαλεί απορρίνιση του μητρικού πετρώματος είτε λόγω της κυματικής δράσης είτε λόγω της διαβρωτικής δράσης των νερών ενός χειμάρρου σε κάποιο χαλαρό σχετικά υλικό (κροκαλοπαγές). Η κυματική δράση επιφέρει σημαντικό βαθμό αποσάθρωσης ακόμη και σε συμπαγή πετρώματα όπως τα μάρμαρα και οι γνεύσιοι.

Στα οριζόντια σπήλαια συμπεριλαμβάνονται οι υπόγειες ανθρωπογενείς κατασκευές, όπως υπόγεια υδρομαστευτικά έργα και μεταλλεία. Στην περιοχή της Κάρλας όπως και του Αγίου Δημητρίου εντοπίζονται υπόγεια έργα υδρομάστευσης. Επίσης από το Καλαμάκι μέχρι και τη Ζαγορά, εντοπίζεται έντονη μεταλλευτική δραστηριότητα. Κατά τη διάρκεια της έρευνας καταγράφηκαν 4 θέσεις με υπόγειες μεταλλευτικές εργασίες. Η εκμετάλλευση ήταν προσανατολισμένη στην εξόρυξη μεταλλεύματος μικτών θειούχων ορυκτών, και κυρίως του αργυρούχου γαληνίτη. Μεταξύ των νεωτέρων υπογείων εκμεταλλεύσεων εντοπίστηκε αρχαία μεταλλευτική στοά νοτιοδυτικά του Ξουριχτίου.

❋ ❋ ❋

Introduction

The contribution of physical-geographical research to the Pelion Cave Project provides basic information about the landscape in which the survey area is situated. Data on the genesis, history and evolution of the present landscape is an aid in the understanding of the constraints imposed and the possibilities offered by the environment to human activities, now and in the past.

Within the framework of the survey, a geological study was carried out on the mountain. Because of the large extent of the survey area, the geological study could only have a general character. The aims of this work were: 1) to study the geology and geomorphology in order to present a description of landscape units, 2) to situate caves and rockshelters as part of the landscape and 3) to understand the genetic relation between the different types of caves and the landscape.

Geological overview

The local and regional geological background for Pelion and Eastern Thessaly has been described in general terms in earlier publications.[1] Mount Pelion constitutes the southern extension of the Olympus–Ossa–Pelion range. As part of the Hellenic orogen, it is closely related to the convergence between the African and Eurasian plates during the Mesozoic

(250-65 my) and Tertiary (65-2.6 my) periods. Pelion is a complex geological structure formed by the rock series of the Pelagonian Zone. The latter represents a continental block (Cimmerian plate) that detached 220 m years ago from the large Gondwana continent (African Plate with smaller plates), and subsequently merged with the southeastern Lavrasia (European plate with Asian and smaller plates) during the late Mesozoic. The result of this continental collision can be found in the large volumes of metamorphic rocks and ophiolitic series that form a nappe pile in Pelion.[2] This nappe stacking started in the middle-upper Jurassic with the obduction of the ophiolites into the Pelagonian carbonate rocks.[3] The absolute thickness of the nappe is estimated at up to 3000 m.

The mountain contains large series of metamorphic rocks such as gneisses, schists and amphibolites (Cimmerian Plate), and carbonate rocks such as marble and limestone. Carbonate sediments were deposited in the ocean basins across the Cimmerian Plate. The ophiolites constitute a series of igneous rocks; they represent the submarine volcanism in these oceans and are accompanied by fine-grain ocean sediments. Currently, ophiolitic rocks are being developed at the Mid-Ocean ridge of the Atlantic Ocean in a 400-km strip of a north–south direction.

---

1 Kilias *et al.* 1995; Caputo & Pavlides 1993, 339-62; Stiros *et al.* 2004.

2 A *nappe* is a mass of rock that is thrust over other rocks by thrust faulting or a recumbent fold or both.
3 Jacobshagen *et al.* 1978, 537-64; Vergely 1984; Mountrakis 1986, 335-47.

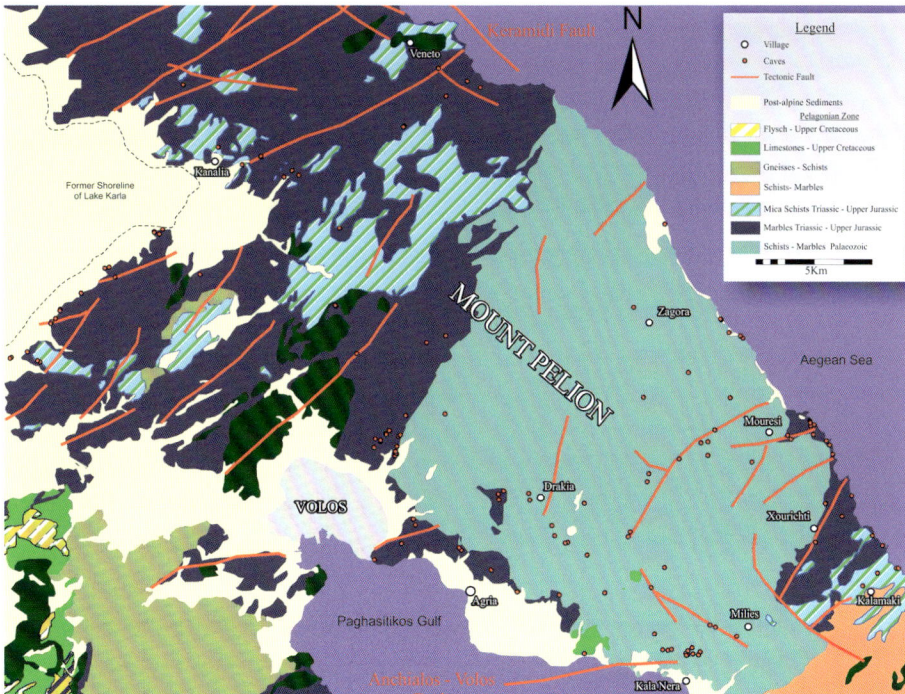

Fig. 2.1.1. Geological map of Pelion.[4] Cave sites are marked with red dots.

The oldest rocks found on Mount Pelion are Palaeozoic schists and gneisses with ages ranging up to the Triassic (200 my). These rocks are overlain by gneisses, schists, marbles, recrystallised limestones and ophiolitic rocks of the Alpine orogenic belt (Fig. 2.1.1). The post-alpine sediments consist mainly of Tertiary and Quaternary alluvial fans and debris cones found principally near the coasts of Pelion, and occasionally at high altitudes.

Lacustrine deposits are found at the northwestern margin of Pelion in the deposits of the former Lake Karla. This lake existed at the southeastern part of the Thessaly plain until 1962, when it was completely drained due to a reclamation project, mainly for agricultural purposes. Lake Karla covered over 160 km^2 in the Neolithic period and 56 km^2 in 1953, with an average depth of 5 metres. The lake's surface area fluctuated from 40 to 180 km^2 because of the low-angle slopes and its recharge regime. The extent of the lake is not well defined because of periodic flooding. Recent observations demonstrate that Mavrovouni and Pelion mountains have played a role in Karla's overflow karstic inlets at the lake's eastern

and southern borders. During flooding, water from the lake was intruding into the karstic aquifer due to Karla's proximity to the marbles. The entrance points, called sinkholes, are now visible as numerous rockshelters filled with silt and clay.

Rock sequences at Pelion have been tectonically deformed and metamorphosed. The African plate on the margin of the European plate has resulted in strong tensions in the area of Pelion and tectonic uplift has led to large normal faults. Pelion is characterised by such rapid tectonic uplift, which leads to exposure of underlying rocks and series (Makrinitsa unit) and subsequent erosion of these. Due to uplift, the peaks of Pelion are characterised by high altitudes (1000-1624 m.a.s.) with a distance from the shoreline of only 7-11 km. Rapid uplift, which is a typical feature of Mavrovouni, Ossa and Olympus, has led to a distinct coastal geomorphology on the western side of Pelion. The eastern coast is steep, comprising marble cliffs that reach a height of 120 m

---

4   Based on IGME 1978 Volos, Zagora-Syki, Ayia Sheets.

Fig. 2.1.2. Fault scarp on the eastern coast of Pelion. The arrows point to the rockshelter complex of Kryfo Scholio and Panagia Megalomata (MOU-10).

and reveal numerous caves inside the karstic rocks. In contrast, the western slopes of Pelion are gentler and there are extensive alluvial fans, with Volos being situated on the largest.

Pelion experiences seismic activity and several earthquakes have affected the broader area. The most significant active fault zone is the Nea Anchialos –Milies fault system. The active Nea Anchialos Fault System is caused by a north–south trending extensional regime, where a 6.5-magnitude seismic event in July 1980 shocked the whole area, generating major damage and numerous ground ruptures.[5] Several earthquakes caused by this fault zone have affected the city of Volos and Pelion with enormous reported damages (1864 and 1955 earthquakes). Based on archaeological data, it appears that the same fault system has been active since at least the third to fourth century AD.[6] Another fault zone located at the eastern area of Keramidi village in a submarine fault scarp generated three main seismic processes in 1905 (magnitude 6.4), in 1911 (magnitude 6.0) and in 1930 (magnitude 6.1). During two major seismic events in 1930, many houses in Pouri, Zagora, Horefto and Makrirachi were de-

stroyed. The earthquakes also destroyed houses in the city of Volos and affected the whole of Thessaly and East-Central Greece.[7]

On the eastern part of Pelion, several ore deposits have been found. Minerals such as pyrite, galena, zinc (P.B.G.) and chalcopyrite were mined in the Zagora–Xourichti area. Small-sheeted veins of ore deposit were found in the schists and the marbles. The P.B.G. appearances were exported from the beginning of the twentieth century until the 1940s, partly for the extraction of silverous galena.[8]

Definitions of cave types

Caves are usually defined in terms of human access rather than geological setting. Caves are natural voids beneath the land surface that are large enough to admit humans.[9] This definition implies that a cavity is connected to the surface through entrances, which is, however, not always the case. Many small

---

5   Papazachos *et al.* 1983
6   Caputo 1996.

7   Papazachos & Papazachou 1989, 317; Spyropoulos 1997, 243.
8   Teller 1879.
9   Palmer 2007.

Fig. 2.1.3. Typical erosion cave in a debris cone at Veneto (KER-9).

underground openings and fissures are formed the same way as caves, but they are too small to be explored and are defined simply as *underground voids* (cavities). Caves and cavities are both characterised as *caveforms*.

Caves are often classified based on their morphological appearance (shafts, rockshelters and horizontal caves), their origin (solution, glacial, volcanic caves) or their host rock (limestone, sandstone, granite). Cave geologists use these speleogenetic characteristics and usually the origin of a cave is the main classification criterion. However, many cavers use the morphological criterion, although this does not constitute the proper orology.

Horizontal caves develop on a horizontal plane and have modest inclinations compared to shafts or pits that have orientations ranging from vertical to horizontal. Small horizontal caves at the base of rock walls with no permanent dark zone are called rockshelters. They are often associated with cultural remains. Shafts or pits are vertical formations of caves and they are usually explored with the single-rope technique.

Solution caves are formed by underground water dissolution of specific rocks that are called carbonate, such as limestone, marbles, dolomite etc. They crop out over approximately 10% of the ice-free continen-

tal areas and underlie much more.[10] The dissolving action takes place in the underground pores, fissures and joints of the carbonate rocks. Rocks dissolved by underground activity are called karstic rocks and the process karstification. The term "karst" was introduced in the Karst area in northwestern Slovenia near to Trieste, where the phenomenon was first studied. The majority of the Pelion caves are karstic rockshelters. Many of these caves are located near the shoreline, where active tectonism has revealed them and the erosive force of the waves has enlarged them further. These caves are called sea caves if they have a direct connection with the sea or littoral caves if they are located a few metres from the shoreline.

Erosion caves result from the mechanic scouring of moving water. They are usually formed by water flowing into unconsolidated rocks causing weathering of the rock across tectonic or stratigraphical discontinuities. They are common in debris cones or debris accumulations (Fig. 2.1.3). Erosion caves are usually related to *talus caves.* These caves consist of interconnected voids between boulders in the talus piles that accumulate at the bases of cliffs and are sometimes related to fault scarps. They can form in all rock types.

―――――――――――
10   Ford & Williams 1989.

Artificial caves constitute another group of underground formations that can be classified as caves although they have a human origin. They include mines, water tunnels and other underground constructions.

Karstic regions on Pelion

Caves and karst areas are abundant throughout most of Pelion (Fig. 2.1.4). The term "karst" refers to the geomorphological and morphological results generated in carbonate rocks (marble and limestone) due to the solvent action of water. It is an international term that describes these specific landscape characteristics and underground water circulation. Marbles and carbonate rocks, in general, are widely distributed all over Pelion, and, therefore, karst phenomena are well developed and present all over the mountain.

Karstic erosion creates both underground forms (caves and other underground voids) and surface karst forms (karrens and sinkholes). The ability of water to dissolute is increased when rainwater engages the atmosphere" CO_2. The latter reacts with the water (H_2O), giving carbonic acid (H_2CO_3). The rain enriched with carbonate acid erodes the sur-

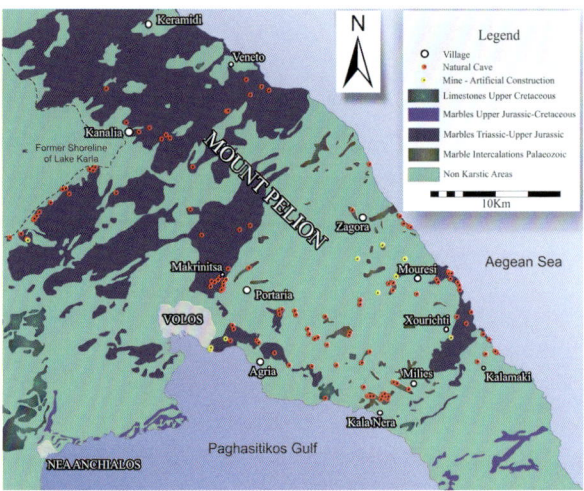

Fig. 2.1.4. Map of karstic areas on Pelion. Cave sites are marked with red dots. Based on IGME 1978 Volos, Zagora-Syki, Ayia Sheets.

face karst areas, and infiltrates the bedrock through tectonic discontinuities (faults and joints), thus creating underground karstic water conduits and cavities. Generally, karst areas (Fig. 2.1.5) appear on the surface as dry landscapes without a significant drainage system because of the vertical movement of water towards the karstic aquifer (a layer of rock or soil that can hold or transmit water). Where the aquifers discharge, the karst water is often of major

Fig. 2.1.5. General view of the karstic landscape.

importance to surrounding villages, monasteries, pastoral facilities, etc.

The northern part of Pelion is mainly made of erosion-resistant marble. However, the major part of the peninsula is dominated by schist and gneiss, which produce more fertile soils than marble. In addition, water here does not disappear into the ground but runs along the surface where it can be more readily exploited. One such example is Timpano above Kissos village, where open conduits, canals and artificial tunnels carry water to the fields and villages.

Factors affecting the development of a cave and the evolution of karst are the bedrock type, the tectonic conditions in the area and the hydrology of the region. The cave genesis follows the oldest tectonic lines (faults, fissures, joints) in the karstic rocks and forms large cavities depending on the amount of water entering the karst system and the solubility of the rocks.[11]

During the Pelion Cave Project survey, 160 localities were explored. Of these, more than 75% are rockshelters and small, horizontal caves.

Of the caves surveyed, 84% were in karst areas, associated with carbonate rocks (Table 2.1.1). Karst regions are typified by the dominant dissolution process, the lack of surface water and the development of stream sinks (dolines), cave systems and resurgences of springs. The main process behind the formation of karstic caves is the process of solution inside tectonic or stratigraphic discontinuities. Inside a cave, there are several generations of tectonic faults. The development of an underground void follows the old cracks and the rock strata, while the younger faults disrupt the continuity of the cave, and sometimes cause the exposure of karstic conduits.

Cave classification based on speleogenetic characteristics	
Solution caves	131
Artificial caves (mines, water tunnels)	9
Erosion caves	18
Cave classification based on the host bedrock	
Caves in marble and limestone	133
Caves in schist and gneiss	14
Caves in conglomerate	11

Table 2.1.1. Classifications of caves encountered during the survey.

The majority of the cavities created in the phreatic phase of speleogenesis are karstic pipes filled with moving water from the aquifer (Fig. 2.1.6). The phreatic phase characterises the speleogenetical phase when the karstic pipe is below water table. The tectonic uplift combined with intense erosion and the presence of faults leads to the formation of the karst cavities. The lowering of the aquifer's surface causes the karst conduits to evacuate the water.

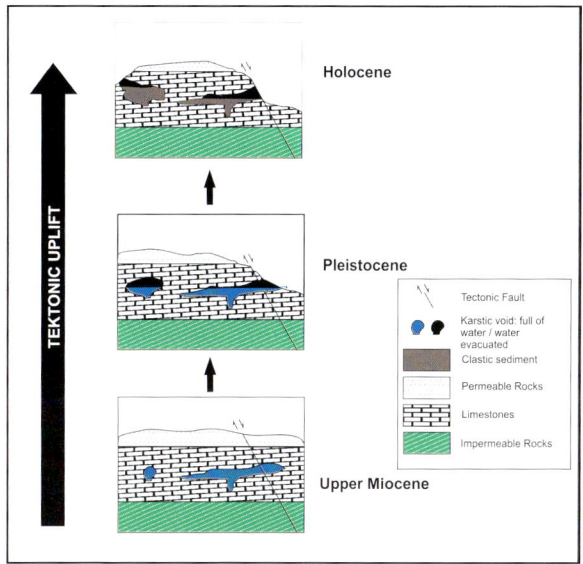

Fig. 2.1.6. Sketch map of the evolution process of the Pelion caves based on other speleogenetical models in Eastern Thessaly (Vaxevanopoulos 2006).

---

11   Gunn 2004, 1421; Vaxevanopoulos 2006; Ford & Williams 2007.

Most karstic caves during their evolution function as karstic springs with several large water supplies. The amount and quality of the water are controlled by the direction of the large penetrating cracks in the underground karst and variations in rock type (e.g. slate). The activation of a tectonic fault is likely to lead to a sudden interruption of a water source, and a subsequent earthquake can lead to its reopening. Caves exposed by surface erosion are often filled with clastic sediments (sand, clay, silt, etc.) after the water has drained them, making them only partly accessible to humans.

Once a karstic conduit evacuates water, chemical sediments are created, such as stalactites, stalagmites, columns or cave corals. The cave of Agios Athanasios (KAR-8) in the area of Karla has more speleothems than any other cave in the Pelion region. Stalactites adorn the ceiling of the cave, these being generated from calcium carbonate ($CaCO_3$), which is deposited by drops of water (Fig. 2.1.9). Where excess calcium carbonate drips off the stalactite, a corresponding stalagmite is formed beneath it.

Fig. 2.1.7. Tectonic fault at the entrance of AFE-6 near Lambinou.

Fig. 2.1.8. Karstic conduits-caves at the tectonic surface in marbles. A fault has revealed the underground voids (Agripou Caves).

Fig. 2.1.9. Speleothems in
Agios Athanasios Cave (KAR-8).

Fig. 2.1.10. Calcitic columns in the cave of Osios
Symeon (KER-7).

Merging stalactites and stalagmites create columns
(Fig. 2.1.10). A drapery (stalactite curtain) is formed
in cases where the ceiling is inclined and deposition
of calcium carbonate takes place along inclined directions. Cave coral is a kind of sediment in karstic
caves, which is usually composed of calcite due to
the condensation of water films on cave walls. They
have a botryoidal shape with concentric layers of
calcite (Fig. 2.1.10).

The temperature inside a cave is usually constant
and, as a rule, equal to the average annual temperature of the region, depending on the specific altitude.
For example, in several caves in Pelion, the temperature stays between 12-18 °C. Near the entrance
of the cave, there is a transitional zone where the
temperature is affected by external weather conditions. This zone reaches a few metres into the cave.
Inside the cave, the area that shows no significant
temperature change is called the homothermic zone.
It is characterised by a significant stability in temperature. In most cases, the caves of Mount Pelion
have no homothermic zone due to their generally
small sizes and low height. Nevertheless, there are

Fig. 2.1.11. Shallow cave (MOU-9) formed in a marble debris cone by lateral erosion.

Fig. 2.1.12. Cave at Milopotamos Beach formed along an axial break and enlarged by wave action (MOU-4).

differences between the inner environment of even small caves and the outside temperature and humidity. The relative humidity in deeper caves (30-40 m) reaches 100% if there is no second entrance.

In addition to the Pelion karstic caves, there are caves near the coasts or in stream beds that have been formed by erosion (Fig. 2.1.11). The active agent is mainly water, which has eroded the parent rock due to wave action or sediment-carrying streams. Wave action can result in significant erosion of even solid rock, such as marble and gneiss (Fig. 2.1.12).

Extractive industries

Many artificial caves on Pelion were developed due to mining. Ore in quartz veins occurs on East Pelion from Zagora to Xourichti. These are acidic quartz infiltrations characterised by the presence of pyrite, sphalerite, galena, chalcopyrite and other minor minerals. In the early twentieth century until the early 1940s, extractive industries exploited these occurrences for argentiferous (silver-bearing) galena, iron ore, antimony, arsenic and iron sulphides.[12]

___
12   Teller 1880; Zachou *et al.* 1965.

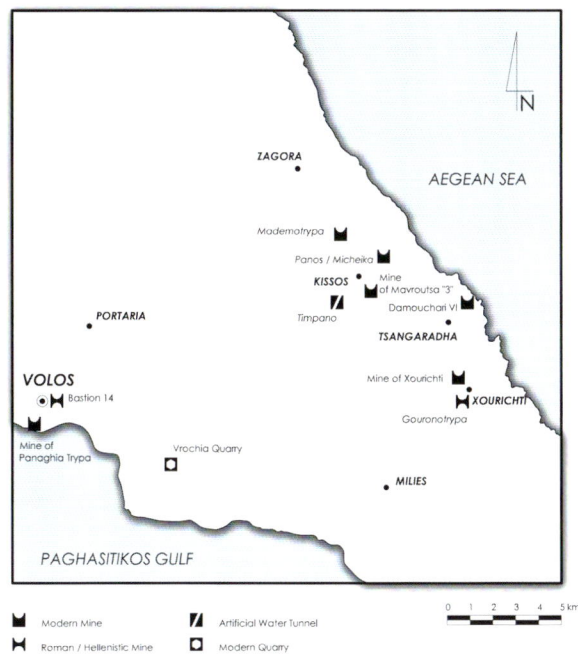

Fig. 2.1.13. Map of part of the survey area showing the location of artificial caves. Drawing by T. Touis-souzoglou.

Abandoned mining tunnels, of which several have now partly collapsed, are the only traces of this industry. These are long, straight tunnels with slightly curving walls and roofs. Remains of internal support structures were only found in one case, where a wooden support beam lay discarded near the entrance. Fig. 2.1.13 shows the location of eight ancient and modern mines as well as a quarry and a water tunnel documented within the survey area.

Mademotrypa (MOU-28) is a mining tunnel near Makrirachi village, which was exploited prior to World War I (Fig. 2.1.14). According to locals, ore was extracted from the mine and carried by mules to near the Taxiarches monastery, where a smelting industry processed the ore. Merchants of Zagora were involved in the exportation of the ore and they used Agioi Saranta as a shipping port. The industrial structures are now completely abandoned and the remaining walls of the complex are used for cultivation terraces.

An underground mining gallery, possibly dating to the Roman period, was found southwest of Xourichti (Gouronotrypa, MOU-2). It was developed in dolomitic marbles. The total length of the mine is 130 m with the main corridor in a north–south direction. At the end of the main gallery, there has been a major collapse blocking a possible continuation. The

Fig. 2.1.14. View into the mine of Mademotrypa (MOU-28). The tunnel is 60 m long and ends in a collapse. At two or three locations inside the tunnel, the roof has collapsed onto the otherwise smooth tunnel floor. Quarried stone has been stacked alongside the tunnel walls near the entrance. The tunnel's dimensions range in width from 1.5-2.2 m and in height from 1.6-2.0 m.

corridors show two faces of exploitation, probably one from the Roman period and one earlier face. Ancient metallurgy on Pelion is mostly unknown and the discovery of the Xourichti mine is expected to provide new information about ancient mining technology in this region.

2.2 Built structures in and around the Pelion caves
Niels H. Andreasen

Greek summary

Ένας σημαντικός στόχος της έρευνας ήταν η τεκμηρίωση, λειτουργική ερμηνεία και χρονολόγηση των κατασκευών που προκύπτουν μέσα και γύρω από τα σπήλαια στο Πήλιο. Σχεδόν το ένα τρίτο των χώρων σχετίστηκαν με ευθείες ή ημικυκλικές κατασκευές ξερολιθιάς στα ανοίγματα των σπηλιών. Οι σπηλιές σε γενικές γραμμές δεν είχαν τροποποιηθεί για να διευκολύνουν την ανθρώπινη κατοίκηση και στις περισσότερες περιπτώσεις χτίστηκαν κατασκευές από αγρότες ή βοσκούς που χρειάζονται να ελέγχουν και να προστατεύουν κοπάδια των οικόσιτων ζώων έξω από τα χωριά. Μερικές φορές χρησιμοποιούνταν άλλα στοιχεία και δομικά υλικά σε συνδυασμό με πέτρινους τοίχους, όπως στέγες, διατηρώντας ή διαχωρίζοντας τοίχους, κουφώματα ή κάθετες επεκτάσεις του τοίχου. Φράκτες που σχετίζονται με σπηλιές ήταν σχετικά σπάνιοι.

Οι επενδύσεις στην κατασκευή και συντήρηση των κατασκευών είναι γενικά χαμηλές σε περιοχές όπου οι άνθρωποι μετακινούνται ελεύθερα μέσα και έξω. Στις περιπτώσεις όπου το ιδιοκτησιακό καθεστώς είναι πιο ελεγχόμενο και η διαμονή στις κατασκευές πιο σταθερή (συνήθως κοντά σε χωριά ή δρόμους), οι κατασκευές είναι πιο πιθανό να διατηρηθούν και να επαναχρησιμοποιηθούν. Η επένδυση ενός ατόμου σε μια κατασκευή δηλώνει σαφώς την ταυτότητά του και την αξίωσή του σε ορισμένα δικαιώματα στην τοποθεσία.

Η συνολική κατάσταση των τοίχων δεν ήταν καλή, μόνο λίγες ξερολιθιές διατηρούνται σε καλή κατάσταση. Ακόμη και εκεί που τοίχοι φαίνονταν να είναι ακέραιοι, μέρος από τα ενσωματωμένα σανιδώματα αλλοιώνονται. Οι τοίχοι μας επέτρεψαν χρονολογική ανάλυση μόνο στις περιπτώσεις όπου δυνητικά χρονολογούμενα οικοδομικά υλικά ήταν ενσωματωμένα στην κατασκευή τους ή όπου πληροφοριοδότες γνώριζαν την ημερομηνία κατασκευής τους.

Το ποσό των δραστηριοτήτων επισκευής και ανακατασκευής αντικατοπτρίζει τους κύκλους της ποιμαντικής ακμής και παρακμής, με τις κατασκευές να ρημάζουν σε περιόδους οικονομικής ύφεσης. Ωστόσο, η εγκατάλειψη και η φθορά των κατασκευών μπορεί να προκύπτει από πολλές αιτίες, συμπεριλαμβανομένης μιας μείωσης του αγροτικού εργατικού δυναμικού, ενός αναπροσανατολισμού των ποιμαντικών συνηθειών, ή μιας αυξημένης προτίμησης για υπαίθριες τοποθεσίες. Οι αιτίες της εγκατάλειψης δεν θα πρέπει να συνάγονται μόνο από τα επιφανειακά αρχαιολογικά δεδομένα, και τυχόν παρατηρήσεις πρέπει οπωσδήποτε να αναλύονται στο πλαίσιο τόσο εθνογραφικών όσο και ιστορικών στοιχείων.

Fig. 2.2.1. Drystone dividing walls built between four adjacent, shallow rockshelters (KER-9). The sidewalls were built to increase the depth of the shelters and improve the protection they provided. Each "compartment" had a different function.

Introduction

An important objective of the archaeological survey was the documentation, functional interpretation and possible dating of structures in and around caves on Pelion. An understanding of the causes and circumstances surrounding the use, modification and abandonment of structures associated with caves was considered important because these processes potentially link use patterns with structural remains and artefacts distributed on the surface. [13] For instance, socially motivated change may often be the cause of frontal stonewalls and other structures falling into disuse.

Structures were defined as constructions consisting of multiple parts that have been built or brought into the cave to facilitate human activities. While a structure usually has a vertical quality to it, we also refer to horizontal constructions, such as stone-paved floors or hearths, as structures.

What is noticeable, when one surveys a large number of caves and rockshelters, is the variability of structures associated with these sites in terms of size, shape and detail. The most obvious explanation for this variety is that structures are fitted to varying topographies and practical requirements at each cave

and that the individual cave user will apply particular technical skills and use his preferred materials. In other words, structures at a cave or rockshelter can be viewed as the outcome of individual, often practical, choices and as site-specific adaptations to certain constraints.

Most structures were situated close to cave entrances. However, the confined spatial limits of caves restrict the possible range of activities and may lead to the formation of open activity areas. In broad terms, divisions of functional space in and around caves can be divided into: 1) separations within the cave/rockshelter, 2) functional separations between two or more adjacent caves (Fig. 2.2.1) and 3) separation between an inner area (cave interior and sometimes a roofed area immediately in front of the cave) and an outer (enclosed) area. Activities in caves and open activity areas may be different although on occasion they are complementary.[14] The open activity area was defined as the area protected by a built structure, such as fences or rock-built perimeter enclosures. Although activity areas may extend beyond such structures, we rarely found evidence of this.

---

13   Cameron 1991, 155.

14   Gorecki 1991, 256.

Figs 2.2.2. & 2.2.3. Small rockshelters (shepherd's shelters) with semi-circular walls of dry-stacked, unshaped limestone (MIL-21 & KAR-35).

Walls

Drystone structures (built without mortar) or remnants of these found during the survey were associated with almost one-third of the caves and rockshelters (Figs 2.2.2 & 2.2.3). A drystone structure was defined as at least three stones or more placed horizontally or vertically in association with each other. This definition, therefore, includes drystone structures which were probably not intended as walls but which cannot be differentiated from wall remnants.

Creation of perimeters by drystone walling is a way of controlling space, which is part of a cultural tradition that extends far back in time. These constructions were mainly built and maintained by poor farmers or landless herders who needed to control and protect flocks of animals outside villages. Aside from keeping the flock together, walls constructed in front of the caves would serve to protect animals and humans from the weather, retain heat in the cave, and deny predators access to the flock at night. Low rubble walls outside rockshelters (but connected to these) were probably used as corrals in the summer.

Drystone walls were mostly straight or semi-circular constructions running close to the drip line of the cave or immediately outside this. Apart from functioning as exterior walls in front of caves, they could also act as internal partitions or as retaining walls close to cave openings.

There are several methods of constructing drystone walls, depending on their function and the quantity and type of stones available. Most walls are constructed from blocks and boulders found around the caves, but some may be from stone quarried in the immediate vicinity. In many places at Pelion, local marble and limestone split naturally into slabs of differing thickness (Fig. 2.2.4). These

Fig. 2.2.4. Straight drystone wall built from shaped and unshaped local limestone slabs (MIL-1). Many stones bear marks from the walling hammer.

are convenient building materials as they can easily be stacked on top of each other. Most dry walls on Pelion are constructed by placing two rows of stones along the boundary to be walled. Walls are built up to the desired height layer by layer and at random intervals, with larger through-stones spanning both faces of the wall. These have the effect of bonding and increasing the strength of the wall.

Three main categories of drystone wall were encountered:

- Coursed random: Squared stonework built of natural, rectangular blocks whose horizontal and vertical joints do not line up. The final layer on the top is usually no different from the rest of the wall (Fig. 2.2.5).
- Stones with pinning: Built of unworked and unsorted boulders. Because many boulders are naturally rounded, packing (or pinning) with smaller stones is required to keep the wall stable (Fig. 2.2.5).
- Random rubble: Built of angular, unshaped and poorly sorted stones. In some cases, rectangular slabs may be built in at corners and entrances to provide more stability. In one case, we observed the use of tilted capstones along the top of a section of a wall (MIL-5).

In most cases, walls had a purposely-built gap functioning as an entrance for both herder and livestock. We did not come across other openings in walls apart from in two cases (a chapel and an animal enclosure), where small timber-framed, glass-less windows were set into the dry wall (MOU-10, MIL-2).

Other elements and building materials were sometimes used in combination with stonewalls, such as roofs, dividing walls, doorframes or vertical extensions of the wall. At MIL-1, five different types of wooden elements were used. Recycled telephone poles in a horizontal position were fastened with metal wire to the top of the drystone wall to support tie beams (slim stems of local trees) extending to the rock wall. Doorways were constructed from square beams while boards and plywood were used for doors and wall partitioning (Fig. 2.2.6). Iron nails and metal wire were used to join wooden components. To supply coverage, tarpaulins or large sheets of plastic would have functioned as roof coverage. Left of the entrance, corrugated iron plates were used to extend the height of the drywall.

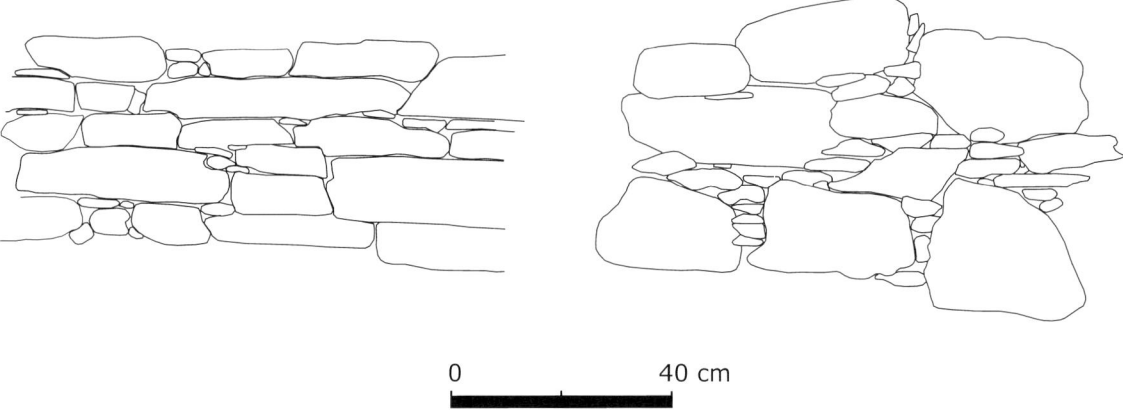

0   40 cm

Fig. 2.2.5. Section samples of drystone walling used at Pelion. Left: Coursed random (MIL-1); right: Stones with pinnings (MIL-4).

Fig. 2.2.6. Telephone poles, local tree stems and plywood as building elements in a covered animal pen near Milies (MIL-1).

Fig. 2.2.7. Pig enclosure with wooden posts and chainlink fence (KAR-1).

Fig. 2.2.8. Pig enclosure with metal gates and metal doors in front of shallow rockshelter (KAR-9).

Fig. 2.2.9. Drystone wall and lath fence with a metal sheet covering (KER-9).

Drystone walls are designed to settle into durable structures without the need for binding substances, such as mortar or cement. Materials and transport costs are also considerably higher when such substances are used and it is therefore no surprise that they are used sparingly outside the villages. At the edge of Lake Karla, cement render was used in two cases to secure now missing roof constructions between the bedrock and the built wall (KAR-2, KAR-9). Mortar was used in two instances to consolidate walls that served non-pastoral purposes. A mortared wall was constructed, probably around 1900, to provide a whistle stop in front of a small rockshelter

next to the Ano Lechonia-Milies train line (MIL-9). At a small cave chapel next to MOU-10, mortar was used (probably relatively recently) to consolidate loose stones in the dry wall.

Mortar and/or cement was also used in association with tunnel constructions (MOU-25, KAR-13) or spring caves (AFE-2, ZAG-6), where the purpose was transporting or manipulating the flow of water. KAR-13 is a metre-wide tunnel built from mortared stonework. It has a slightly pointed arched entrance (Fig. 2.2.10). *Voussoirs* (wedge-shaped stones used in the construction of an arch) are made from slabs of unequal size and there is no central keystone at

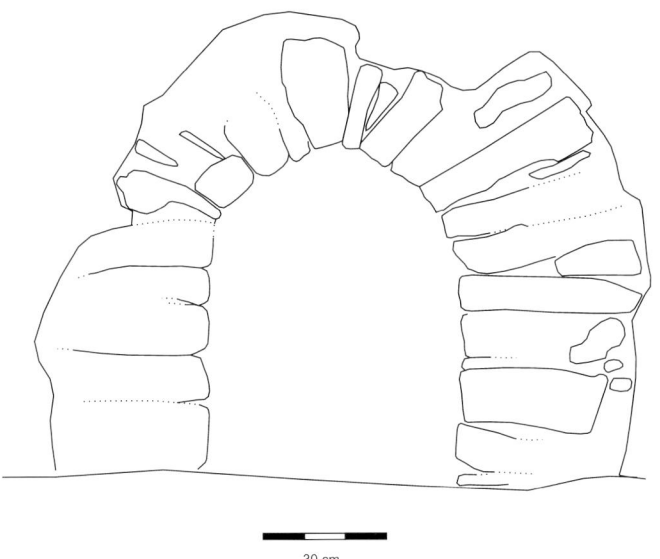

Fig. 2.2.10. Entrance to a hand-dug tunnel at Lake Karla with mortar-reinforced walls (KAR-13). Age unknown.

30 cm

the apex. The tunnel, which runs more than 30 m back into the hillside, might have been a semi-public opening to a spring or well.

"Timpano" is the name of a water tunnel in active use above the fields of Kissos. It was carved through the rock face with hammer and chisel in the beginning of the twentieth century and leads part of the discharge from a tributary to Mega Revma through a concrete channel down to the village of Agios Dimitrios for irrigation purposes. Inside the tunnel is a narrow walkway and a flow channel partly constructed from stone and mortar. The tunnel is about 1.75 m wide and reaches a maximum length of 8.5 m.[15]

These examples contain evidence of economic variation in site structure. Mortar and cement, for instance, is mainly reserved for "public" or shared structures around cave springs or for features connected to water tunnels. In contrast, mortar and cement were rarely used for pastoral structures, and

only as render in small amounts to fasten roof constructions.

Fences

Fences were surprisingly rare and occurred only at eight sites. They could be related to animal enclosures in all cases but one (Table 2.2.1). Chainlink fencing (wire mesh) was used at KAR-1 in combination with wooden pallets, round and square wooden posts as well as metal irrigation pipes, to construct a durable pig enclosure in front of a cave. A little further north at KAR-9, an enclosure was constructed from an assortment of discarded iron gates and doors tied together and attached to iron posts with wire. The sturdiness of the construction and the relatively low height of the enclosure indicate a pig enclosure.

At the shallow rockshelter of MOU-9, a goat-track leads immediately past the shelter and a partly ruined chicken-wire fence runs alongside the sloping floor of the ledge in front of the site. A disused wooden trough indicates that the shelter was used as a watering station, probably for goats.

---

15   Similar tunnel/channel constructions exist at Chalkidiki and at Kridines at Kavala (Vaxevanopoulos, pers. comm. 2006).

Fence type	Materials	Function	Site
Chain link	Chainlink, wooden pallets, round and square posts, iron water pipes	Pig enclosure	KAR-1
Iron	Iron gates, iron doors, iron posts	Pig enclosure	KAR-9
Chicken wire	Chicken wire	Goat enclosure	MOU-2
Lath w. metal sheet covering	Laths, zinc-plates, hammered-out oil barrels	Goat enclosure	KER-9
Lath	Laths	Animal enclosure	MOU-7
Lath	Laths, metal wire	Animal enclosure	MIL-3
Wire?	Iron posts, iron pipes, metal wire	Goat enclosure	MAK-3
Metal	Wooden posts, hammered-out oil barrels	Storage	ART-2

Table 2.2.1. Types of fencing encountered during the survey.

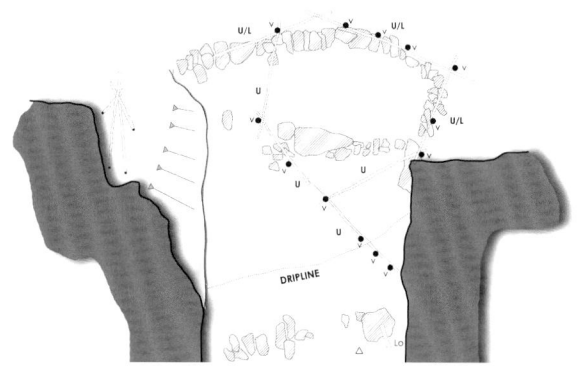

Fig. 2.2.11. Entrance section of MOU-7 with remains of drystone walls and lath fence (V=vertical post; U/L=upper/lower horizontal beam).

A short, robust drystone construction at KER-9, extending from one side of the shelter, was used to support a mobile lath fence frame covered with scrap zinc-plates and hammered-out, flattened oil drums. Laths consist of wood from young trees growing in the vicinity.

At MOU-7, a stationary lath fence encloses an area of 16 m² immediately in front of the drip line (Fig. 2.2.11). Apart from vertical posts, the fence consists of horizontal upper and lower laths. There was no sign of wire mesh or any other materials that would be supported by the frame. Part of the fence runs on top of a retaining drystone wall along the ledge in front of the cave.

Remains of a completely ruined fence at MAK-3 consist of iron posts (both in situ and ex situ) and entangled bundles of metal wire.

ART-2 is a storage cave where the opening is completely sealed off by hammered-out oil drums, supported by vertical wooden beams.

Hearths

A hearth was defined as the floor of a fireplace, which can be open or stone- or brick-lined. Hearths were used for cooking, heating and processing of faunal and vegetal resources, but no concentrations of food debris were observed around hearths. Only a single fire-affected bone was recovered (Panagiotidou, this volume) from one site.

Definable hearths were present at only three rockshelters and two small caves. These were either open hearths (placed on areas paved with flat stones) or loosely stone-lined hearths. Both types are markedly low-investment features in terms of energy. Apart from the five identified hearths, possible remains of hearths were identified at at least 30 sites as discrete areas that contained charred fragments of wood, almond shells, individual charcoal pieces or diffuse ash-deposits. Some of these caves had distinct fire- or smoke-affected rock walls. While expedient hearths or isolated fires are the most likely source of cultural charcoal in the caves, it is also known that periodical burning of stable waste can produce visible ash deposits and fine charcoal (see Chapter 4).

Despite the presence of fire-affected wood and other indications of kindling fires within the caves, the scarcity of formal hearths was surprising. While site formation processes may quickly deform or disperse built hearth features and thus make them difficult to identify, it is clear that hearths in the Pelion caves were mostly expedient features. Ethnographic investigations from caves outside Greece show that there is little relationship between factors such as length of occupation or the degree of mobility and specific types of hearths.[16]

While brief stays in the cave do not correlate directly with expedient hearths, the scarcity of hearths might reflect brief occupations during mainly the warmer part of the year when it was necessary only occasionally to build a fire.

Construction and maintenance of structures

Apart from expedient hearths, only in a few cases did we observe structures built specifically to facilitate human short- or long-term habitation (MOU-10, MIL-3, MIL-9). This does not mean that people did not spend time in the caves, only that they were generally not modified for this use and that in most cases, construction materials were for building and maintaining animal enclosures.

Materials necessary for the construction and maintenance of structures had to be obtained from different sources and it can be difficult to determine

exactly where they came from. Heavy, not readily transportable materials of modest value, such as stone and most types of wood were probably obtained in the immediate vicinity (Table 2.2.2). Originally built with what was found around the cave, no dry walls were important enough to warrant transport of stones over any great distance, whereas easily transportable items and materials of greater value could have originated from far away.

Material	Ease of procurement	Speculated cost
Drystone	On-site	Gratis
Tree stems	On-site	Gratis
Telephone poles	Vicinity	Gratis
Plywood	Imported to site	Gratis
Square beams	Imported to site	Low cost
Planks	Imported to site	Low cost
Nails	Imported to site	Low cost
Metal wire	Imported to site	Low cost
Tarpaulin or plastic sheets	Imported to site	Medium cost
Corrugated iron	Imported to site	Medium cost

Table 2.2.2. Construction materials used at MIL-1 related to ease of procurement and cost.

Whichever method is used to build a dry wall, some skill is required. Selection of the correct stone for every position in the wall makes an enormous difference to the lifetime of the finished product, and an experienced builder will take time making the selection. While most dry walls are expedient constructions made with random rubble, the exceptions are interesting because they suggest an investment of resources and skill. There would have been plenty of stonemasons around on Pelion during the eighteenth and nineteenth centuries, when many stone-paved paths, mansions and bridges were constructed. How-

16   Galanidou 2000, 247.

ever, such professional labour was expensive and would hardly have been employed to build stone walls in front of caves. All shepherds and members of local households are assumed to have been capable of building simple walls of random rubble, and it is likely that even some of the better walls (perhaps except mortared structures) were within the range of capabilities of more accomplished shepherds.

Work rates are difficult to specify, as these depend on various factors including skill and experience of the builder, type of stone, type of wall and site and weather conditions. On average, an experienced builder may complete about 2.7-4.6 m per 8-hour day.[17] It follows from this that one person working for a couple of days would have had no difficulty constructing a linear wall in front of a small rock-shelter or cave opening. Obviously, caves or rock-shelters with wider openings needed longer walls and, therefore, required more work and building material. The longest wall encountered during the survey was approximately 20 m and it would have been possible to construct it in less than a week, perhaps more if the stone had to be quarried from the bedrock close to the site.

With the advent of modern wire fencing, areas around caves could be fenced in much less time and at less expense; however, the initial expense of building drystone walls is offset by the fact that they are sturdy, low maintenance constructions and are easy to repair. Stability of dry walls depends on the quality of the construction and the amount of maintenance the structure receives. Leaping goats and sheep can quickly degrade a wall and stones can easily topple if they are not secured by wall plates or protected by brushwood.

The survey results show that investments in construction and maintenance are generally low in areas where individuals or family groups move in and out freely. In situations where property ownership is more structured and occupancy of struc-

tures more stable (typically close to villages or near roads), structures are more likely to be maintained and reused. Establishment of structures in front of caves can be a means of nonverbal communication that transmits a message about rights of access.[18] A person's investment in a structure clearly announces his identity and his claim to certain right over the location. The widespread phenomenon of cave graffiti is a further emblem of identity that can establish such claims.

Condition of structures

Drystone walls were assessed in the field as being in one of six condition categories (Table 2.2.3). Preservation of drystone walls varied greatly from complete to severely damaged, but the overall condition of walls is poor, with about 8 out of 10 caves falling in the bottom three categories (D–F). Few drystone walls have survived in good preservation, except for those that have been maintained until recently. Even where walls appeared to be undamaged, some of the integrated woodwork was deteriorating, particularly doors and their frames.

A	Stock proof and in excellent condition	6%
B	Sound & stock proof with minor defects	12%
C	Major signs of advancing deterioration	4%
D	Not stock proof, becoming derelict	2%
E	Derelict	31%
F	Remnant	45%

Table 2.2.3. Condition of drystone structures associated with caves on Pelion. The categories only refer to structures of stone and not to associated woodwork, etc. Note that Category F includes intentionally piled stones with unknown function.

---

17 Brooks & Adcock 1999, 34.

18 Galanidou 2000, 269.

Fig. 2.2.12. Remains of drystone wall in front of MIL-4.

Fig. 2.2.13. Collapsed storage facility in Kanalia village (KAR-31).

Fig. 2.2.14. Remains of cement render above small cave (KAR-2).

Fig. 2.2.15. Drystone wall with a wood-framed window (MIL-2).

Fig. 2.2.16. *Mandri* below Makrinitsa built against two shallow rockshelters (MAK-18). The interior of the *mandri* is equipped with electrical lighting, dividing walls and feeding troughs.

The few cases falling in the intermediate categories C and D could indicate that the process of deterioration from stock proof to derelict/remnant is relatively fast, which is why most walls remain only as derelict or remnants. Removal of structural components when buildings are abandoned speeds up deterioration and decreases potential for later reuse. Especially salvageable timber may be removed after abandonment to be used elsewhere or as firewood. The most significant threats to drystone walls are their demolition for reuse of the stones elsewhere. The remnant category in Table 2.2.3 includes walls where only a row of half-buried wall foundation stones could be seen or where a few stones on a line or in a pile could be interpreted as possible remains of a wall. Such remains tend eventually to become overgrown with grasses and other plants, and the stones themselves become dispersed or buried in the ground. It was common to find stones and other construction materials ex situ in front of the caves, particularly at Lake Karla.

An understanding of the causes and circumstances surrounding abandonment of cave structures was important to the project because abandonment links the use of a structure with the structural remains and artefacts found inside and around the caves.

All structures are thought to follow a trajectory from initial construction to total collapse. Structure use-life can be defined as the period from initial construction to final abandonment. Within this period, structures undergo change, both from natural deterioration processes and through the activities of the occupant.[19] Stone wall structures probably had functional use-lives of more than fifty years. While functional disuse quickly leads to structural decay, the stability of stone walls depends on the quality of their construction, the amount of maintenance the structure receives and the massiveness of construction.[20]

Structure abandonment is the result of a series of decisions to maintain, repair, rebuild, reuse or dismantle cave structures. The functional lifecycle of individual structures is connected both to processes of structural decay and to changing social conditions. The amount of repair and rebuilding activity would have closely reflected cycles of pastoral prosperity and decline, with structures falling into disrepair in times of economic depression. A large number of derelict and remnant structures on Pelion may be interpreted as caused by a downturn in pastoralism in this region, but it is important to note that abandonment of structures results from a wide range of causes. Structures may be abandoned and allowed to decay because of a declining rural workforce, because of reorientation of pastoral routines or because of an increased preference for open-air locations. Most dry walls are no longer needed for animal control, and even when they are needed for this purpose, fences can replace them more cheaply. Caves may become unfavourable because of insect infestations or trash, or locations that are more favourable may be found near a newly constructed water tank or fountain. MIL-4 near the bottom of a ravine below Vyzitsa may have fallen out of regular use after a fatal flash flood drowned a shepherd with his flock inside the shelter in 1913.[21] River sediments within the area, protected by a crumbling drystone wall, show that the shelter may have been inundated several times. Other catastrophic events such as the earthquake of 1930 may cause damage to caves and rockshelters and furthermore shorten structure use-life.[22]

An obvious obstacle to shedding light on the reasons for structure abandonment was the fact that stone walls usually do not allow a precise chronological resolution. Drystone walls associated with caves were classified according to their condition in order to establish an abandonment sequence. However, the general lack of chronological anchor points made it

---

19   Cameron 1991, 157.
20   Cameron 1991, 161.

21   PCP ethnographic field notes 2007.
22   Vaxevanopoulos, this volume.

difficult to use that information towards establishing a chronology of cave abandonment. Based on this, we acknowledge that it is not feasible to identify causes of abandonment from surface data alone. In other words, the validity of inferences from the survey must necessarily be analysed in the context of both ethnographic and historical data (see Chapter 4.8 for further discussion of this point).

2.3 Vertebrate faunal remains from the Pelion caves[23]

Theodora Panagiotidou

Greek summary

Την σύγχρονη εποχή έλαβαν χώρα έντονες αλλαγές στην φύση της σχέσης μεταξύ ανθρώπου και ζώου, οι οποίες περιελάμβαναν την ανάπτυξη νέων ειδών ζώων, την εισαγωγή νέων μορφών αγροτικών μηχανημάτων και την εκτεταμένη χρήση τεχνητής τροφής και αχύρου, που αποσυνέδεσαν την σχέση ανάμεσα στην εποχικότητα και τους φυσικούς βιολογικούς κύκλους. Επιπλέον, η αστική δημογραφική έκρηξη διευκόλυνε την εκβιομηχάνιση της παραγωγής κρέατος και γαλακτοκομικών προϊόντων και την εμφάνιση διηπειρωτικών δικτύων εμπορίου τροφίμων.

Μία συλλογή 236 οστών σπονδυλωτών της σύγχρονης περιόδου συνελέγη από 23 σπηλιές. Περισσότερη από τη μισή συλλογή συνελέγη από τις MOU-16, KAR-8 και KAR-29. Αναγνωρίστηκαν τουλάχιστον 14 ήδη αμφιβίων, πουλιών και θηλαστικών. Τα ευρήματα ανήκουν σε οικόσιτα ζώα (196 δείγματα), ακολουθούμενα από παρεισφρητικά είδη (24 δείγματα) και άγρια είδη (17 δείγματα). Η πιο κοινή κατηγορία ζώων είναι αιγοπρόβατα (69,9%). Υπάρχουν επίσης βοοειδή και χοίροι, και ένα δείγμα αναγνωρίστηκε ως γάιδαρος.

Δύο δείγματα χοίρων και 12 αιγοπροβάτων αναγνωρίστηκαν ως νεογέννητα. Τα περισσότερα οστά νεογέννητων αιγοπροβάτων βρέθηκαν στην MOU-16 και πιθανώς άνηκαν στο ίδιο ζώο. Στοιχεία σχετικά με την φθορά των επιφύσεων και των κάτω γναθών των αιγοπροβάτων, υποδεικνύουν την παρουσία μικτών στρατηγικών εκμετάλλευσης των ζώων. Ενώ τα περισσότερα ζώα σφαγιάζονται στην ηλικία των 1-3 ετών για το κρέας τους, κάποια θανατώνονται νέα για την εξάλειψη των ανταγωνιστών στην παραγωγή γάλακτος, και κάποια κρατούνται εώς την ενηλικίωσή τους για το μαλλί τους.

Βρέθηκαν λίγα ίχνη ανθρώπινης παρέμβασης. Τρία οστά είχαν κοπεί πιθανώς για την αφαίρεση του μυελού, όπως για παράδειγμα τα βραχιόνια οστά βοοειδών και κόκκινων ελαφιών. Άλλα δύο έφεραν ίχνη διαμελισμού (για τον διαχωρισμό διαφόρων ανατομικών μερών), ενώ σε τρία αναγνωρίστηκαν σημάδια τεμαχισμού σε φιλέτα, με σκοπό τον διαχωρισμό του κρέατος από το υπόλοιπο οστό. Η σφαγή των ζώων θα είχε συμβεί μακριά από την στάνη. Επομένως, όπως είναι φανερό από την συλλογή του οστεολογικού υλικού, συγκεκριμένα ανατομικά μέρη, όπως πυρήνες κεράτων, μέρη του κρανίου και την άνω οδοντοστοιχίας και κάτω τμήματα των μπροστινών και πίσω άκρων, απουσίαζαν.

❋ ❋ ❋

───────────

23   I would like to thank Sherry Fox, the former director of the Wiener Laboratory of the American School of Classical Studies, for allowing access to the laboratory's equipment and reference collection.

Introduction

Animal bones are ubiquitous archaeological finds and their detailed study can shed light on a diverse range of past human activities. These include identification of subsistence strategies, economic regimes, use of animals and animal parts in craft and industry, attitudes to animals and the role of animals as food items. The importance of these lines of investigation, together with the abundance of faunal remains on many sites, has meant that analyses of animal bone are now routinely undertaken as part of the post-survey/excavation process. Despite their centrality, however, zooarchaeological studies deriving from the Ottoman and post-Ottoman periods in Greece have remained undervalued, though archaeologists frequently encounter Modern materials.

This oversight is surprising when one considers the profound changes in the nature of human–animal relationships that took place in the Early Modern period. In domestic livestock husbandry, for example, major technological changes occurred, which included: the development of new breeds of livestock, the introduction of new forms of agricultural machinery and the expanded use of artificial feed and hay, which decoupled the relationship between seasonality and natural biological cycles. Social change also profoundly affected agricultural practice. The urban population boom facilitated the industrialisation of meat and dairy production and the emergence of transcontinental food-trade networks, while the drive to increase output resulted in major changes to the conformation and appearance of domestic livestock through breeding programmes. Profound changes in attitudes to animals also occurred in the Modern period in Greece, with the adoption of pets, formalisation of veterinary care and appearance of organisations dedicated to protecting animals.

The goals in this chapter are: a) to describe and analyse the vertebrate remains collected during the survey at the Pelion caves, and b) briefly to address questions about ecology and animal consumption

Fig. 2.3.1. Skull of red fox in situ from the inner chamber of Tsounaga (MOU-1).

and production in and around caves in the recent past.

Methodology

The survey produced a small but diverse assemblage of vertebrate remains (Fig. 2.3.1). The finds comprised 236 animal bones, hand-collected from 23 caves (Table 2.3.1). More than half of the assemblage was recovered from MOU-16, KAR-8 and KAR-29. The bones were all scattered on the surface of the cave floors in a fragmentary condition. They were collected by hand and no excavation or sieving of the material was involved. Most bones were plotted on plan drawings prior to their collection.

The faunal remains were transferred to the Wiener Laboratory at ASCSA where they were recorded and analysed. All bone fragments were numbered separately and due to their good state of preservation, they could be easily identified as regards species. Apart from species identification, other information such as the side of the body, presence/absence of proximal or distal parts of the skeleton, state of epiphyseal fusion and dental development, fragmentation, sex, evidence for burning and butchery marks and metrical data were recorded, where appropriate.

Caves	NISP
MOU-16	72
KAR-8	33
KAR-29	28
MOU-1	18
AFE-7	15
KAR-23	13
MIL-3	12
KER-3	7
MOU-24	5
KAR-6	4
MOU-28	4
MAK-23	4
MOU-7	3
ART-9	3
KER-1	3
KAR-16	3
VOL-4	3
VOL-3	1
VOL-1	1
ART-1	2
AGR-3	1
MOU-12	1
MAK-7	1
Total	**237**

Table 2.3.1. Caves from Pelion revealing animal bones.

Identifications of remains were based on reference specimens[24] and relative atlases.[25] The distinction between sheep and goat was based on Boessneck for postcranial bones, on Payne for deciduous teeth and on Halstead *et al.* for permanent teeth.[26] Age at death of domestic animals was estimated from the state of eruption and mandibular tooth wear (sheep and goat: Payne,[27] cattle: Grigson and Halstead,[28] pig: Bull & Payne[29]), and from the epiphyseal fu-

sion of postcranial bones.[30] Butchery marks were identified according to Binford,[31] while bone measurements were taken following von den Driesch.[32] Finally, quantification in terms of Minimum Number of Individuals (MNI) was avoided, because the bone sample from each cave was too small. Animal species had to be treated as deriving from one individual area and so only the Number of Identified Specimens (NISP) was used.

Species composition and age profiles

Faunal material involves both postcranial bones and teeth (mandibular and maxillar). Three main categories were identified (Table 2.3.2): "Domestic", "Wild" and "Intrusive" animals (Fig. 2.3.2). Intrusive are the so-called burrowing animals that tend to enter caves, disturbing deposits and sometimes adding their own and other intrusive bones.[33]

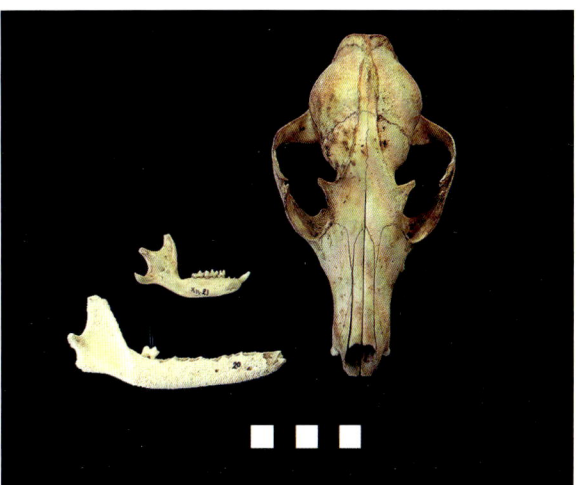

Fig. 2.3.2. Red fox skull (MOU-1) and mandible and hedgehog mandible (KAR-8), as examples of wild and intrusive animals.

---

24   These are housed at the Wiener Laboratory, The American School of Classical Studies at Athens.
25   Schmid 1972.
26   Boessneck 1969; Payne 1985; Halstead *et al.* 2002.
27   Payne 1973.
28   Grigson 1982; Halstead 1985.
29   Bull & Payne 1982.

30   Silver 1969.
31   Binford 1981.
32   Driesch 1976.
33   Payne 1985.

	Species	NISP	%
domestic	*Equus asinus* (donkey)	1	0.4
	Bos taurus (cattle)	11	4.6
	Sus scrofa (pig)	7	3
	Ovis aries (sheep)	26	11
	Capra hircus (goat)	32	13.5
	Ovis/Capra (sheep/goat)	108	45.6
	Gallus gallus (chicken)	11	4.6
wild	*Cervus elaphus* (red deer)	2	0.8
	Vulpes vulpes (fox)	14	6.0
	Sus scrofa (wild boar)	1	0.4
intrusive	*Lepus europeus* (hare)	8	3.4
	Erinaceous europeus (hedgehog)	1	0.4
	Martes sp. (marten)	9	3.8
	Testudo sp. (turtle)	5	2.1
	Anura sp. (frog unspecified)	1	0.4
	Total	**237**	**100.0**

Table 2.3.2. Species of domestic, wild and intrusive animals found during the survey.

At least 14 different species of amphibians, birds and mammals were identified. The majority of the faunal assemblage was comprised of domestic taxa (196 specimens), followed by intrusive (24 specimens) and wild species (17 specimens). The most common taxonomic category is sheep/goat (69.9% of the total assemblage). This category includes the majority of remains that could not be identified with certainty as either sheep or goat (N=107, 45.3%), as well as those remains identified as sheep (N=26, 11%) and goat (N=32, 13.6%). Goat specimens tend to be more numerous than sheep because of the mountainous character of the area. Sheep mainly graze, while goats like climbing on rocky places and even on trees to chew small branches and leaves easily found on

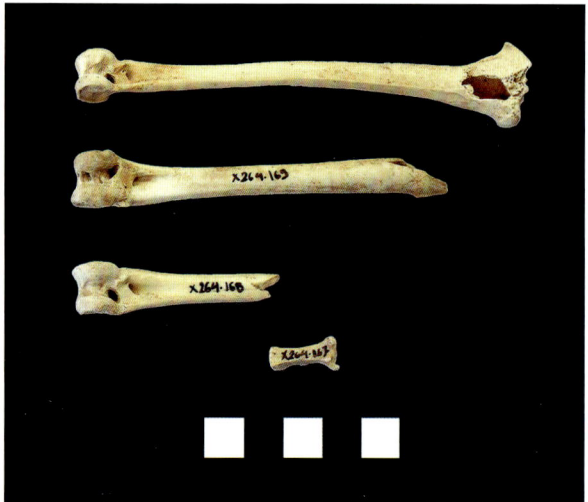

Fig. 2.3.3. Chicken bones collected during the survey.

Pelion. Cattle and pigs are also present in the faunal assemblage, and one specimen was identified as a donkey (Table 2.3.2).

The next most common species after sheep and goat are red fox (5.9%) and chicken (4.7%; Fig. 2.3.3).[34]

Because the faunal sample was quite small and most specimens did not provide ageing criteria, only domestic animals were further analysed for identification of the age at death. According to the data for epiphyseal fusion for postcranial bones (Table 2.3.3) and the states of eruption and wear of mandibular teeth (Table 2.3.4), many age categories are present: from newborn to fully adult animals.

Fourteen specimens of the total assemblage were identified as newborn (2 derived from pig and 12 from sheep/goat) indicating probably infant mortality, either because of a disease or delivery problems (especially pigs, which when newborn are smaller and weaker). Most newborn sheep/goat bones were found at MOU-16 and probably belonged to the same animal.

34   I sincerely thank Dr Katerina Trantalidou for identifying the chicken bones.

Species				Fusion			Total
				Fused	Unfused	Indet.	
cattle	FUS group	stage 1		0	1	0	1
		stage 2		1	0	0	1
		stage 3		0	0	0	0
		stage 4		1	3	0	4
	Total			**2**	**4**	**0**	**6**
	% within FUS group			33.3	66.6	0.0	100.0
pig	FUS group	stage 1		0	1	0	1
		stage 2		0	0	0	0
		stage 3		0	1	0	1
	Total			**0**	**1**	**0**	**1**
	% within FUS group			0.0	100.0	0.0	100.0
sh/gt	FUS group	stage 1		2	5	1	8
		stage 2		0	2	0	2
		stage 3		0	11	0	11
		stage 4		1	21	0	22
	Total			**3**	**39**	**1**	**43**
	% within FUS group			7.0	90.7	2.3	100.0
sheep	FUS group	stage 1		1	1	0	2
		stage 2		3	0	0	3
		stage 3		3	0	0	3
		stage 4		5	2	0	7
	Total			**12**	**3**	**0**	**15**
	% within FUS group			80.0	20.0	0.0	100.0
goat	FUS group	stage 1		5	1	0	6
		stage 2		3	0	0	3
		stage 3		1	0	0	1
		stage 4		7	3	0	10
	Total			**16**	**4**	**0**	**20**
	% within FUS group			80.0	20.0	0.0	100.0

Table 2.3.3. Epiphyseal fusion for mortality in cattle, pig, sheep/goat, sheep and goat (Silver, 1969).

Teeth age	Sheep	%	Goat	%
6 months	2	40.0	0	0
1-2 years	0	0	2	33.3
3 years	0	0	1	16.7
3-4 years	0	0	1	16.7
4-6 years	2	40.0	2	33.3
6-8 years	1	20.0	0	0
Total	**5**	**100.0**	**6**	**100.0**

Table 2.3.4. Age profiles in sheep and goats based on mandibular tooth wear stages.

Ages at the death of cattle and pigs according to epiphyseal fusion (Table 2.3.3) indicate that mainly young animals (1-3 years) were slaughtered for their meat. In modern societies, animals such as cattle are no longer useful for traction and are slaughtered before reaching adulthood for meat and dairying management.

The case of the sheep/goats provides a clearer picture of ageing because mandibular eruption and wear are added to the data for epiphyseal fusion (Table 2.3.3 & 2.3.4). According to the epiphyseal fusion data, sheep and goats are mainly slaughtered between 2.5-3 years, when they reach the optimum point in weight gain. Nevertheless, younger ages, such as 6-10 months, are also present. This is also obvious from teeth ages (Table 2.3.4). Both young mandibles (6 months) and adult mandibles (4-6, 6-8 years) are present.

	Species	Fragmentation				
		5/5	4/5	3/5	2/5	1/5
domestic	*Equus asinus*					1
	Bos taurus	1	1	2	2	5
	Sus scrofa	3		1		3
	Ovis aries	11	6	2	2	5
	Capra hircus	14	9	3	5	1
	Ovis/Capra	18	19	14	17	40
	Gallus gallus	2	2	4	2	1
wild	*Cervus elaphus*	1				1
	Vulpes vulpes					1
	Sus scrofa	13	1			
intrusive	*Lepus europeus*	2		3		3
	Erinaceous europeus	1				
	Martes sp.	1	3		2	3
	Testudo sp.					5
	Anura sp.			1		
	Total	**67**	**41**	**30**	**30**	**69**

Table 2.3.5. Bone fragmentation data for domestic, wild and intrusive animals.

Epiphyseal fusion and mandibular wear data indicate for sheep/goats the presence of mixed strategies of animal exploitation.[35] While most of the animals are slaughtered at the age of 1-3 years for their meat, some, especially the lambs/kids surplus to breeding requirements, are killed quite young to eliminate competitors for milk production, and some are kept until adulthood for their wool.

Bone preservation and taphonomy

Most specimens recovered in Pelion caves were in quite good preservation. From Table 2.3.5 it is obvious that especially smaller animals such as sheep, goats and foxes tend to be fragmented less and in some cases found complete. On the other hand, bigger animals such as donkeys, cattle and red deer, because of their bone size, break more easily into smaller splinters and as a result, they were found incomplete.

The recovery technique used during the survey was a source of bias that negatively affected recovery of bones from smaller animals and smaller bones of larger animals.[36] As is apparent from Tables 2.3.5-2.3.12, smaller bones, such as astragalus, calcaneum, phalanges (especially second and third) ulna and caudal vertebrae are underrepresented in the bone sample. Furthermore, all teeth were found attached to mandibular or maxillar bones and no loose teeth were recovered. A reliable retrieval of these smaller animal remains would require the application of specific collection methods, such as water flotation or water sieving.

Despite the fact that some bones suffered from recovery biases, none of the rest bore gnawing traces from smaller carnivores or rodents and only seven bones bore traces of alteration and erosion. This

Element	Side			Fragmentation					Modifications		NISP
	L	R	Unid.	5/5	4/5	3/5	2/5	1/5	Cut/chop marks	Alterations	
skull			1					1			1
upper teeth	1			1							1
vertebra, lumbar			1			1					1
ribs			1		1						1
humerus, shaft			1				1	1			1
radius, distal			1			1					1
ulna, proximal			1				1		1		1
pelvis		1					1				1
femur, proximal	1	1			1		1				2
phalanx 1			1	1							1
Total	**2**	**2**	**7**	**1**	**1**	**2**	**2**	**5**	**1**	**1**	**11**

Table 2.3.6. Number of Identified specimens (NISP) of cattle anatomical parts.

35   Payne 1973.

36   Payne 1972.

Fig 2.3.4

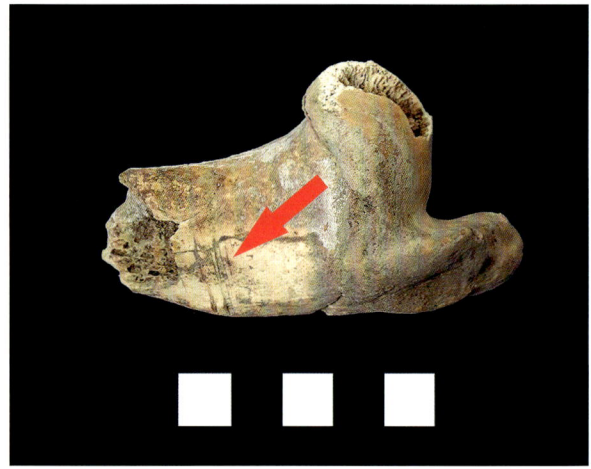

Fig. 2.3.7. Filleting butchery marks on sheep/goat bone from KAR-8.

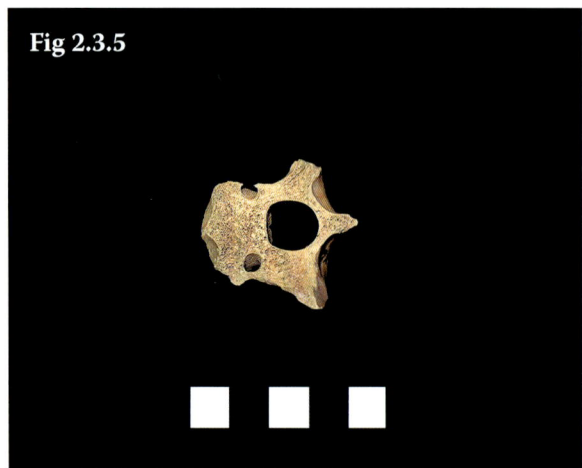

Fig 2.3.5

Element	Side		Fragmentation			NISP
	L	Unid.	5/5	3/5	1/5	
skull		1			1	2
ribs		4	1	1	2	4
radius, proximal/ distal	1		1			1
carpal/ tarsal		1	1			1
Total	**1**	**6**	**3**	**1**	**3**	**7**

Table 2.3.7. Number of Identified specimens (NISP) of pig anatomical parts.

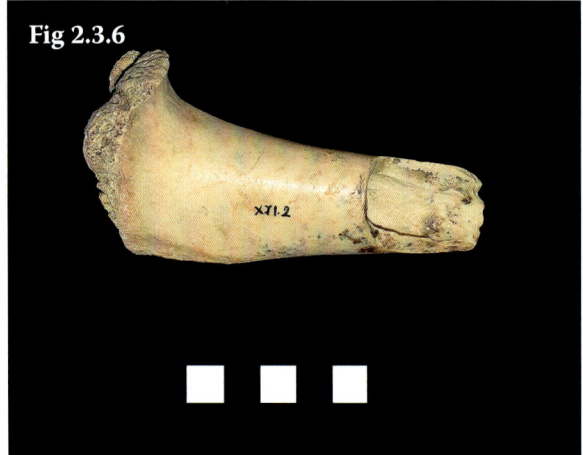

Fig 2.3.6

Figs 2.3.4.-2.3.6. Chopping butchery marks for marrow extraction. Bone samples from MOU-12 and MIL-3.

clearly shows that the faunal assemblage from the Pelion caves is quite modern. Otherwise, most of the superficial bones would have been found eroded because of the humid environment of the caves, and they would have been gnawed and trampled by humans and other animals. Carnivores, and especially dogs, often create bone cylinders by chewing animal bones in order to reach the marrow. The fact that none of the bones were gnawed by dogs indicates that dogs were generally kept outside the caves.

Element	Side			Fragmentation					Modification	NISP
	L	R	Unid.	5/5	4/5	3/5	2/5	1/5	Cut/chop marks	
lower teeth	2	3	1	6						6
atlas			1	1						1
axis			2		1	1				2
humerus, proximal		1						1		1
humerus, distal	1							1		1
radius, distal		2				1		1		2
ulna, proximal		1			1					1
metacarpus, proximal		2			2					2
metacarpus, distal			1				1			1
pelvis	1							1	1	1
femur, proximal		1						1		1
femur, distal		1					1			1
calcaneus, proximal		1			1					1
metatarsus, proximal	1				1					1
phalanx 1			1	1						1
phalanx 2			2	2						2
phalanx 3			1	1						1
Total	**5**	**12**	**9**	**11**	**6**	**2**	**2**	**5**	**1**	**26**
%	19.2	46.2	34.6	42.3	23.1	7.7	7.7	19.2	3.8	100

Table 2.3.8. Number of Identified Specimens (NISP) of sheep anatomical parts.

Human users of the Pelion caves modified animal bones by inflicting butchery marks.[37] Three bones were chopped (Figs 2.3.4-2.3.6), probably for marrow extraction, such as the cattle and red deer humeri. Two others bore traces of dismemberment (for the separation of different anatomical parts), while three, including a sheep humerus (Fig. 2.3.7), were identified with filleting marks, made in order to separate meat from the rest of the bone.

---

37   Binford 1981.

Animals and animal use at the Pelion caves

Palaeoecology

The vertebrate remains provide relatively little palaeoecological information. A species like the pine marten prefers relatively closed-canopy forests. All remaining taxa would do quite well in forests with the odd meadow and some additionally in the drier, open land on the northwest foothills of the mountain around Lake Karla (tortoise and fox). The faunal

Element	Side			Fragmentation					Modification		NISP
	L	R	Unid.	5/5	4/5	3/5	2/5	1/5	Root-etching	Alteration	
horn core	1					1					1
lower teeth	2	4	1	7							7
axis			2		2						2
ribs			1					1			1
scapula, proximal	3			1	2						3
humerus, proximal	2	1			2	1					3
humerus, distal		2					2				2
radius, proximal		1			1						1
radius, distal		1					1				1
ulna, proximal	1	1		1	1						2
femur, proximal		1					1				1
femur, distal		1			1						1
calcaneus, proximal	1	1		1	1						2
metatarsus, proximal	1						1		1	1	1
metatarsus, distal		1		1							1
phalanx 1			2	2							2
phalanx 2			1	1							1
Total	**11**	**14**	**7**	**14**	**9**	**3**	**5**	**1**	**1**	**1**	**32**
%	34.4	43.7	21.9	43.7	28.1	9.4	15.7	3.1	3.1	3.1	100

Table 2.3.9. Number of Identified Specimens (NISP) of goat anatomical parts.

remains do not indicate any variation in the presence or abundance of these types of habitat. Turning to the microfauna, the single frog from KAR-29 was probably present within the cave itself.

The domestic fauna reveal what animals were exploited near the caves. All of the caves where goat/sheep remains were collected are in areas where goats are (or recently were) herded. Exceptions to this are perhaps MOU-28 and AFE-7.

Food consumption and production

Many age categories, from newborn to adult animals, were present in the bone sample, indicating mixed strategies of animal exploitation for milk, meat and probably wool. Intentional culling of animals at a young age, shortly after birth, makes sense if herders are interested in eliminating competitors for milk produced by the goats and the sheep. Alternatively, neonatal bones, especially those of pig, could indicate infant mortality either due to weakness from birth or a disease.

Element	Side			Fragmentation					Modification			NISP
	L	R	Unid.	5/5	4/5	3/5	2/5	1/5	Fire 300-350^{0} C	Cut/chop marks	Alteration	
skull			3					3				3
maxilla			1					1				1
upper teeth	3	1	1	4			1					5
atlas			1					1				1
vertebra			3					3				3
vertebra, cervical			6		3	2		1		1		6
vertebra, thoracic			7			4	2	1				7
vertebra, lumbar			2					2				2
sacrum			1			1						1
ribs			37	3	11	5	10	8		2	1	37
humerus, proximal	1			1								1
humerus, shaft			6					6				6
humerus, distal	1						1					1
radius, proximal	2			1	1							2
radius, shaft	1							1				1
radius, distal			1					1				1
ulna, proximal			1					1				1
metarcarpus, proximal			4	3		1						4
pelvis	2	2	1				1	4				5
femur, proximal	2			2								2
femur, shaft			1					1				1
tibia, proximal	2	2				1	1	2				4
tibia, shaft		1	2			1		2	1	1		3
tibia, distal	3	3	1	4	3							7
metatarsus, shaft			1					1				1
phalanx 1			1					1				1
phalanx 2			1		1							1
Total	**17**	**9**	**82**	**18**	**19**	**14**	**17**	**40**	**1**	**4**	**1**	**108**
%	15.8	8.3	75.9	16.7	17.6	12.9	15.8	37	0.9	3.7	0.9	100

Table 2.3.10. Number of Identified Specimens (NISP) of sheep/goat anatomical parts.

Element	Side		Fragmentation		Cut/chop marks	NISP
	L	Unid.	5/5	1/5		
humerus, proximal	1			1	1	1
phalanx 2		1	1			1
Total	**1**	**1**	**1**	**1**	**1**	**2**
%	50.0	50.0	50.0	50.0	50.0	100

Table 2.3.11. Number of Identified Specimens (NISP) of red deer anatomical parts.

Only a few traces of human modification (butchery marks) were identified, indicating animal slaughtering and consumption of bone marrow by the cave users. This is further supported by the absence of bone pathologies. Studies have showed that animal butchering occurred in areas distant from the animal fold.[38] Therefore, as is apparent in the faunal assemblage, specific anatomical parts, such as horn cores, parts of the skull and upper dentition and lower parts of the front and hind limbs (phalanges, astragals, metapodials) were absent,[39] and there was little or no evidence of butchered remains in the immediate vicinity of the caves.

Both age profiles and patterns of human modification on bones give us the opportunity to propose different ways that bones of domesticated animals were deposited in the Pelion caves (Table 2.3.13). Bones from newborn and senile animals may indicate that shepherds used caves as temporary pens, especially during summer. Weaker or diseased animals could have died naturally, with the shepherds leaving the carcasses behind to decompose. In such cases, bones can be in a good state of preservation, with possible traces of gnawing from rodents and smaller carnivores. Older and diseased animals might also have entered caves not used by shepherds to die there.

Element	Side			Fragmentation		Modification	NISP
	L	R	Unid.	5/5	4/5	Alteration	
upper teeth		2		2			2
lower teeth	1			1			1
atlas			1	1			1
vertebra, cervical			1	1			1
vertebra, lumbar			3	3			3
humerus, proximal	1	1		2			2
radius, proximal	1	1		2		1	2
ulna, proximal	1	1		1	1	1	2
Total	**4**	**5**	**5**	**13**	**1**	**2**	**14**
%	28.6	35.7	35.7	92.8	7.2	14.3	100

Table 2.3.12. Number of Identified Specimens (NISP) of red fox anatomical parts.

---

38   Chang 2000.
39   Brain 1981.

	Rodent or carnivore gnawing	Burning	Cut marks	Age profiles	Identified skeleton part
Newborn animals, because of delivery problems or disease, died in the cave, which is used as a pen.	possible	not possible	not possible	newborn	whole carcass
The animal (possibly sick or old) entered the cave by itself and died there.	possible	not possible	not possible	senile	whole carcass
The animal was killed in the cave by humans, then butchered and filleted for immediate consumption.	possible	possible	possible	young	bone elements with more meat
Humans brought selected parts of meat (with bones) to the cave for consumption; the bones were subsequently discarded.	possible	possible	possible	young	bone elements with more meat

Table 2.3.13. Conceivable interpretations for the presence of bones of domesticated animals in caves.

In cases where caves were used in the recent past as a temporary shelter for humans who slaughtered and consumed animals for food, finds are more likely to include bones from young animals (1-2 years). The bones are more fragmented and usually suffer from burning and butchery marks, as they are food waste. In this case, specific parts of the skeleton are expected to be found. Unwanted anatomical parts, such as horn cores, upper dentition and lower parts of the front and hind limbs, would either be discarded outside the caves or, if the animal was slaughtered somewhere else, would never be brought to the cave.

The finds of chicken bones from six caves probably represent either butchered individuals for human consumption or refuse from predatory animals that have ventured near human habitation to eat waste from meals or to raid chicken coops. The fieldwork on Pelion recorded several ethnohistorical narratives of wolf and fox in connection to caves and shepherd stories.

The relatively low quantity of bone remains and the low minimum number of individuals can perhaps be explained by the fact that carcasses are rarely left in regularly used caves. At the same time, it shows that the domestic use of caves was occasional or limited to a few shepherds, who did not make use of their herd as a meat resource.

Hunting

There is limited evidence of large game hunting, while the taphonomic status of the small game (red fox, marten and rabbit) is uncertain. This is perhaps not surprising, as one would expect that hunting always was a relatively unimportant subsistence activity relative to herding. The presence of fired birdshot shells at some caves (e.g. ART-1, MAK-7, AGR-8) indicate their use as occasional hunting-stands for game bird shooting.

The only possible indication of hunting within the bone assemblage is a red deer humerus from MIL-3 identified with chopping marks, probably for marrow extraction. In this case, either hunting was not performed by Pelion cave users on a regular basis or wild animals were hunted and slaughtered far from the immediate vicinity of the caves. However, only dismemberment and filleting marks can indicate animal slaughter. Chopping marks are only inflicted on bone elements when the meat has already been cooked.

2.4 Marine invertebrates from the Pelion caves

Niels H. Andreasen

Greek summary

Πέντε σπηλιές του Πηλίου συνέβαλαν στην συγκέντρωση των 32, ως επί το πλείστον πλήρων, δειγμάτων τριών ειδών θαλάσσιων ασπόνδυλων (Patella spp. (Πεταλίδες), Monodontinae (Top Shells), Pinna spp. (Fan mussels), ένα καβούρι, και ένα sp Helix (Βρώσιμο σαλιγκάρι της γης). Όλα μπορεί να ερμηνευθούν σαν υπολείμματα τροφίμων, ίσως με εξαίρεση τα Fan mussel. Το μοναδικό βρώσιμο σαλιγκάρι που βρέθηκε είναι πιθανώς μία νεότερη εισβολή.

Τόσο οι Πεταλίδες όσο και τα Top Shells αναμφίβολα συγκεντρώθηκαν με το χέρι (ίσως με ένα μαχαίρι) από τις βραχώδεις ακτές του Ανατολικού Πηλίου – ίσως κατά τη διάρκεια του καλοκαιριού, όταν τα οι τροχοί (Top Shells) μεταναστεύουν στο ανώτερο τμήμα της ζώνης παφλασμού. Αν και τα μαλάκια μπορούν να καταναλωθούν ωμά, η μεταφορά τους πίσω στις σπηλιές δηλώνει πρόθεση να μαγειρευτούν ή να διατηρηθούν. Η υπόθεση ότι οι χρήστες σπηλαίων ταξίδευαν, ίσως τακτικά, μεταξύ της ακτής και μεγαλύτερων υψομέτρων πάνω στο βουνό, καταδεικνύει μια τοπική κινητικότητα που ξεπερνά τις παραδοσιακές ιδέες των μετακινήσεων χωριού/υψιπέδων.

Τα όστρακα από την ART-9 μπορούν να συνδεθούν με έναν μεσαιωνικό (τέλη του Μεσαίωνα/αρχές μετά τον Μεσαίωνα) οικισμό κοντά στη βραχοσκεπή που βρέθηκαν τα κελύφη. Τα άλλα όστρακα συλλέχθηκαν σε σπήλαια που περιέχουν τόσο ελληνιστικά-ρωμαϊκά αντικείμενα, καθώς και αποδεικτικά στοιχεία για την χρήση τους στη μετά-Μεσαιωνική περίοδο (19ος-20ος αιώνας). Η πιθανότητα ότι τα οστρακοειδή που βρέθηκαν μέσα στις σπηλιές ανήκουν σε μία ιστορικά πρώιμη χρονική περίοδο, την ελληνιστική-ρωμαϊκή, αυξάνει, εάν υποθέσουμε ότι το έδαφος έχει διαταραχθεί. Αν συμβαίνει αυτό, τότε κανένα από τα υπολείμματα των οστρακοειδών από την έρευνα, δεν μπορεί να χρησιμοποιηθεί για να αποδειχτεί η χρήσης τους ως συμπληρωματική πηγή τροφής σε σπήλαια, στη σύγχρονη εποχή. Σε κάθε περίπτωση, συγκεντρώσεις μαλακίων όπως η τρέχουσα, είναι πιο κατατοπιστικές σχετικά με τις διαδικασίες σχηματισμού του χώρου και τις ανθρώπινες δραστηριότητες σε σπήλαια, παρά σχετικά με τοπικά ή περιφερειακά παλαιο-περιβάλλοντα.

❋ ❋ ❋

Introduction

Five Pelion caves (MOU-1, MOU-7, MOU-16, KAR-23, and ART-9) contributed a small archaeomalacological assemblage of 32 mostly complete specimens (Table 2.4.1). They divide into two main categories: 1) marine molluscs and crustaceans carried by humans to the caves for consumption, and 2) edible terrestrial molluscs, which either are natural intrusions or were collected by humans.

Due to the character of the survey, none of the shells are from closed contexts, and their tentative dating is only suggested by the associated ceramics. The specimens from ART-9 can be associated with a late Medieval / early post-Medieval settlement close to the rockshelter where the shells were recovered. In this case, the shells could have been secondarily deposited within the shallow shelter along with

English name	Latin name	Cave	NISP
Crab (front chelae)	Brachyura (suborder)	MOU-1	1
Limpet	*Patella* spp.	MOU-1, MOU-16, KAR-23, MOU-7, ART-9	25
Top Shell	*Monodonta turbinata* Born	MOU-1, MOU-7	5
Fan mussel	*Pinna* spp.	ART-9	1
Edible land snail	*Helix* sp.	MOU-16	1

Table 2.4.1. Taxonomic representation of invertebrates (Crustacea and Mollusca) found during the survey (NISP=number of identified specimens).

numerous ceramic sherds and tile fragments. The other shells were collected in caves that contain both Hellenistic–Roman artefacts as well as evidence of use in the post-Medieval period (nineteenth and twentieth centuries).

Ecology and assemblage composition

Patella spp. (Limpets)

Limpets prefer rocky, coastal environments, where they cluster around the splash zone to water of a few metres in depth.[40] They avoid shallow bays and estuaries where discharging rivers cause fluctuations in salinity.[41] Sub-fossil limpet shells are difficult to identify to species based on their shell alone, but it is possible that at least two species of limpets are present in the Pelion assemblage, of which *P. caerulea* (Mediterranean Limpet) makes up the majority (Fig. 2.4.1). Other species represented are probably either *P. rustica* L. (Rustic Limpet) or *P. ulyssiponensis* (Rough/China Limpet). All are adult specimens (except a premature limpet from ART-9) with lengths of between 22-50 mm (adult limpets usually range from 20-66 mm in length).

Limpets are common on archaeological settlement sites and have traditionally been exploited as a snack or as bait for fishermen and other coastally based communities.[42] In the same manner, limpets could have served as an occasional snack during stays in the caves, where they could have been eaten raw or smoked, dried, boiled, roasted or steamed.

Monodontinae (Top Shells)

Like the limpets, the little top-shell of the Mediterranean, *Monodonta turbinata*, is a common species in the rocky intertidal and sublittoral zone and they sometimes share the same habitat (Fig. 2.4.2). *Turbinata* and the related species *Monodonta articulate* can nowadays be found in large numbers along the coast of the Pagasitikos gulf. Top shells are regularly consumed in Mediterranean countries after being boiled,[43] and can be found on today's shellfish markets.

Pinna spp. (Fan mussels)

The Fan mussel, a large marine bivalve mollusc of the Pinnidae family, is endemic to the Mediterranean Sea. It lives offshore among sea grass meadows on loose substrates down to a depth of 60 m.[44] It can grow to 60 cm or even larger and is relatively

---

40   Delamotte & Vardala-Theodorou 1994, 204.
41   Hayward *et al.* 1995, 502.

42   Shackleton 1968, 129; Forbes 1976, 134-5.
43   Davidson 1981, 191.
44   Katsanevakis 2007, 1320.

Fig. 2.4.1. & 2.4.2. Limpets (*Patella caerulea*?) from KAR-23

Fig. 2.4.2. Top Shells (*Monodontinae*) from MOU-7.

fragile. The flesh of the Fan mussel is edible and, due to its peppery flavour, it is even reputed to be an aphrodisiac.[45] Otherwise, *Pinna nobilis* is best known as the origin of sea silk, made from the byssus of the animal.

Helix sp. (Edible land snail)

Helix species have a European-wide range.[46] They thrive in cool, moist and shady environments and are known to take refuge in caves during their dor-

mant periods. There is a long history of land snail consumption in the Mediterranean, but few comprehensive data are available on their use as food and their nutritional value.[47]

Taphonomy

Taphonomic processes that are active on individual shells in terrestrial contexts are dissolution, chemical conversion, heating and trampling.[48] Acid environments can dissolve shells and in contrast to large shell aggregations, thin scatters or individual shells are susceptible to leaching of their $CaCO_3$. In addition, burned shell fractures more easily than unburned shell.[49] If the caves were used as animal pens, then the deposits of dung and urine would have decreased the pH level of the soil and increased degradation of the shells. The effects of animal trampling would have further decimated the assemblages.

Some degree of dissolution is apparent on most, if not all, shells in the assemblage. Periostracum (the outer organic layer) is absent from the limpets from

---

45   www.sardolog.com/bisso/english/quoi.htm
46   Kerney & Cameron 1979.

47   But see Lubell 2004.
48   Claassen 1998, 54.
49   Claassen 1998, 61.

ART-9, MOU-16 and KAR-23, but the growth lines are still well defined. There are slight variations in the stage(s) of deterioration between limpets found at ART-9, and the poorest preserved shells have taken on a slightly "chalky" appearance and have damaged edges. The snail shells of *Monodontinae* are quite durable, and animal trampling or intentional crushing of the spire to facilitate extraction of snail meat are the most likely causes for fragmentation of two top shells from MOU-7.[50] Although the surface of the *Helix* sp. shows modest degradation and light flaking, it appears to be of a more recent date than the marine molluscs. Brown banding on the body whorl is still visible.

Exploitation

Invertebrate remains are common on both Prehistoric and Historic sites, and may either be food remains and/or have served various secondary uses. All invertebrate remains found during the survey can be interpreted as food waste, perhaps with the exception of the fragmented Fan mussel, which is not a frequent food species. Although *Helix* species were widely exploited from Prehistory into the Historical period, the single edible land snail could well be intrusive, especially since no other *Helix* sp. were found nearby.

Limpets and top shells were undoubtedly gathered from the rocky shores of East Pelion where the only influx of freshwater is from seasonal torrents, allowing marine salinity to be maintained. Shellfish gathering does not require special techniques or skill and the limpets from the Pelion caves were gath-

ered by hand or by using a lever, possibly a knife, to dislodge them. This must be done with a quick movement since the limpet will secure a firm grip once it has been touched. Both limpets and top shells can be plucked from rocks during the summer when the top shells migrate to the upper part of the splash zone.[51]

The shellfish and crab are easily accessible coastal resources and may have been recovered in small amounts on a regular basis and transported back to the caves by their users. Although molluscs can be eaten raw, their transfer back to the caves indicates an intention to cook or preserve them (as opposed to eating them raw on the spot). The shells may not represent immediate consumption by the collectors or consumption at all. Molluscs could have been dried for delayed consumption or the flesh could have been used for bird bait.[52]

With the possible exception of ART-9, it is probable that individual herders/cave users extracted small amounts of coastal invertebrates directly rather than obtaining them from intermediaries. The suggestion that cave users travelled, perhaps regularly, between the coast and higher altitudes on the mountain demonstrates a local mobility that transcends traditional ideas of village/upland movements. The four caves are situated at relatively low altitudes (<350 m) but at different distances from the sea. The closest is MOU-16 and the farthest is KAR-23. Several questions arise from these observations. At what occasions did cave users collect shellfish at the coast? Why were shellfish transported back to the caves (rather than consumed on the spot)?

We know little about shell accumulation at contexts such as those investigated by the Pelion Cave Project. The potential association of shellfish and Hellenistic–Roman occupation in all cases but one raises the possibility that the shellfish belong to dis-

---

50 Four shells from MOU-7 were found inside the cave in an area where the concentration of ceramics and animal bones were higher than in the rest of the cave. This points either to the presence of a more intensely occupied area of the cave or to a disturbance in this area, which has brought artefacts to the surface from underlying levels.

51 Çakırlar 2009, 62.
52 Claassen 1998, 187.

turbed Early Historical facies within the caves. If this is the case, then none of the shellfish remains from the survey can be used to demonstrate their use as a supplementary food source in caves in the Modern period.

In either case, mollusc assemblages like this one are more informative about site formation processes and human activities at caves than about local or regional palaeoenvironments. Small-scale shellfish exploitation in pre-industrial communities has been infrequently investigated despite 150 years of archaeological interest in shells. Ethnography undoubtedly offers the most direct avenue to understanding local use of molluscs as a food source.

2.5 Human skeletal remains from the Pelion caves[53]

Eleanna Prevedorou

Greek summary

Η συλλογή επιφανειακών ευρημάτων από τις σπηλιές του Πηλίου περιλαμβάνει συνολικά 24 ανθρώπινα οστά, προερχόμενα από τέσσερις σπηλιές. Η πλειοψηφία του ανθρωπολογικού υλικού προέρχεται από τη Σπηλιά του Αγίου Αθανασίου, ενώ ένα οστό βρέθηκε στην Τσούναγκα, δύο οστά βρέθηκαν στη Γκόριτσα II και τέσσερα οστά στη Νταμούχαρη IV. Τα ανθρωπολογικά κατάλοιπα βρέθηκαν εκτεθειμένα στην επιφάνεια σε αποσπασματική κατάσταση. Η παρουσία μικρών οστών, όπως μίας φάλαγγας, ενός μετακαρπίου και ενός μεταταρσίου, δηλώνει πρωτογενείς αποθέσεις. Ωστόσο, δεν μπορεί να προσδιοριστεί εάν αποτελούν κατάλοιπα ταφικών πρακτικών, απλών εναποθέσεων, ή ατόμων που πέθαναν μέσα στις σπηλιές. Στην περίπτωση της Σπηλιάς του Αγίου Αθανασίου, τα οστά ενδεχομένως να προέρχονται από ταφή της Μυκηναϊκής περιόδου που είχε συληθεί στο παρελθόν.

Στη Σπηλιά του Αγίου Αθανασίου βρέθηκαν 17 οστά, κυρίως μακρά οστά, σπόνδυλοι και πλευρά.

Ένα τμήμα αριστερού πλευρού φέρει τραύμα μερικώς επουλωμένο, το οποίο επήλθε μερικές εβδομάδες πριν από το θάνατο. Η κάτω γνάθος ανήκει σε ενήλικα πιθανώς άνδρα και παρουσιάζει προθανάτια απώλεια δοντιών, περιοδοντίτιδα και έντονη φθορά στις μασητικές επιφάνειες. Ενδιαφέρον παρουσιάζει η παρουσία μιας μεσοδόντιας αύλακας στον αυχένα της άπω επιφάνειας του δεξιού πρώτου γομφίου, αποτέλεσμα επανειλημμένης χρήσης εργαλείου για τον οδοντικό καθαρισμό ή της παρουσίας ημιέγκλειστου δεύτερου γομφίου σε πρόσθια κλίση. Στην Σπηλιά Τσούναγκα αναγνωρίστηκε μία μη συνοστεωμένη κεφαλή μηριαίου οστού νεαρού εφήβου. Στη Σπηλιά Νταμούχαρη περισυνελέγησαν μία αριστερή και μία δεξιά άνω γνάθος σε άρθρωση με τα αντίστοιχα ζυγωματικά. Παρά τις γενικές ομοιότητες σε μέγεθος και διατήρηση, η ασυμμετρία στη μορφολογία και δομή των δύο πλευρών, καθώς και η μη άρθρωση των δύο άνω γνάθων, δηλώνουν ότι οι δύο πλευρές προέρχονται από διαφορετικά άτομα. Τα σωζόμενα δόντια φέρουν έντονη τερηδόνα, ενώ η δεξιά άνω γνάθος παρουσιάζει σοβαρές αλλοιώσεις στο ιγμόρειο ως αποτέλεσμα χρόνιας φλεγμονής, η οποία πιθανόν να οφείλεται σε ακρορριζικό απόστημα στην περιοχή των προγομφίων.

❊ ❊ ❊

53 I would like to thank the Wiener Laboratory of the American School of Classical Studies at Athens, and especially former director Dr Sherry Fox for her constructive comments and for allowing access to the laboratory's equipment (stereoscope, X-ray facilities). Also, I would like to thank Dr Rozalia Christidou and Dr Sireen El Zaatari for their input on the interproximal groove.

Introduction

The finds from the Pelion Cave Project included twenty-four human bones, collected from four caves in total (Table 2.5.1). The majority of the human bones came from KAR-8 (Agios Athanasios; Fig. 2.5.1), while only a few bones were recovered from the caves of MOU-1 (Tsounaga; Fig. 2.5.2), VOL-3 ("Goritsa II"; Fig. 2.5.3), and MOU-16 ("Damouchari IV"; Fig. 2.5.4). The human skeletal remains were all found scattered on the surface of the cave floors and were in a fragmentary condition. The recovery of small bones such as a manual phalanx and a metacarpal in KAR-8, a metatarsal in VOL-3 and an unfused femoral epiphysis in MOU-1 indicate depositions of complete bodies. However, whether or not these represent intentional burials, disposal of bodies, or people who died within the cave cannot be determined given their disturbed context. In the case of KAR-8, the human remains might belong to a Mycenaean burial that was robbed in the recent past. In terms of dating, the ceramic sherds at KAR-8 represent a variety of periods spanning from the Late Bronze Age to Roman times; however, the proximity of the cave to a Mycenaean cemetery and the presence of the disturbed burial might suggest a Mycenaean date for the human remains. At MOU-1, the ceramic assemblage dates to Hellenistic and Roman times, while at MOU-16 the assemblage dates to the Late Hellenistic/Early Roman period. At VOL-3, the small number of recovered ceramics is less diagnostic, including both Late Bronze Age and Roman sherds. Nevertheless, given that none of the assemblages are derived from closed contexts, any dates for the human remains have to remain tentative, while disturbance and/or disposals in later periods cannot be excluded. An inventory of the recovered human skeletal remains is included (Table 2.5.1) and a description of the most important finds by cave follows.

KAR-8 (Agios Athanasios)

In KAR-8, seventeen human bones were recovered in total (Table 2.5.1; Fig. 2.5.1). The lower arm bones probably belong to the same individual. Even though

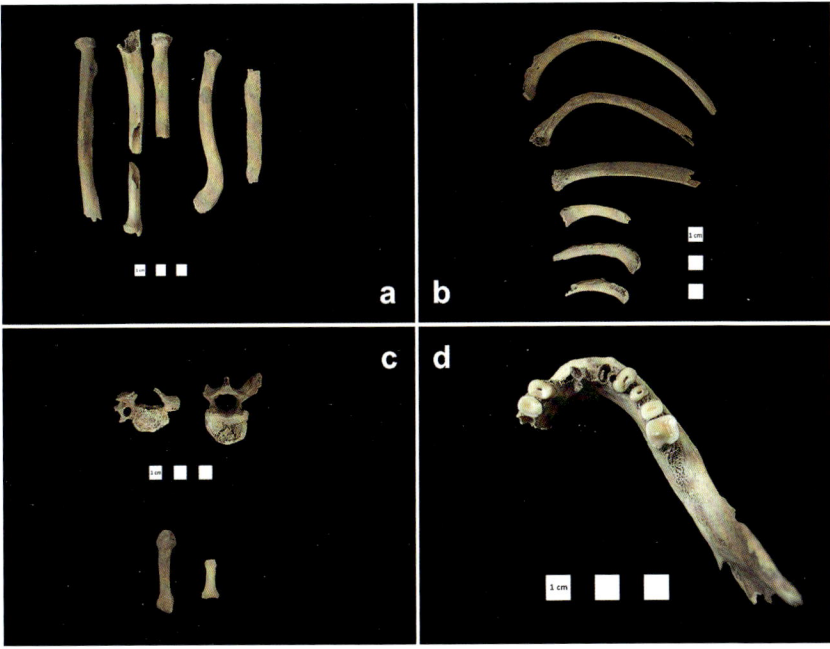

Fig. 2.5.1. Human remains from KAR-8: (a) right radius, left ulna, left radius, left clavicle and femoral fragment (from left to right); (b) ribs (superior aspect); (c) inferior aspect of cervical and thoracic vertebrae (upper row), dorsal aspect of left third metacarpal and manual proximal phalanx (lower row); (d) mandible (occlusal aspect).

ID #	Bone	Side	Comp/ness	Cave	Provenance
PCP1	3rd metacarpal	Left	1	KAR-8	X13
PCP2	clavicle	Left	1	KAR-8	X21
PCP3	femur (?)	Unidentified	3	KAR-8	X16
PCP4	rib (3-10)	Left	2	KAR-8	X5
PCP5	rib (3-9)	Left	1	KAR-8	X3
PCP6	rib (3-10)	Left	2	KAR-8	X2
PCP7	11th rib	Right	3	KAR-8	X26
PCP8	12th rib	Left	2	KAR-8	X26
PCP9	radius	Right	1	KAR-8	X8
PCP10	thoracic vertebra (2-10)		1	KAR-8	
PCP11	cervical vertebra (3-7)		2	KAR-8	
PCP12	ulna	Left	3	KAR-8	
PCP13	manual proximal phalanx (2-5)	Unidentified	1	KAR-8	
PCP14	rib (11-12)	Right	3	KAR-8	
PCP15	radius	Left	2	KAR-8	
PCP16	ulna	Left	2	KAR-8	
PCP17	mandible		2	KAR-8	X1
PCP18	femur	Right	3	MOU-1	
PCP19	5th metatarsal	Right	1	VOL-3	X97
PCP20	1st rib	Left	2	VOL-3	X97
PCP21	maxilla	Left	1	MOU-16	X251 AS.534
PCP22	zygomatic	Left	1	MOU-16	X251 AS.534
PCP23	maxilla	Right	1	MOU-16	X251 AS.534
PCP24	zygomatic	Right	1	MOU-16	X251 AS.534

Table 2.5.1. Inventory of the human remains recovered from the Pelion caves. Completeness is coded according to the following: 1 = >75% present; 2 = 25-75% present; 3 = <25% present.

the left proximal radius (PCP15) is slightly larger than the right proximal radius (PCP9), the difference is not considered enough to suggest a different individual, and it is probably the result of bilateral asymmetry (Fig. 2.5.1a). The measurements of the maximum radial head diameter of both radii suggest a male individual.[54] While no duplication of skeletal elements is observed for the assemblage, the fragmentary ribs PCP8 and PCP14 are smaller than the remaining ones and could suggest the presence of a second, smaller and/or younger individual (Fig. 2.5.1b). Nevertheless, it should be noted that floating ribs can be highly variable.

The two vertebrae preserved are relatively small and show signs of osteoarthritis, which is a common find in archaeological human remains (Fig. 2.5.1c).

54   Berrizbeitia 1989; Charisi *et al.* 2011; Mall *et al.* 2001.

Fig. 2.5.2. Unfused femoral head from MOU-1.

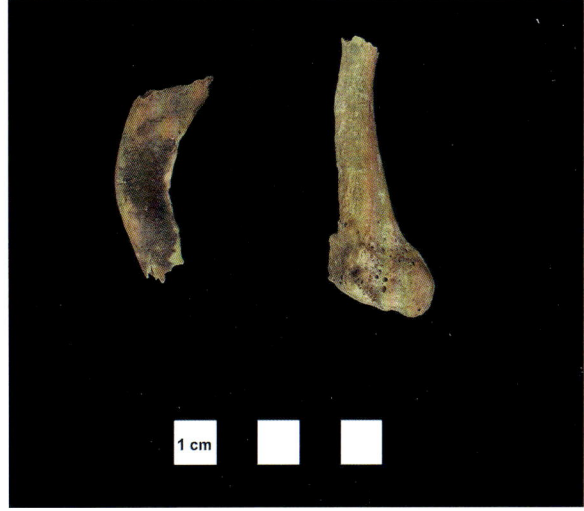

Fig. 2.5.3. Left first rib and right fifth metatarsal (from left to right) recovered from VOL-3.

The body of the cervical vertebra (PCP11) shows porosity and lipping on both anterior and inferior surfaces, though more pronounced on the inferior. In addition, a small facet-like porous lesion is observed on the inferior aspect, between the body and the left transverse foramen. However, in the absence of the adjacent vertebrae, it is not possible to determine its exact nature. The thoracic vertebra (PCP10) also shows slight porosity on both superior and inferior surfaces of the body, while marginal lipping is observed on the inferior surface.

Among the six fragmentary ribs from KAR-8, a left rib segment (PCP6) showed an antemortem fracture at the lateral shaft (Fig. 2.5.5). The preserved rib segment can be identified as a typical middle rib (fourth to ninth), belonging to the category of the ribs most commonly fractured.[55] The fracture was still in the process of healing at the time of death, as can be seen from the newly deposited woven bone and the recently formed callus that is observed on the visceral surface. The periostitis observed on the anterior surface is probably secondary to the injury. Considering the variation in the duration of fracture healing and its multifactorial nature (including se-

verity of fracture, type of bone, the age of the individual, etc.), the presence of the active bony callus suggests that the injury occurred at approximately 3 to 9 weeks before the time of this individual's death.[56] The location of the fracture near the costal angle suggests either anteroposterior compression,[57] or an impact from the front.[58] Unfortunately, the postmortem breakage on the site of the callus does not allow for a complete evaluation of the fracture type (e.g., transverse, oblique, greenstick). Rib fractures can result from direct or indirect blunt force chest traumas (e.g., blow or fall against a hard object) and biomechanical stress, and can be associated with accidents, interpersonal violence or underlying pathological conditions such as persistent coughing, osteoporosis and osteomalacia.[59] However, specifying the etiology for rib fractures can be rather challenging, especially when the adjacent ribs and other associated skeletal elements are missing,[60] which in

---

55   Brickley 2006; Lovell 1997.

56   Lovell 1997; Ortner 2003.

57   Galloway 1999.

58   Lovell 1997.

59   For an extended bibliography on the etiology and complications of rib fractures see Brickley 2006; Matos 2009; Sirmali *et al.* 2003.

60   Matos 2009.

the KAR-8 specimen is further complicated by the postmortem breakage. In addition, a second left rib segment (PCP4) shows signs of periosteal reaction with sclerotic margins at the area of the tubercle and the neck, while the tubercle appears to be displaced inferiorly. Due to the postmortem fracture at this location, it is not possible to determine whether or not this reaction is associated with the ligaments that attach to this area or with a traumatic event. In any case, the reaction is completely healed in contrast to the active lesion observed in the rib PCP6.

Furthermore, a mandible (PCP17) showing unusual oral pathologies was recovered in KAR-8 (Fig. 2.5.1d). The mandible was partially preserved, missing the superior portion of the right body and the majority of the left side. Based on morphological and developmental features, the mandible probably belongs to a male, adult individual.[61] Six teeth were present out of thirteen tooth sockets observable (left first premolar, left and right canines, right first and second premolars, right first molar). The right second and third molars and the left lateral incisor were lost antemortem. The alveolar remodelling and rough porosity present in those areas with signs of an active inflammation suggest that the teeth were lost relatively recently before the time of death.

The dentition shows severe wear exposing the pulp cavity in three out of the six preserved teeth (left and right canines, right first premolar; Fig. 2.5.1d). Marked alveolar resorption with root exposure and generalised inflammation is evident throughout the preserved mandible, indicative of periodontal disease, and is more pronounced at the area of the right molars. A periapical abscess associated with alveolar and root resorption, and reactive tissue are associated with the left canine, indicating active inflammation and that the individual was in the process of losing the tooth; a large periapical abscess is also present in the area of the missing central incisors

(but it is unclear whether or not the central incisors where lost antemortem). A deep opening (cloaca) to drain the pus is located on the mesiolabial margin of the right first premolar, associated with marked root exposure and a large lytic lesion at the apex of the root. This individual shows no signs of caries or calculus formation, suggesting that the antemortem tooth loss, the periapical abscesses, and the inflammation of the surrounding tissues are the result of the observed heavy wear. Pulp exposure due to wear can lead to infection of the pulp cavity and consequently of the supporting alveolar bone.[62] In addition, slight antemortem enamel chipping is observed on the mesiolingual and distolingual corners of the left first premolar. The severity of the tooth wear is indicative of a highly abrasive diet including rough-textured, fibrous and/or gritty foods.[63]

The distal surface of the right first molar shows a non-carious interproximal groove-like depression, situated at the dental neck and extending over the tooth root (Fig. 2.5.6). The modified area measures 6.50 mm in length (maximum buccolingual dimension), and 4.30 mm in height (maximum apicocervical dimension). The groove is oblique and asymmetrical, oriented buccolingually, with the lingual margin being longer, sharper and deeper than the buccal margin, the latter being shorter and smoother. The deepest region measures 1.40 mm and is located at the lingual margin of the cervico-enamel junction. In addition, there is an occlusal notch at the centre of the distal aspect of the crown, measuring 2.66 mm (Fig. 2.5.1d). The edges of the notch are smooth while faint vertical striae are evident when observed microscopically.

A known interproximal dental modification is interproximal grooving, a phenomenon with a long history in dental anthropology from both bio-archaeological and paleoanthropological contexts. Interproximal grooves are located at the tooth

---

61   Bass 1995; Buikstra & Ubelaker 1994; Eliopoulos 2006.

62   Ortner 2003.

63   Powell 1985.

cervix, on the mesial and distal surfaces of mostly posterior teeth, especially molars. The grooves have smooth and even polished surfaces, and often show a tubular appearance involving both the mesial and distal surfaces of two adjacent teeth.[64] Generally, the grooves on mandibular teeth are oriented lingually, while those on maxillary teeth are oriented buccally.[65] Higher frequencies of interproximal grooving in maxillary than in mandibular dentitions, as well as in male than in female skeletons, have been reported for some skeletal assemblages.[66]

Interproximal grooving, first identified by A. Siffre in 1911, shows wide chronological and geographic distributions. The occurrence of interproximal grooves has a remarkable temporal range from 1.84 million years ago to the present, spanning a variety of time periods and groups.[67] Besides anatomically modern humans, the grooves have been observed in several extinct species including *Homo habilis*, *Homo erectus* and *Homo neanderthalensis*.[68] Interproximal grooving has been identified in North America,[69] South America,[70] Africa,[71] Pakistan,[72] Uzbekistan,[73] China,[74] and Australia.[75] In Europe, interproximal grooving has been reported,

among other countries, in Denmark,[76] Germany,[77] Switzerland,[78] Spain,[79] France,[80] Italy[81] and Croatia.[82] With regard to Greece, dental grooves have been identified at the Early Bronze Age cemetery (3300/3100-2300/2200 BC) of Koilada in Western Macedonia.[83] The majority of the grooves were located at the lingual surfaces of maxillary anterior teeth, possibly related to some task activity. Triantaphyllou also reported the presence of an interproximal groove on the distal surface of a maxillary second molar.[84]

The literature on the etiology of interproximal grooving has been not only extensive but also quite controversial. A variety of explanations has been hypothesised, including chemical erosion,[85] grit-laden saliva from sand and soil in the food and drinking water[86] and task activity involving the stripping of animal sinews or fibrous materials between the teeth.[87] The etiology generally accepted, however, consists of abrasion due to the habitual use of wooden or bone dental probes (toothpicks) for oral hygienic purposes, to remove impacted food, or for therapeutic and/or palliative purposes, to alleviate pain and discomfort resulting from inflammation and irritation of the gingiva.[88] Based on ethnograph-

---

64 Berryman *et al.* 1979; Formicola 1991; Ubelaker *et al.* 1969.

65 Frayer 1991.

66 Berryman *et al.* 1979; Ubelaker *et al.* 1969; Wallace 1974; Willey & Hofman 1994.

67 Ungar *et al.* 2001.

68 See Ungar *et al.* 2001 for a review of interproximal grooves on fossil hominin teeth.

69 Berryman *et al.* 1979; Schulz 1977; Ubelaker *et al.* 1969; Willey & Hofman 1994.

70 Eckhardt & Piermarini 1988; Lessa & Guidon 2002.

71 E.g. Bermúdez de Castro & Arsuaga 1983; Bermúdez de Castro & Pérez 1986; Boaz & Howell 1977; Bonfiglioli *et al.* 2004; Ungar *et al.* 2001; Wallace 1974.

72 Lukacs & Pastor 1988.

73 Turner 1988.

74 Weidenreich 1937.

75 Brown & Molnar 1990; Campbell 1925.

76 Alexandersen 1978.

77 Frayer 1991.

78 Alt & Koçkapan 1993; Alt & Pichler 1998.

79 Bermúdez de Castro *et al.* 1997; Puech & Cianfarani 1988.

80 Bermúdez de Castro & Pérez 1986; Hartweg 1945; Siffre 1911.

81 Formicola 1988.

82 Frayer & Russell 1987.

83 Triantaphyllou 2001, 88-90.

84 Triantaphyllou 2001, 88.

85 E.g. Brothwell 1963; Campbell 1925.

86 E.g. Wallace 1974.

87 E.g. Brown 1991; Brown & Molnar 1990; Eckhardt 1990; Schulz 1977.

88 Bermúdez de Castro & Arsuaga 1983; Berryman *et al.* 1979; Bonfiglioli *et al.* 2004; Eckhardt & Piermarini 1988; Frayer & Russell 1987; Frayer 1991; Siffre 1911; Tratman 1956; Ubelaker *et al.* 1969.

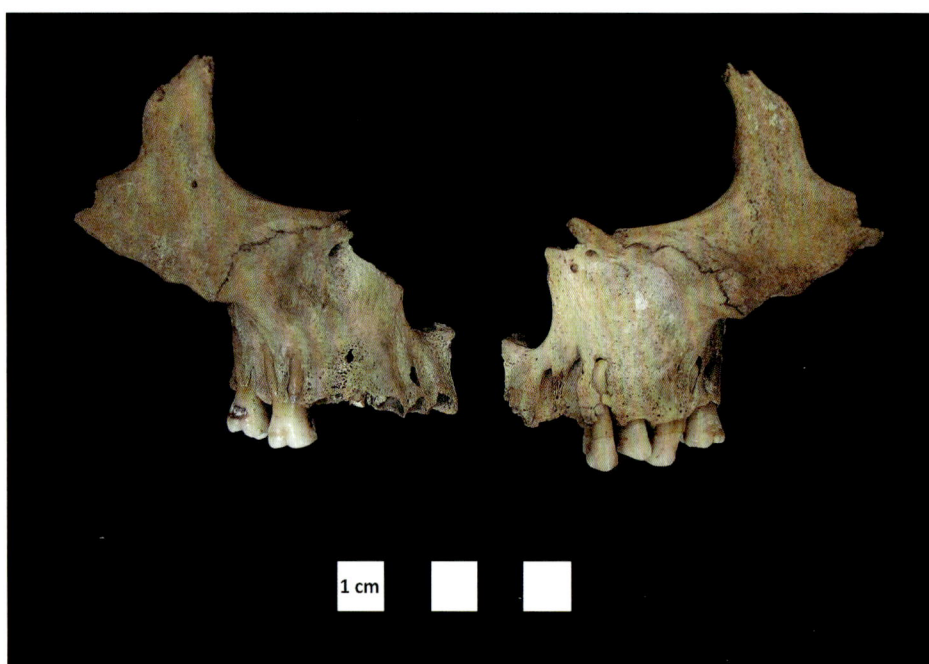

Fig. 2.5.4. Left and right maxillae and zygomatics recovered from MOU-16.

ic examples, some scholars have proposed the use of plants with analgesic and anesthetic properties as probes causing the interproximal grooving, including the black Sampson in North America,[89] and cactus thorns in Brazil.[90] In the absence of associated oral and dental pathological lesions, idiopathic, nonfunctional, cultural and/or psycho-cultural motives for the habitual probing causing interproximal grooving have also been proposed.[91] Interestingly, in their review of the fossil record, Ungar and colleagues suggested that tooth picking reflects a behaviour unique to the genus *Homo*, which in early *Homo* was probably palliative, related to a dietary shift to include the consumption of significant quantities of vertebrate animal protein.[92] Nevertheless, a number of scholars have emphasised that the aforementioned hypotheses are not necessarily mutually exclusive

and that different grooves can have different etiologies.[93]

In the case of the mandible from KAR-8, the shape of the groove does not match exactly the typical cases of interproximal grooving reported in the literature. The tubular form of the groove involving the opposing mesial surface of the adjacent tooth cannot be evaluated, given that the second molar was lost antemortem. The morphology generally suggests formation by a relatively hard object with buccal insertion and a lingual direction restricted by the cheeks, with the sharp lingual border representing the stopping point at the dental neck. However, the groove from Pelion is less confined, with clear superior and lingual margins, but with the inferior border extending over most of the exposed root. In addition, the modified area shows a W-shaped inferior edge that is not consistent with the use of a rounded instrument. The shape of the inferior

---

89   Willey & Hofman 1994.

90   Lessa & Guidon 2002.

91   E.g. Berryman *et al.* 1979; Formicola 1988; Willey & Hofman 1994.

92   Ungar *et al.* 2001.

93   Bermúdez de Castro & Arsuaga 1983; Bermúdez de Castro *et al.* 1997; Berryman *et al.* 1979; Formicola 1991; Frayer 1991; Ungar *et al.* 2001; Willey & Hofman 1994.

Fig. 2.5.5. Visceral surface of the fractured left rib (PCP6) showing bony callus (posterior is to the left; KAR-8).

border is formed by the removal of micro-flakes due to the induced pressure. Interproximal grooves have been reported to appear more confined in early stages, and deeper and continuous in more advanced ones, with different features reflecting different stages in the formation process.[94] In that case, the groove could have been extended inferiorly later on, given the resorption of the supporting alveolar bone and the antemortem loss of the second molar, and considering also that Frayer has reported interproximal grooves only on the molar root surfaces.[95] The occlusal involvement could be an old fracture that was worn smooth over time, forming a notch. The heavy tooth wear and the oral pathologies, including antemortem loss of the adjacent tooth, alveolar resorption and active inflammation, may suggest some type of interdental probing to alleviate gingival irritation, discomfort and pain for the Pelion mandible, possibly in combination with a grit-laden diet that would cause further abrasion. Of course, an etiology involving a dental practice or tooth use previously undescribed cannot be excluded.

Nevertheless, an alternative explanation for the interproximal depression on the right first molar

might be the former presence of an impacted second molar. Specifically, the enamel and root defect could have resulted from a partially impacted second molar in a mesioangular position (mesially inclined), pressing up against the adjacent first molar at the cervico-enamel junction for a long period of time. Impaction is a common anomaly in tooth eruption, often caused by a physical obstacle in the eruption path of a tooth, such as an adjacent tooth or bone; it may also result from the abnormal position of the tooth within the alveolar bone.[96] Even though impaction typically implies that the tooth is completely enclosed within the alveolar bone, there can be many variations in its form and degree. Impaction can be complete or partial, and a tooth can be partially or fully erupted sideways with only the side of the crown or a number of cusps showing uppermost.

The teeth most commonly impacted are the maxillary and mandibular third molars, followed by the maxillary canines. Impaction of the second molar, however, is a rare dental anomaly. The prevalence of impacted mandibular second molars has been

94   Bonfiglioli *et al.* 2004.
95   Frayer 1991.
96   Andreasen & Kurol 1997.

reported to be between 0.6/1000 and 3/1000.[97] Impacted second molars have been reported to occur more frequently in males than in females, unilaterally (on the right side) rather than bilaterally.[98] In a modern Greek sample recruited at the School of Dentistry at the University of Athens, out of the 940 impacted teeth observed (in 425 patients with at least one impacted tooth), 91.6% were third molars, whereas only 1% were second molars.[99] A slightly higher frequency was observed in a modern Northern Greek sample at the School of Dentistry at the Aristotle University of Thessaloniki: out of the 225 impacted teeth in total (in 170 patients with at least one impacted tooth), 3.6% were second molars.[100] If that is the case for the Pelion mandible, the antemortem loss of the second and third molars may be associated with an abnormal position; eruption disturbances of permanent molars can result in clinical problems such as periodontal disease, caries, and bone and root resorption, sometimes resulting in the loss of neighbouring teeth.[101] However, extraction of the problematic teeth may also be plausible.

MOU-1 (Tsounaga Cave, inner chamber)

The only human skeletal element recovered in MOU-1 consists of the unfused head of a right femur (Table 2.5.1; Fig. 2.5.2). The epiphysis is fully developed with no signs of fusion, indicating the presence of a subadult individual, probably a young adolescent. Time of fusion for the femoral head is reported to occur between 12-16 years in females and 14-19 years in males.[102]

---

97   Grover & Lorton 1985; Johnsen 1977; Varpio & Wellfelt 1988.
98   Varpio & Wellfelt 1988.
99   Gisakis *et al.* 2011.
100   Fardi *et al.* 2011.
101   Gisakis *et al.* 2011; Raghoebar *et al.* 1991.
102   Scheuer & Black 2000.

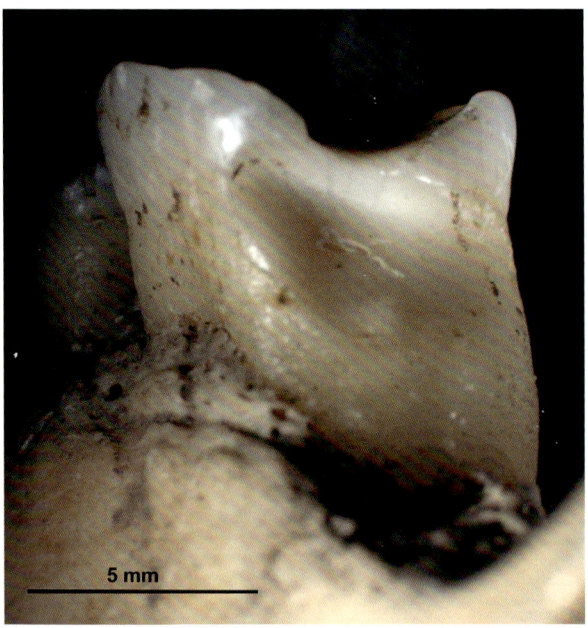

Fig. 2.5.6. Magnified view of the interproximal groove on the distal surface of the mandibular right first molar (KAR-8).

MOU-16 ("Damouchari IV")

In MOU-16, scattered bones were present on the cave floor, but only four bones were collected due to time and space constraints. The bones collected consist of a left maxilla and a left zygomatic bone in articulation, and a right maxilla and a right zygomatic bone in articulation (Table 2.5.1; Fig. 2.5.4). The overall size and preservation of the two sides are similar; nevertheless, the sutures of the two maxillae do not articulate. Moreover, after close inspection, the two sides show significant differences in structure and morphology. First and foremost, the shape of the nasal aperture is completely different: the inferior margin of the nasal aperture of the left maxilla is narrow and deep, whereas on the right maxilla it is wide and shallow, showing marked guttering. The left canine fossa is large and deep, whereas the right one is much shallower, a contrast associated with observed differences in the morphology of the two maxillary sinuses and the structure of the antral walls. Dissimilarities in the bones of the two

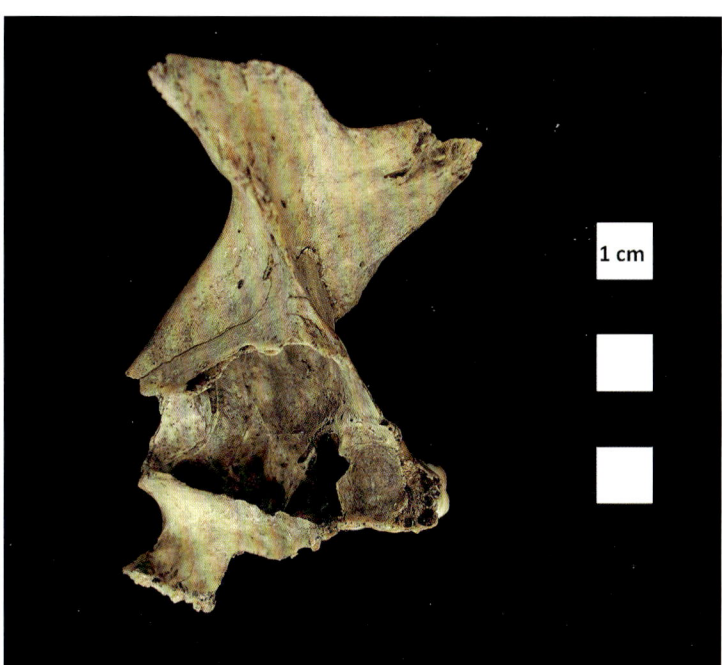

1 cm

sides are also observed in the size and shape of the temporal processes of the zygomatic bones, the infraorbital, zygomaticofacial, zygomaticoorbital and zygomaticotemporal foramina, as well as in the zygomaticomaxillary and temporozygomatic sutures. As a final comment, the tooth wear of the dentitions of the two sides does not match, with the left side showing more pronounced wear, suggesting an older age and/or different diet; however, given the absence of the associated mandibular dentition, the etiology for the dental wear dissimilarities remains inconclusive. Thus, the bones of the two sides are considered to represent two different (but perhaps related) individuals.

Turning to the dentitions, on the left maxilla (PCP21) four teeth were present (first and second premolars, first and second molars) out of the seven tooth sockets observable, while on the right maxilla (PCP23) three teeth were present (first premolar, and first and second molars) out of the seven tooth sockets observable (Fig. 2.5.4). The remaining teeth were lost postmortem. Agenesis of the third molar is observed on the left side, given that the alveolar

process ends at the second molar. On the right side, the socket of the third molar is not observable. A tooth root remnant 7.0 mm in length is present in the intermaxillary suture of the left maxilla.

The dentitions of both maxillae show severe dental caries. On the left maxilla, the first and second molars show large dental caries with destructive lesions penetrating the pulp cavity, resulting in infection of the supporting tissues and in periapical lytic lesions of the alveolar bone, more pronounced on the palate. In addition, the left first premolar shows interproximal caries on the distal surface. On the right side, the first and second molars show caries on the mesial and the buccal surfaces, respectively. The second molar also shows initial stages of fissure caries occlusally. The crown of the right first premolar is completely destroyed by caries, leading to the infection of the supporting alveolar bone, associated with a large lytic lesion and marked porosity in the premolar area. Dental caries is a common oral pathology characterised by the progressive destruction of dental calcified tissues (enamel, dentin, cement) caused by microbial activity in the dental plaque.[103] The occurrence of dental caries in archaeological specimens is widely used for paleodietary reconstruction, indicating consumption of refined foods that are high in carbohydrates, particularly sugars.[104] High rates of caries have been reported for various Bronze Age samples from Greece, particularly from Mycenaean contexts.[105]

The maxillary sinus of the right maxilla (PCP23) shows large remodelled lobules of bone and large bony spicules indicative of severe maxillary sinusitis (Fig. 2.5.7). Maxillary sinusitis is an inflammation of the mucosa lining the maxillary sinuses, which are the largest of the paranasal sinuses.[106] Maxillary sinusitis is a commonly diagnosed disease worldwide today,[107] and it has been observed and studied in a number of bioarchaeological samples.[108] Based on duration, sinusitis is generally classified as acute, subacute or chronic; however, it is the chronic form that is diagnosed paleopathologically, given the time needed for bone involvement.[109] The major causes of maxillary sinusitis are rhinogenous (viral upper respiratory tract infections, poor air quality or smoke from fuel burning), odontogenic (dental disease), asthma, immunodeficiency disorders, infections such as leprosy and allergies.[110] Even though the radiographic examination showed no evidence for oro-antral fistulae penetrating the sinus, the presence of a periapical abscess in the premolar region and the observed lytic lesion at the root apex of the first premolar may suggest an odontogenic origin for the maxillary sinusitis of the right maxilla from MOU-16.[111] Finally, the maxillary sinus of the left maxilla (PCP21) shows some evidence of pitting that could be associated with the hypervascularisation occurring in the initial stages of infection.[112]

---

103   Hillson 1996; Ortner 2003; Powell 1985.
104   Hillson 1996; Powell 1985.
105   e.g., Angel 1944; Lagia *et al.* 2007; Schepartz *et al.* 2009; Triantaphyllou 2001; Tsivilakos *et al.* 2002.
106   Boocock *et al.* 1995a; Brook 2009.
107   Brook 2009; Slavin *et al.* 2005.
108   Boocock *et al.* 1995a,b; Lewis *et al.* 1995; Merrett & Pfeiffer 2000; Panhuysen *et al.* 1997; Roberts 2007; Wells 1977.

109   Boocock *et al.* 1995a; Roberts 2007; Tovi *et al.* 1992.
110   Boocock *et al.* 1995a; Lewis *et al.* 1995; Roberts 2007.
111   Mehra & Murad 2004; Merrett & Pfeiffer 2000.
112   Merrett & Pfeiffer 2000.

2.6 Pottery from the Prehistoric to Roman periods[113]

Ioannis Voskos

Greek summary

Στο παρόν υποκεφάλαιο εξετάζονται τα 293 πήλινα αντικείμενα (283 όστρακα και 10 τμήματα κεραμίδων) από τους Προϊστορικούς και Ιστορικούς χρόνους, έως την Υστερορωμαϊκή ή Πρωτοχριστιανική περίοδο (6ος-7ος αι. μ.Χ.). Οι ιδιαίτερες συνθήκες διατήρησης αρχαιολογικών καταλοίπων εντός των καρστικών εγκοίλων, και κυρίως η δράση μετα-αποθετικών παραγόντων, συνέβαλαν στον εντοπισμό και την περισυλλογή ως επί το πλείστον διαβρωμένων ή πλήρως καλυμμένων από ιζήματα οστράκων. Ως εκ τούτου, η μακροσκοπική μελέτη του υλικού επικεντρώθηκε στα στοιχεία που θα μπορούσαν να οδηγήσουν στη σχετική χρονολόγησή του. Ο κατάλογος που προέκυψε από την ανάλυση των οστράκων περιλαμβάνει στοιχεία προέλευσης, διαστάσεις, λεπτομερείς πληροφορίες για την κεραμική ύλη, επεξεργασία επιφάνειας και διακόσμηση (όπου υπάρχει) και ακόμη την όπτηση, την κατάσταση διατήρησης και την πιθανή χρονολόγησή τους.

Από την ανάλυση αυτή προκύπτουν εξαιρετικά ενδιαφέροντα δεδομένα για τη χρήση των σπηλαίων του Πηλίου διαχρονικά. Παρά την απουσία νεολιθικών οστράκων που προκαλεί κάποια ερωτήματα, αντιπροσωπεύονται με σημαντικούς αριθμούς η Εποχή του Χαλκού και κυρίως η Ελληνιστική και Ρωμαϊκή/Υστερορωμαϊκή περίοδος. Εντύπωση, από την άλλη, προκαλεί η μικρή παρουσία οστράκων των Αρχαϊκών και Κλασικών χρόνων. Το γεγονός αυτό ίσως οφείλεται σε παράγοντες αναγνωρισιμό-τητας και όχι απαραίτητα στη μη χρήση των σπηλαίων κατά την περίοδο αυτή.

Όσον αφορά στα όστρακα της Εποχής του Χαλκού ξεχωρίζουν τα μυκηναϊκά και μία ντόπια κατηγορία χειροποίητης στιλβωμένης κεραμικής. Η ύπαρξη οστράκων με τη χαρακτηριστική μυκηναϊκή διακόσμηση επιβεβαιώνει τις έντονες επαφές της ευρύτερης περιοχής της Μαγνησίας με το Μυκηναϊκό πολιτισμό της Νότιας Ελλάδας. Η παρουσία, ωστόσο, και της στιλβωμένης κεραμεικής οδηγεί στο συμπέρασμα της ύπαρξης και δράσης ακόμη μίας πολιτισμικής ομάδας που διατήρησε εν μέρει τη ντόπια παράδοση της χειροποίητης κεραμεικής.

Από εκεί και πέρα, κατά την Ελληνιστική και Ρωμαϊκή/Υστερορωμαϊκή περίοδο, παρατηρείται εντατικοποίηση της χρήσης του Πηλίου με άφθονο υλικό προερχόμενο από πολλά σπήλαια. Η αναγνώριση, μεταξύ άλλων, αρκετών οστράκων από αμφορείς (Κωακούς, Ροδιακούς, Ιταλικής προέλευσης κ.ά.), αλλά και κάποιων εισαγμένων καλής ποιότητας αγγείων σερβιρίσματος, φανερώνει τις διευρυμένες επαφές των κοινωνικών ομάδων του Πηλίου που συμμετείχαν δυναμικά στην εντεινόμενη εμπορική δραστηριότητα των περιόδων αυτών. Η παράλληλη χρήση των σπηλαίων για αποθηκευτικές, κτηνοτροφικές και άλλες δραστηριότητες ενισχύει το σημαντικό ρόλο τους στο πολιτισμικό περιβάλλον της εκάστοτε περιόδου.

❊ ❊ ❊

---

113 I owe many thanks to Dr Dimitris Grigoropoulos for his valuable comments concerning the Roman sherds and to Anthoula Tsarouha, Anthi Balitsari and Elli Tzanni for their help and comments.

Introduction

The ceramic assemblage of the Pelion Cave Project numbers 536 sherds and 55 roof tile fragments, collected from the surface of 45 caves and rockshelters. In most cases, the sherds are weathered and eroded. As a result, nothing more than rough speculations, mostly based on fabric, firing and surface treatment, can be made regarding their date. Another problematic issue is the lack of relevant bibliography for Pelion. Indeed, even though Magnesia and Thessaly, in general, flourished during the Neolithic, Mycenaean, Hellenistic and Roman periods, Mount Pelion has been archaeologically neglected. In this chapter, I present a brief analysis of the sherds from the Prehistoric and the Historical periods (until the sixth to seventh centuries AD), focusing mainly on datable features. The assemblage includes 283 sherds (33 of which are unidentifiable) and 10 roof tile fragments. A unique catalogue number (Cat. No.) has been assigned to every sherd.

The pottery of the Neolithic to Bronze Age periods

Nine caves yielded 51 prehistoric sherds. Many of these are difficult to identify with certainty and some may belong to eroded wheelmade wares of the Historical periods. For some of the smaller and poorly preserved fragments, estimations of chronology are based entirely on fabric and surface finish. For the decorated Mycenaean sherds, I have followed the standard typology and terminology established by Furumark (i.e. "FS" for Furumark's shape and "FM" for Furumark's motif).[114]

The Neolithic period

Only a single sherd from MOU-1 (Cat. No. 423) can be dated to the Neolithic period with some cer-

tainty. This contrasts with the fact that the Neolithic in Thessaly is well investigated and that some of the most important settlements were excavated in Magnesia (e.g. Sesklo and Dimini). The decoration and surface treatment of Cat. No. 423 are reminiscent of blacktopped ware. Such vessels were fired upside-down with part of the pot buried in the ground. The black colour of the interior and the buried part is a result of the reduced atmosphere during firing, whereas the rest of the vessel displays a clay colour (usually reddish or brownish). Blacktopped ware is believed to appear in Northern Greece during the Middle Neolithic and becomes abundant in Greece during the Late Neolithic.[115] The discovery of only one tentative Neolithic sherd could suggest that the caves of the Pelian landscape did not constitute part of the social environment of Thessalian Neolithic groups. Based on many examples from southern, central and insular Greece we know that this is not generally the case and caves were periodically utilised for various purposes.[116] Tomkins, for instance, recently proposed a ritual model of cave use, a pattern that also applies to some isolated mountainous and semi-mountainous areas.[117] Concerning the Pelion caves, therefore, the absence of contemporary material suggests that any Neolithic layers might be well buried and that their discovery will require systematic excavations.

The Bronze Age periods

Cat. Nos 127 and 203 are coarse wares with a thin dark wash on both inner and outer surfaces (Figs 2.6.1 & 2.6.2). This particular surface finish is typical of handmade wares in Macedonia and inland Thessaly during the EBA and MBA.[118] The tech-

---

114   Furumark 1941; see also Mountjoy 1986.

115   See for example Koukouli-Chrissanthaki 1996, 114.

116   See for example Papathanassopoulos 1996, 39-40.

117   Tomkins 2009. For example, Zas cave at Naxos, Skoteini cave at Tharrounia Euboea, etc. See Zachos 1999; Sampson 1992.

118   Feuer 1992, 287.

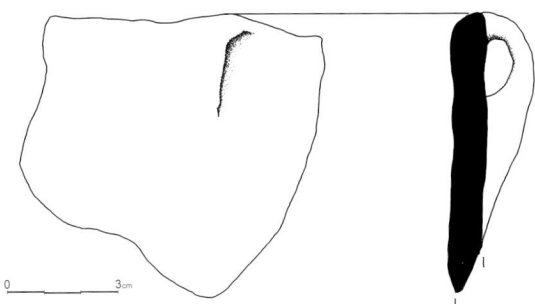

Fig. 2.6.1. Fragment of coarse ware vessel from KAR-8 (Cat. No. 127).

nique is also similar to so-called "Barbarian Ware" or Handmade Burnished Ware (HBW), which has been identified at sites all over Greece. [119]

HBW appears in Greece, Troy and later in Cyprus from LH IIIB and onwards. Even though its provenance and chronology are still being debated, a probable date of the two fragmentary vessels from KAR-8 (Cat. No. 203) and VOL-3 (Cat. No. 127) is LH III, judging by their shape, fabric and finish, which are almost identical to ceramics recorded at Sparta[120] and elsewhere. Alternatively, these two

fragmentary vessels could be local imitations of Mycenaean shapes or, more likely, products of local handmade techniques that survived through the LBA and later.[121]

A fragmentary rim from a handmade burnished vessel is reminiscent of LBA shapes[122] or MH local wares (Cat. No. 255; Fig. 2.6.3).

The absence of clearly datable features (e.g. the typical plastic or impressed decoration on these vessels) makes a more precise dating impossible. It probably comes from a medium-sized open vessel for storage. The presence of mica suggests local manufacture.

More problematic is a group of sherds from KAR-15 and KAR-16 that share some common characteristics, such as the high polished light-coloured slip applied to their outer (and in some cases inner) surfaces (Cat. Nos 354-358, 378-379, 388-389, 397-403 and 405; Figs 2.6.4-2.6.6). With the exception of Cat. No. 378, which is from a bowl, the other sherds are from closed, small to medium-sized vessels.

However, only a handle (Cat. No. 379) and a fragment of a closed vessel with part of a handle (Cat.

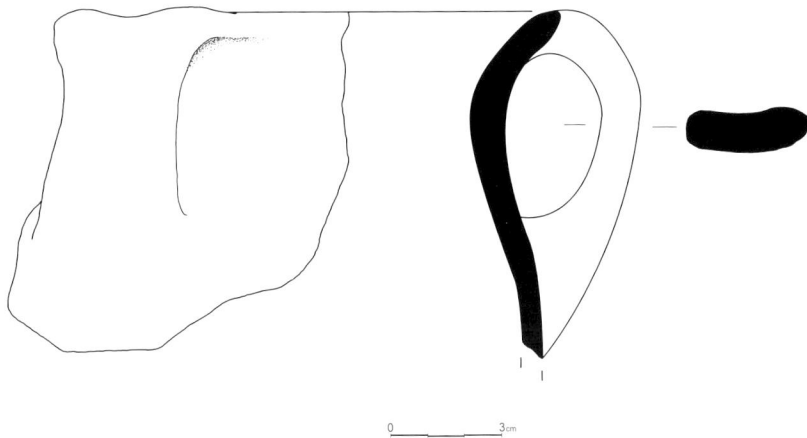

Fig. 2.6.2. Fragment of coarse ware vessel from KAR-8 (Cat. No. 203).

---

119 Rutter 1975; Small 1990.
120 See Small 1990, 4, fig. 1, no. 1.

121 Verdelis 1958, 93.
122 See for example Small 1990, 4, figs 1, 2 and 6.

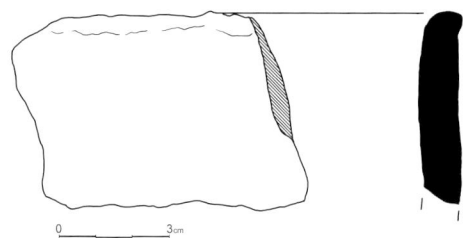

Fig. 2.6.3. Rim from handmade, burnished vessel (Cat. No.. 255).

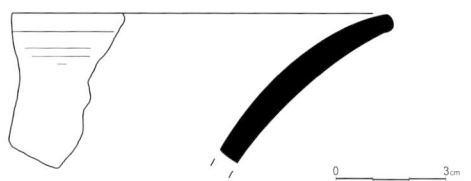

Fig. 2.6.4. Fragment from bowl with high polished light-coloured slip (Cat. No. 378).

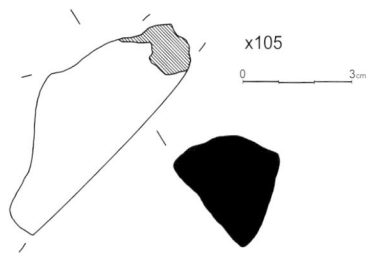

Fig. 2.6.5. Handle from closed vessel (Cat. No. 379).

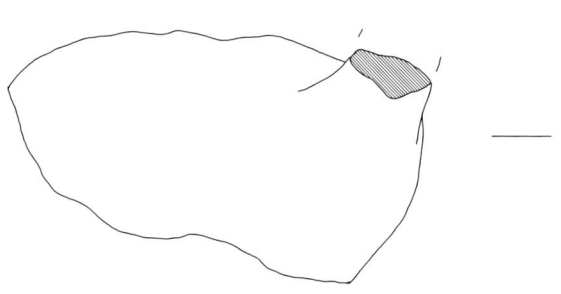

Fig. 2.6.6. Fragment from closed vessel with part of a handle (Cat. No. 405).

No. 405) provide some clues about their shapes. The surface treatment, quality of slip and particularly the fact that they all are from handmade vessels suggest a date within the BA. They could either be products of the local Thessalian-Macedonian handmade burnished wares or of a local workshop attempting to imitate Grey-Yellow Minyan pottery. An even earlier date within the EH cannot be excluded.

Another sherd from KAR-8 (Cat. No. 235) is possibly from a small open drinking vessel (cup?). The surface treatment is typical of Yellow Minyan Ware, a type traditionally dated to the later part of the MH. The fact that Yellow Minyan Ware was still in use at least in the first phases of LBA makes a date to the LH most probable for Cat. No. 235.

The most securely datable sherd from the survey is the upper part of a small stirrup jar (Cat. Nos 207, 248, 249 and 254) from KAR-8 (Fig. 2.6.7).

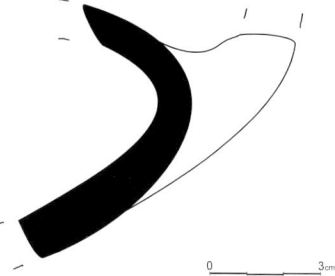

This type of vessel, which has a distant antecedent in the Middle Minoan repertory, is quite popular from LH IIIA and onwards, especially in tombs.[123] Indeed, the cave appears to contain a disturbed burial,[124] and the presence of this stirrup jar would date it to LH III. What is more, the vessel itself offers a good number of features that could give a more exact date. Judging by the shape of its upper part, the vessel belongs to the globular type (possibly FS 173), with a slightly curviform shoulder. The false

---

123 Mountjoy 1986, 105; for general remarks on the development of this shape see Mountjoy 1986, 203.

124 Prevedorou, this volume.

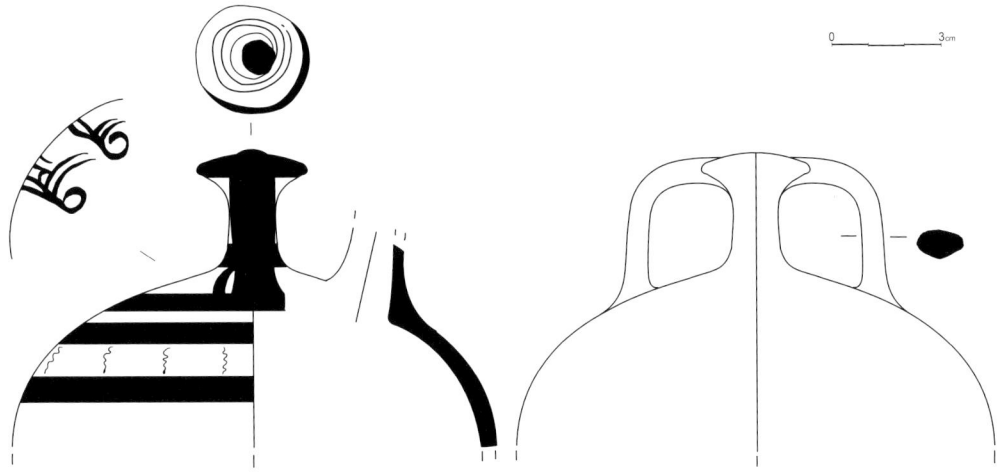

Fig. 2.6.7. Upper part of a small stirrup jar (Cat. Nos 207, 248, 249 and 254).

neck and the cylindrical spout are both relatively long and narrow. The spout's rim is eroded and it provides no information regarding its decoration or the form of its "mouth". The disc above the painted handles (only one of which is preserved) is wide and slightly convex. The vessel's decoration consists of 1) a painted band that connects the lower part of neck with the lower part of the spout, 2) a spiral motif (FM 46) that covers the upper part of the handle disc and 3) a zone on the shoulder with a repeated motif that resembles a schematic version of the horizontal voluted flower (FM 18A). The vessel's body is covered with a succession of horizontal wide bands and thin lines. The second zone of decoration on the vessel's "belly" includes the repeated motif of a single vertical wavy line (FM 53). The form of neck and spout, the type and arrangement of decoration, the existence of the second zone of decoration and the absence of the plastic band in the neck's bottom date the stirrup jar to LH IIIB1 or LH IIIB2.

Other sherds from closed vessels appear to be LH. Cat. Nos 208 and 236 have traces of black paint on their outer surface. The colour of clay and this type of paint are strongly reminiscent of Mycenaean decorated pottery. They could belong either to the same stirrup jar (see above) or a similar LH vessel.

Cat. No. 244 is similar to 208 and could be from the same vessel. Finally, Cat. No. 128 is too eroded and fragmented to provide information on dating and vessel shape. Its pale-yellowish coloured, very fine fabric with no visible inclusions is indicative of LH wares. Especially its shape is reminiscent of FS 230 or a similar small drinking vessel. However, the clay is also reminiscent of Early-Middle Roman lamps.[125]

Cat. No. 252 is a fragmentary plain ware open vessel. Its shape and profile are reminiscent of some variants of LH bowls (or cups). Two marks, one near the rim and another near the vessel's "belly", suggest the existence of a missing handle. Overall, the bowl's shape and its combination with one vertical handle are strongly reminiscent of LH dippers.[126] The regular association of dippers with burials suggests that this vessel could have been an offering as part of the presumed LH burial in KAR-8.

Two more sherds (Cat. Nos 212-213) from KAR-8 are both characteristic of LH decorated pottery. Cat. No. 213 is from a small closed vessel. On its outer

---

125   See for example Petropoulos 1999, 57.

126   For parallels see AD 52-54 in Catling 2009, Vol. I, 221, 415-6 and Catling 2009, Vol. II, 266, fig. 270.

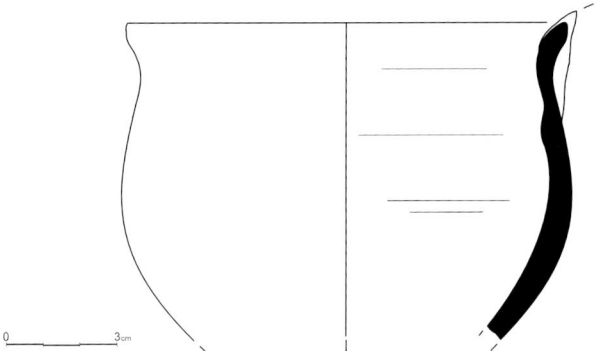

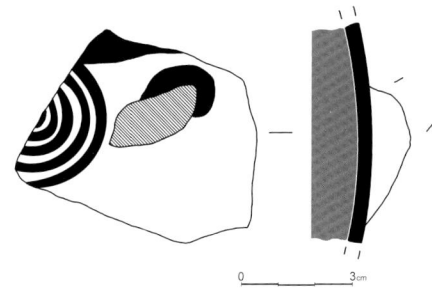

Fig. 2.6.8. Fragment of plain ware open vessel. Late Helladic dipper? (Cat. No. 252).

Fig. 2.6.9. Possibly Late Helladic decorated vessel (Cat. No. 212).

surface, there is typical Mycenaean polished paint in the form of painted reddish and brownish bands and thinner lines. However, these characteristics are not enough to suggest a more precise LH chronology.

Cat. No. 212 includes more datable features. Its inner surface is covered with a dark brown paint and in the outer surface is either a fragmentary spiral motif (FM 46, 51), or most probably a motif of isolated semi-circles (FM 43; Fig. 2.6.9).

A fragmentary horizontal handle (of circular cross-section) leads to a shape like FS 282 (ring-based krater) or FS 284 (deep bowl). Given the fact that FS 282 usually has very thick walls, Cat. No. 212 most probably belongs to a deep bowl (FS 284). A date around LH IIIB2 or early LH IIIC would be suitable for this type of vessel and decoration.

Cat. Nos 421 and 424 from MOU-1 probably belong to the same vessel. Their fabric, firing and especially the type of slip and surface finish (burnishing) of both sherds are all indicative of prehistoric techniques, but there are no clear, datable features, such as any type of decoration (Fig. 2.6.10). On the other hand, nearly all the sherds collected at this cave are from Hellenistic and Roman vessels.

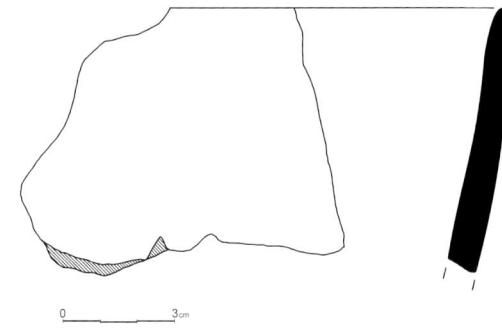

Fig. 2.6.10. Undated fragment of cooking ware (Cat. No. 424).

Thus, Cat. Nos 421 and 424 could also belong to eroded cooking wares of the Historical period.[127] The same situation is also valid for Cat. Nos 540 and 552 from MOU-7 and 566 from MOU-16.

Lastly, the enigmatic vessel from MIL-13 could also be Prehistoric. Only four body sherds (Cat. Nos 334, 336-338) and one rim fragment (Cat. No. 335) were collected from this vessel, but they are enough to reveal a good part of a medium-sized vessel, possibly for short-term storage of liquids. Its coarse fab-

---

127   Rotroff for example, refers to Hellenistic cooking fabrics that contain mica, schist and quartz. These inclusions resemble the fabric of Cat. Nos 421 and 424. See Rotroff 2006, 39-45.

ric includes some large grits (mostly quartz), and the fact that most of its surface is uneven, smoothed and burnished indicates prehistoric techniques.

Notes on Bronze Age

The settlement nucleation that was observed during the Final Neolithic and EBA in southern Thessaly is said to be partly the result of trade/exchange development.[128] Though there are few excavated EH and MH settlements in the region,[129] it seems that this process further intensified during these periods. By the LBA, there were numerous flourishing small and larger settlements in Magnesia, mostly connected with the material culture of southern "Mycenaean" Greece.

It is impossible to reach any conclusions regarding Pelion during the BA based on so few and such eroded sherds. A possible exception is a burial at KAR-8, clearly connected with the nearby LH settlements of Lake Karla. This case and the random discovery of handmade wares in other caves suggest that BA inhabitants of Magnesia were familiar with the Pelian landscape. Whether the mountain and especially its caves were used primarily as an integral part of a mixed agro-pastoral economy or solely for ritual purposes is not clear,[130] but in any case, it seems that during the BA Pelion was part of the social space and the existential world of nearby groups. Concerning the identity of these people, despite the long-standing discussions about barbarian or Indo-European invasions in Greece, the direct equation of pottery with specific ethnic groups is at best problematic. On the other hand, the presence of wheelmade Minyan painted Mycenaean

and handmade burnished wares is noteworthy. It seems that Magnesia was a meeting point of different ceramic production techniques, a fact that suggests the existence of active social groups that maintained contacts with southern "Mycenaean" and other areas of mainland Greece.

The Pottery of the Classical and Hellenistic periods

The Classical period

This period approximately extends from the end of the LBA (last quarter of the eleventh century BC) to the beginning of the Hellenistic period (i.e. 323 BC). After the collapse of the Mycenaean administrative system (c. 1200 BC), mainland Greece saw a population decrease and relocation of many sites. The so-called "Dark Ages" and the Geometric period in Magnesia[131] are well attested at a number of sites. Indeed, after the abandonment of Mycenaean Iolkos and Magoula Pevkakia, archaeological remnants at Palia Volou and especially the cemetery of Nea Ionia indicate continuity until the Geometric period.[132] Even though it covers at least three centuries, no sherds from the PCP survey can be dated to this period. Possible exceptions are the handmade coarse wares that continue the local tradition,[133] but they are most probably dated to the BA (see above).

The lack of identifiable PCP sherds also applies to the Archaic period. Two possible exceptions are Cat. Nos 343 and 377 (Fig. 2.6.11). While Cat. No. 343 is too eroded for precise dating, the very fine fabric and whitish-pale clay suggest either a date in

---

128   Gallis 1996, 37.
129   With the exception of Magoula-Pevkakia, see Maran 2007.
130   For a full discussion on pastoralism, prehistoric cave use in the Aegean and the possible economic or ritual role of caves see Cherry 1988; Halstead 1996a; Halstead 1996b; Tomkins 2009; Tomkins 2010, 33-4.

---

131   For general remarks on this period in Thessaly, see Snodgrass 1971 and Desborough 1972.
132   Batziou-Efstathiou 1999, 117.
133   See for example Heurtley & Sceat 1930-31, 41 and Verdelis 1958, 91-3, 98 for the Marmariani cemetery.

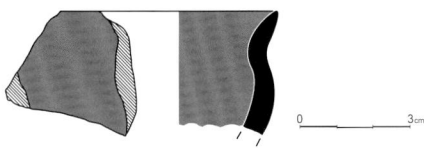

Fig. 2.6.11. Rim fragment of a small vessel with reddish paint, possibly Archaic or Classical (Cat. No. 377).

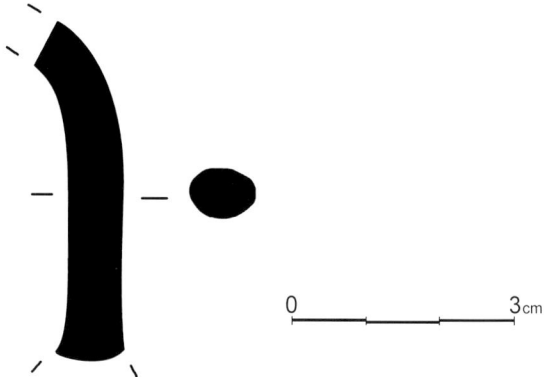

Fig. 2.6.12. Part of a handle with black polished paint, possibly from Classical drinking vessel (Cat. No. 246).

the LBA or the Archaic period.[134] The type of red paint observable on the exterior supports the latter date. Cat. No. 377 is a rim fragment of a small vessel. The fabric, shape and mainly the presence of reddish paint, possibly applied with a brush or a similar tool in the outer surface, suggest a date in the Archaic (or Classical) period.

More problematic are Cat. Nos 386-387 from KAR-15. Their fine fabric and especially the presence of a thin crust of red slip or paint point to a date around the Archaic or Classical period. Both sherds are too fragmentary to reveal the shape and use of the vessels. The only securely observable features in Cat. No. 386 are the mark of a missing handle and traces of red paint on its inner surface suggesting an open vessel, possibly for serving food or drink.

The number of sherds and roof tile fragments that can be dated to the Classical period (i.e. 490-323 BC) is small. One example is Cat. No. 246, a small part of a handle covered with black polished paint (Fig. 2.6.12). A Hellenistic date should not be excluded, but given the fact that the equivalent Hellenistic black paint is usually duller, this handle probably belongs to a small Classical drinking vessel.

More sherds from KAR-8 appear to belong to Classical period vessels, though none are from datable fine wares. Cat. Nos 224-231 come from one or two similar closed vessels, possibly for storing small quantities of liquids (jars or jugs). Cat. Nos 221-222 are body sherds possibly from the same vessel. Their fabric, surface treatment and the fact that they are from a wheelmade vessel indicate a Classical date. Cat. No. 223 is similarly problematic concerning its date. The link between these sherds is that they do not belong to either Prehistoric or Roman wares. A Classical or Hellenistic date seems most suitable.

Some additional eroded and fragmentary sherds (Cat. Nos 327, 352) could be from Classical vessels but cannot be dated with any accuracy. The nine roof tile fragments (Cat. Nos 114-122) from VOL-1 are probably Hellenistic (see below).

The Hellenistic period

The Hellenistic period begins with the death of Alexander the Great (323 BC) and finishes at 31 BC ("battle of Actium") or 27 BC ("beginning of post-Republican period").[135] In Magnesia and Thessaly in general, this period is marked by the increasing

---

134 Dr Dimitris Grigoropoulos suggested a Corinthian provenance based on the clay colour, fabric and firing of this sherd. I thank him for bringing this to my attention.

135 Hayes 1991, 183; Rotroff 2006, 3-4.

power of the Macedonians, the competition between Macedonians, Aetolians and later the Romans for the control of the region and the continuous struggle of the locals for the independence of their cities.[136] The most important turning point for this region was the foundation of Dimitrias (294-292 BC), a city that succeeded ancient Pagasai and would become an important centre and port in central Greece for the next centuries. A good number of PCP sherds may belong to the Hellenistic period. Most of them are from coarse and plain wares for cooking, serving food, transportation and storage activities and thus are difficult to date or even to distinguish from equivalent sherds of the Roman or Classical period in some cases. There are almost no decorated wares (mouldmade bowls with relief decoration, West Slope or Hadra pottery, etc.), with the exception of a few painted sherds, one black painted lamp fragment and some painted roof tiles. The existence of local workshops[137] complicates the chronology since there are currently too few well-dated ceramic assemblages from Thessaly.[138]

Nine roof tiles fragments were collected at VOL-1 (Cat. Nos 114-122). Most are from pan tiles and the curvature observed in Cat. Nos 116, 117, 120 and 121 suggests that the tiles may belong to the Laconian type. What is more, the presence of red or black paint in at least three of them suggests a Classical or Hellenistic date. The latter is more probable since the cave is located immediately beneath a Hellenistic fort, a fact that also explains the discovery of roof tiles inside this cave.

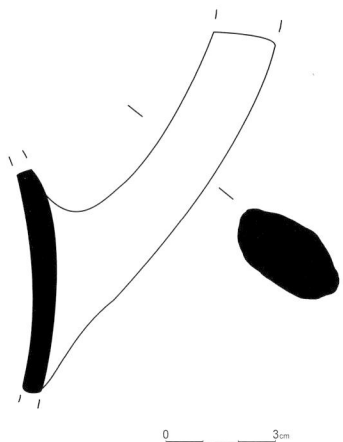

Fig. 2.6.13. Hellenistic ridged handle from KAR-8 (Cat. No. 210).

Another roof tile fragment, possibly a cover tile, was collected at KAR-8 (Cat. No. 204). The presence of red paint on its outer surface suggests a Classical or Hellenistic date.

Of Hellenistic date is also a ridged handle (Cat. No. 210) from KAR-8 (Fig. 2.6.13). It is reminiscent of handles commonly appearing on so-called "Macedonian" amphorae.[139]

"Macedonian" amphorae were traditionally dated to the end of the third and mainly the second century BC.[140] Recently, Papaconstantinou, based on a well-dated local assemblage from Lamia, argued for a date in the first half of the third century BC.[141] This not only means that the type was already common by the beginning of the third century, but also that there were numerous variants produced at various local workshops. Consequently, even though a date around the 3rd or 2nd centuries BC is secure for the handle fragment from KAR-8, it is difficult to esti-

---

136   For a brief but comprehensive review of the Hellenistic history of the region, see Intzesiloglou 2000, 11-5.

137   Doulgeri-Intzesiloglou for example refers to a whole potter's quarter at Pherai (see Doulgeri-Intzesiloglou 1992), and there would definitely be more workshops at Dimitrias.

138   Due to this fact, the main source of parallels for this research was the equivalent ceramic material from the Athenian Agora.

---

139   I thank Anthoula Tsarouha for identifying this type of handle.

140   Drougou 1988, 86; Drougou & Touratsoglou 1994, 135.

141   See Papaconstantinou 2000, 193-203.

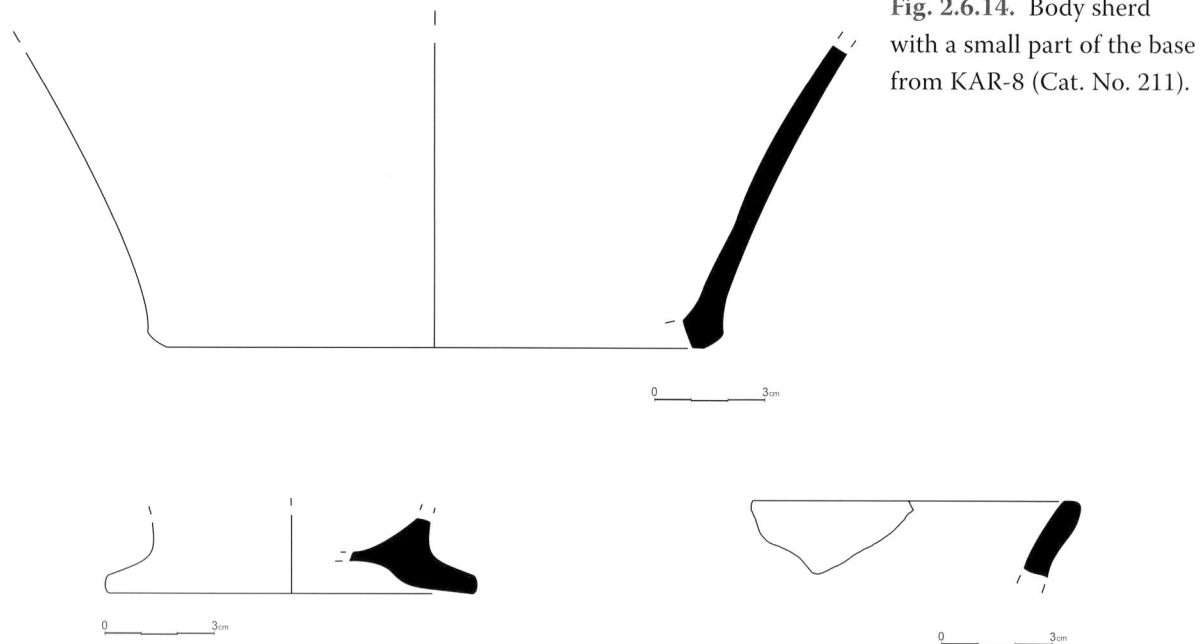

Fig. 2.6.14. Body sherd with a small part of the base from KAR-8 (Cat. No. 211).

Fig. 2.6.15. Fragment from Hellenistic or Roman vessel (Cat. No. 214).

Fig. 2.6.16. Rim from an open vessel, Hellenistic or Roman (Cat. No. 215).

mate the vessel's provenance without further information regarding its shape.

Cat. No. 211 from the same cave is a body sherd with a small part of the base (Fig. 2.6.14). The shape is reminiscent of medium-sized *lekanai* with disc foot from the Athenian Agora,[142] dated between the end of the third and the mid-second century BC. Cat. Nos 214-215 could be either from Hellenistic or Roman vessels. Cat. No. 214 resembles vessels with projecting foot or broad base (Fig. 2.6.15),[143] and Cat. No. 215 is a typical rim from an open vessel (possibly bowl) with no other datable features (Fig. 2.6.16).

Cat. No. 326 from MAK-12 might be from a Hellenistic vessel. The very fine fabric and especially the existence of a band of red gloss are reminiscent of closed vessels with similar decoration from the

Athenian Agora.[144] Based on its fabric and firing, a similar or even earlier (Classical) date is tentatively proposed for Cat. No. 327.

Cat. Nos 450-451 are from open vessels for food serving. Black painted decoration is observable on both surfaces of Cat. No. 451. The presence of a fragmentary ring foot points to a shape like the shallow plates with rolled rim discovered at the Hellenistic cemetery of Echinos in Thessaly.[145] Based on Papaconstantinou's observations, the small diameter of base, thin walls and lack of decoration on part of the surface are features that date these plates to the second half of the second century BC or beginning of the first century BC,[146] a date that fits with most of the sherds collected at MOU-1. Cat. Nos 445-449

---

142   Rotroff 2006, fig. 46, no's 269-70.
143   See Rotroff 1997, fig. 1, nos 12-13.

144   For possible examples, see Rotroff 1997, fig. 20, nos 125-126.
145   See Papaconstantinou 2000, 211-2, pls 11-12. For more parallels, see also Rotroff 1997, 142-50.
146   Papaconstantinou 2000, 210-1.

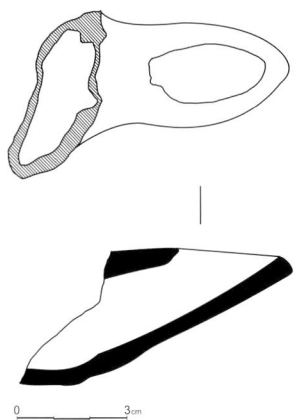

Fig. 2.6.17. Fragment of a lamp with a single nozzle of oval termination (Cat. No. 434).

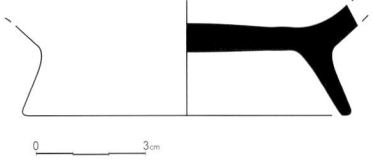

Fig. 2.6.18. Ring foot with black glazed slip, possibly Hellenistic (Cat. No. 454).

are too eroded to date securely, but judging by their similarity in fabric and clay colour and the presence of shallow circular lines on their inner surfaces, they are probably from similar Late Hellenistic vessels. Alternatively, they could belong to lids. Cat. No. 455 dates to the mid-second century BC or slightly earlier. This rilled rim sherd is probably from an unglazed open vessel for serving food, such as the plates and saucers from the Athenian Agora.[147]

The typical Hellenistic glazed slip is preserved in only a few occasions. One is Cat. No. 434, a fragmentary lamp with a single nozzle of oval termination (Fig. 2.6.17). This thick and fat nozzle type is reminiscent of lamps described as having a "spoon-shaped" wick-hole and a collar added to their globular type.[148] The closest parallel is the Attic type 37 and especially 37C.[149] Additional parallels from Thessaly suggest a date around the end of the

second or beginning of the first century BC.[150] Cat. No. 454 is a base sherd (ring foot) with eroded black glazed slip on all surfaces (Fig. 2.6.18). It is possibly from an open vessel for serving food or drink.[151] The existence of black slip suggests a Hellenistic date.

The existence of a thin coating of whitish slip and the pinkish clay of Cat. Nos 430 and 441 resemble amphorae of the Rhodian type.[152] The sherds are too fragmentary to provide a more accurate date than around the second century BC or a bit later.

Other sherds collected at MOU-1 fall within a broad Late Hellenistic–Early Roman chronological range. Cat. Nos 419-420 is a bowl fragment with slightly out-turned rim and a flat base (Figs 2.6.19 & 2.6.20).

While the whole profile of the vessel is preserved, there are no exact parallels for its shape, suggesting a local provenance. On the other hand, its fabric and firing point to a Hellenistic-Roman date. Similarly, Cat. Nos 422, 426, 429, 431, 433, 437, 438 and 443 provide few or no datable features but share some common characteristics that suggest a date around

---

147   See Rotroff 1997, 151-2. According to Rotroff, however, this type is exclusively Attic and was not exported or imitated elsewhere.

148   Howland 1958, 121.

149   Howland 1958, 121, pl. 45, no. 504.

150   See BE 8617 in Doulgeri-Intzesiloglou 1994, 371-2, pl. 278a–b; also BE 16027 in Nikolaou 2004, pl. 12a and BE 9335 in Malakasioti 2004, 95-7, pl. 19.

151   This type of ring foot for example is reminiscent of the two-handled cups from the Athenian Agora dated to the second and first centuries BC. See Rotroff 1997, 117-9.

152   For a Rhodian-type amphora example, see Joncheray 1976, 22, pl. V, nos 52-53.

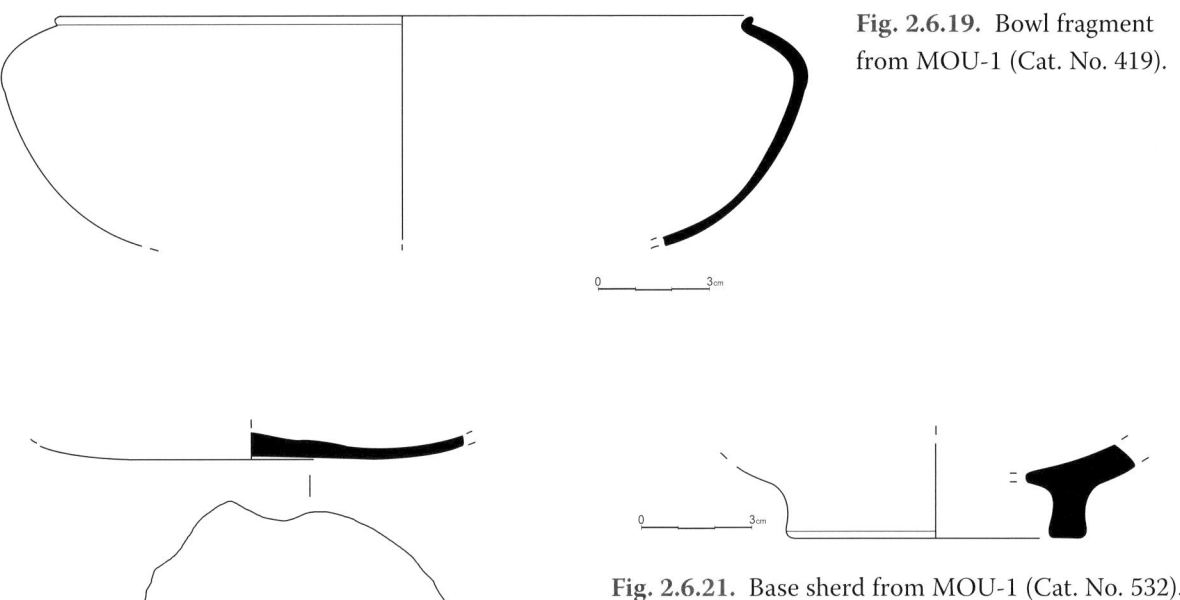

Fig. 2.6.19. Bowl fragment from MOU-1 (Cat. No. 419).

Fig. 2.6.20. Bowl fragment from MOU-1 (Cat. No. 420).

Fig. 2.6.21. Base sherd from MOU-1 (Cat. No. 532).

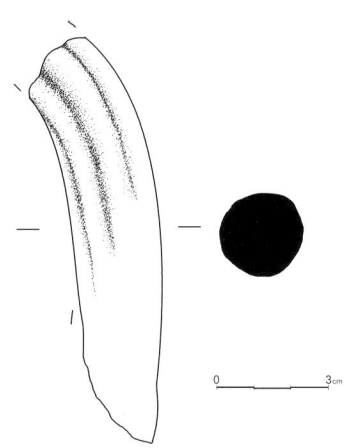

the second or first century BC. All are uncharacteristic body sherds from wheelmade vessels and nearly all are closed shapes. Lastly, a Late Hellenistic date is suggested for Cat. No. 432, a small fragmentary flat base with circular traces of the wheel on both surfaces.

Another fragmentary base sherd (Cat. No. 532) from MOU-7 resembles typical Hellenistic ring foots but the range of vessels it covers leaves no space for speculations concerning its exact date and shape (Fig. 2.6.21). Of Late Hellenistic date is also the fragmentary grooved (and slightly twisted) handle Cat. No. 545, most probably from a closed vessel (possibly *lagynos*, jug or *chytra*; Fig. 2.6.22). Lastly, based on fabric, firing and surface treatment, a Hellenistic date is also suggested for Cat. Nos 537, 544, 548 and the fragmentary flat base 549 (Fig. 2.6.23).

Fig. 2.6.22. Late Hellenistic grooved and slightly twisted handle from closed vessel (Cat. No. 545).

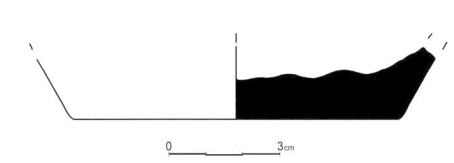

Fig. 2.6.23. Hellenistic flat base (Cat. No. 549).

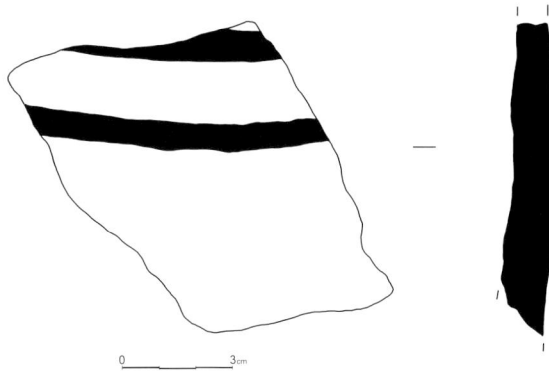

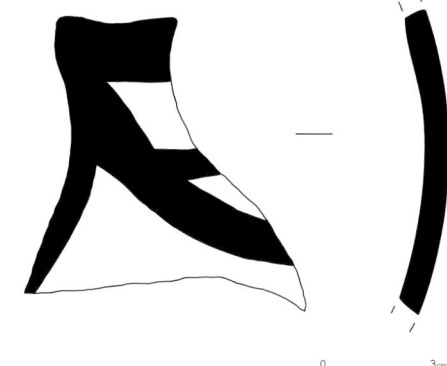

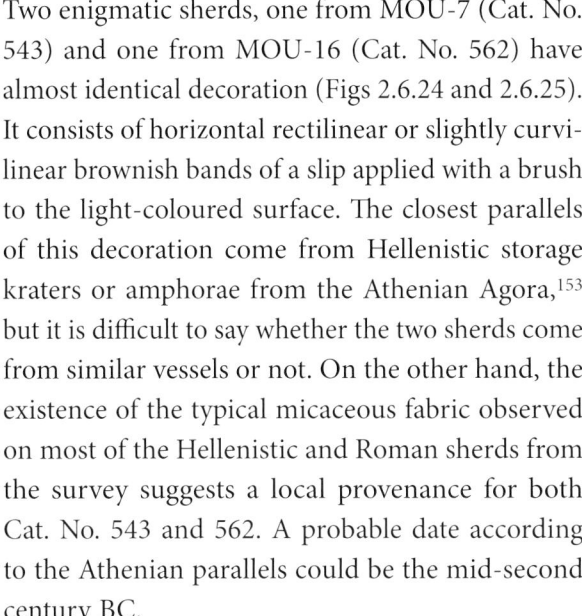

Fig. 2.6.24. Decorated sherd from MOU-7, possibly Hellenistic (Cat. No. 543).

Fig. 2.6.25. Decorated sherd from MOU 16, possibly Hellenistic (Cat. No. 562).

Two enigmatic sherds, one from MOU-7 (Cat. No. 543) and one from MOU-16 (Cat. No. 562) have almost identical decoration (Figs 2.6.24 and 2.6.25). It consists of horizontal rectilinear or slightly curvilinear brownish bands of a slip applied with a brush to the light-coloured surface. The closest parallels of this decoration come from Hellenistic storage kraters or amphorae from the Athenian Agora,[153] but it is difficult to say whether the two sherds come from similar vessels or not. On the other hand, the existence of the typical micaceous fabric observed on most of the Hellenistic and Roman sherds from the survey suggests a local provenance for both Cat. No. 543 and 562. A probable date according to the Athenian parallels could be the mid-second century BC.

The Roman period sherds of PCP[154]

The Roman period in Greece extends from the late first century BC to the early seventh century AD.[155] A further division followed in this chapter is: Early Roman (first century BC – first century AD), middle Roman (circa second century AD – beginning of fourth century AD) and Late Roman (circa fourth – early seventh centuries AD). During the Roman period, Thessaly was part of the province of Achaea.[156] Dimitrias continued to be the centre of Magnesia, even after its destruction by the Pontic fleet (between 88-85 BC).[157] Most of the sherds analysed in this chapter belong to one of these sub-periods. The percentage of datable sherds is again relatively small. On the other hand, Roman wares, fabrics, techniques of surface finish and decorative features are easier to identify or at least classify to one of the Roman periods. More tentative is the classification to specific centuries, and it is even more difficult to be accurate concerning the transitional phase between

153   Rotroff 2006, 89, 105-7.

154   I am grateful to Dr Dimitris Grigoropoulos for his valuable comments and for identifying Cat. No. 439 and most of the amphora sherds.
155   Robinson 1959, 1.
156   Intzesiloglou 2000, 15.
157   Intzesiloglou 2000, 14.

the Hellenistic and Early Roman periods, to which a large number of sherds from MOU-1, MOU-7 and MOU-16 seem to belong.

It is important to note that with only a few exceptions (e.g. Cat. No. 439), most of the Roman sherds are from coarse cooking wares or vessels for short-/long-term storage and from amphorae for the transportation of liquids. Of special importance is the increased presence (compared to the Archaic–Classical periods) of Late Hellenistic and mainly Roman amphora sherds in the caves of eastern Pelion. Magnesia, with Dimitrias as its centre and main port, was naturally part of the dense trade network of these periods. The existence of imported vessels (e.g. Cat. No. 439 from NW Asia Minor) and also of Italian (Dressel 1), Koan, Rhodian and other amphora types suggest that the coastal areas of eastern Pelion were active participants in these long-distance exchanges. It is risky to suggest that the main use of caves was for storage activities based on a few amphora sherds, but what needs to be emphasised here is that there is no sign of the isolation we would expect for an area so difficult to approach from land.

From ART-9, only one rim sherd is probably of Late Roman date (Cat. No. 3). Based on its fabric, firing and surface treatment, it is a sherd from a coarse cooking pot, approximately of the sixth to seventh centuries AD.

The ceramic material from KAR-8 seems to include a few more Roman sherds. Cat. No. 202 is a base and rim sherd from an open vessel (possibly a small plate). Cat. No. 247 most probably belongs to a Roman amphora. Two other sherds collected at the inner part of chamber 1 (Cat. Nos 205-206) are from closed vessels (possibly jugs). The existence of wheel ridges suggests a Roman date for both. Lastly, Cat. Nos 216-217 are two small and thin-walled sherds of the same vessel. Wheel-ridges cover their outer surfaces. They could belong either to a small jar or most probably to a small trefoil-mouth jug.

An approximate date according to parallels from the Athenian Agora is the mid-third century AD.[158]

The three caves around Koukourava village (MAK-12, MAK-15 and MAK-16) mostly include Roman sherds. Only Cat. Nos 324-325 from MAK-12 are datable, and they come from the same vessel. Even though the sherds are eroded, their fabric and the presence of six parallel, incised lines on Cat. No. 324 suggest a Roman amphora (second to third centuries AD). The three sherds from MAK-15 (Cat. Nos 328-330) are also very eroded but their fabric is also reminiscent of Roman pottery. Cat. No. 331, a fragmentary base from MAK-16, is from a closed shape, possibly a jug with round mouth of Early Roman (or Late Hellenistic) period. Based on fabric and firing, a similar date is suggested for the remaining two sherds of this cave, even though they are badly eroded (Cat. Nos 332-333).

The presence of grog and the absence of any paint or other decorative features on a roof tile fragment from VOL-3 suggest a Post-Hellenistic and possibly Roman date (Cat. No. 344).

Only one sherd (Cat. No. 409) was collected at KAR-22 and its chronology is problematic. The existence of three curvilinear incised lines is reminiscent of Late Roman grooved wares (fourth to seventh centuries BC), but a later date cannot be excluded.[159] On the other hand, the finds from KAR-23 include two possible Roman sherds (Cat. Nos 410-411). Both are weathered and eroded but their fabric points to a Middle or Late Roman date. However, given the find context, which includes post-Medieval sherds, a later date should not be excluded.

Another securely dated fragmentary vessel was collected at KER-3. Cat. No. 418 is the upper part of a medium-sized trefoil-mouth jug with visible wheel-ridges on both surfaces (Fig. 2.6.26).

---

158   See for examples Robinson 1959, pl. 13, K69-72 and pl. 16, L4 and L6.

159   See Vroom, this volume.

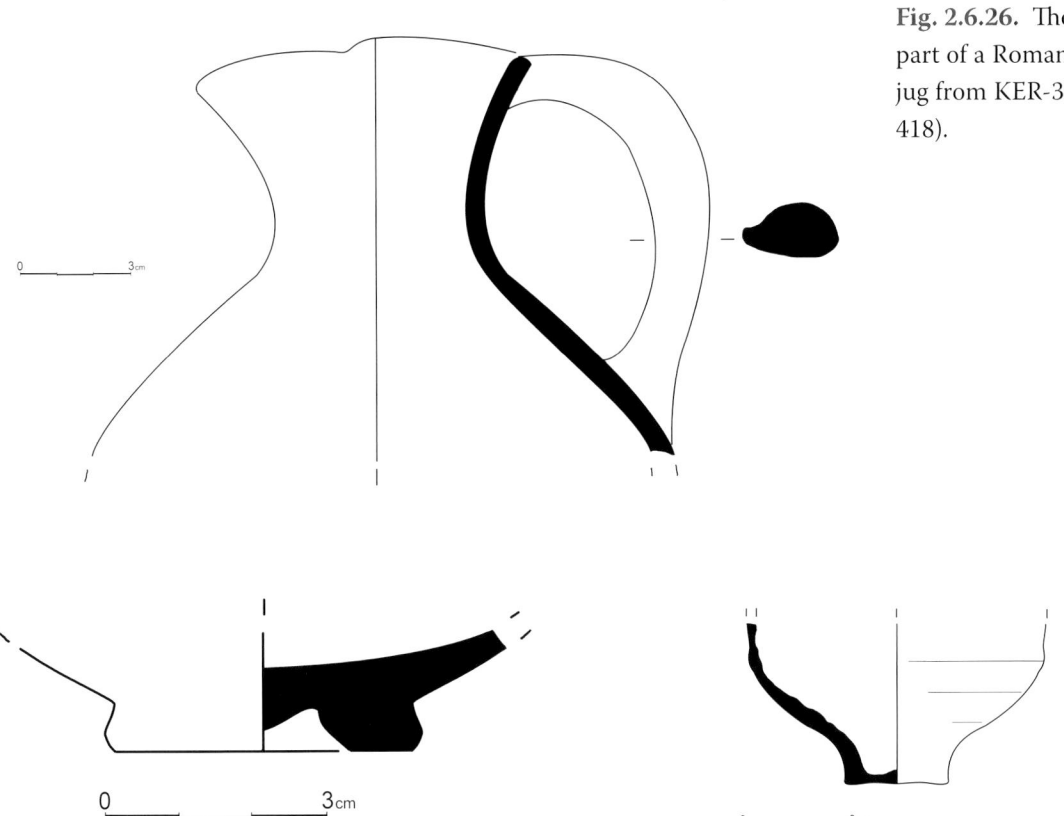

Fig. 2.6.26. The upper part of a Roman, trefoil jug from KER-3 (Cat. No. 418).

Fig. 2.6.27. Disc foot from Early Roman red-gloss "Pergamene Ware" (Cat. No. 439).

Fig. 2.6.28. Lower part of a vessel with flat-based, high foot (Cat. No. 436).

The trefoil-mouth jug is ubiquitous in the Classical, Hellenistic and Roman periods of the Aegean but the fabric, firing and shape of this vessel are indicative of the Middle Roman period,[160] most probably between the second and third centuries AD.

Cat. No. 439, from MOU-1, is one of the few sherds from fine and datable Roman wares (Fig. 2.6.27). It is a base sherd (disc foot) from an Early Roman red-gloss ware, usually termed "Pergamene Ware" or "Eastern Sigillata C". The source of production was the coastal area around ancient Pitane (or modern Çandarli) in West Turkey.

Cat. No. 439 is almost identical to a small-sized bowl with flanged rim (Forma Loeschke 19),[161] dated between 75-150 AD. Similar but slightly later is Hayes form 3, dated to the third century AD.[162] A date in the first or early second century AD seems more probable for 439.

Four more sherds collected at MOU-1 are from Roman vessels. Cat. No. 436 is the lower part of a vessel resting on a high foot with a flat base (Fig. 2.6.28). The type is reminiscent of ovoid or globular jugs dated to the end of the third and the beginning of the fourth centuries AD.[163]

---

160   For parallels, see Robinson 1959, pl. 7, G188 and pl. 23, M101.

161   See Hayes 1985, 76-7, pl. XVII, nos 5-7.
162   See Hayes 1972, 321; also Kallintzi & Chryssaphi 2010, 387-8, pl. 1a.
163   For examples, see Robinson 1959, 101-2, M190-M194.

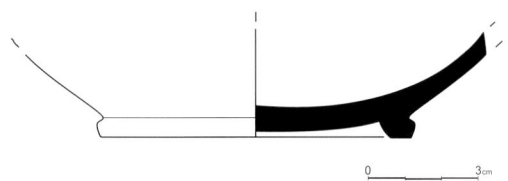

Fig. 2.6.29. Fragment from an open vessel with low projecting foot from MOU-1 (Cat. No. 456).

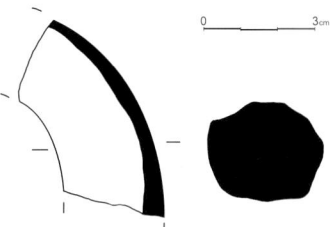

Fig. 2.6.30. Fragmentary amphora handle with a longitudinal ridge from MOU-11 (Cat. No. 521).

Cat. No. 440 has identical fabric and clay colour and it could be from the same vessel. Cat. No. 456 is possibly from an open vessel with low projecting foot (Fig. 2.6.29). Its fabric, clay colour and the existence of shallow grooves and wheel-ridges suggest a Roman date, possibly around the first century BC or AD. A similar date is also suggested for Cat. No. 457, which is a body sherd from a closed vessel, most probably from an Early Roman amphora.

Cat. No. 428 from MOU-1 is a small sherd with pale brown slip on its outer surface. The clay and slip colour and the existence of three small curvilinear incisions are reminiscent of Late Roman grooved wares and thus a date between the late fourth and seventh centuries AD is suggested for it.

At MOU-11, 61 out of the 63 collected sherds are from the same vessel (Cat. Nos 460-520). These are body sherds and most have the characteristic horizontal grooves of Late Roman wares. The shape is identified as a Late Roman I amphora,[164] a vessel used for the transportation of olive oil. It is normally dated between the fifth and early seventh centuries BC. The fragmentary handle with a longitudinal ridge Cat. No. 521 could be from the same amphora, as it was found together with the body sherds (Fig. 2.6.30).[165] Of earlier date but possibly from another Roman amphora of unidentifiable type

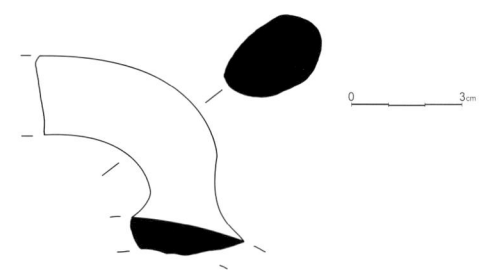

Fig. 2.6.31. Roman handle from MOU-11 (Cat. No. 522).

is handle Cat. No. 522 collected at the same cave (Fig. 2.6.31).

The single sherd from AFE-5 (Cat. No. 523) and eight more from MOU-2 (Cat. Nos 524-531) are uncharacteristic and eroded, but their fabric and firing suggests a possible Roman date. Similarly, for the uncharacteristic sherd Cat. No. 556 from MOU-20 an Early or Middle Roman date (first to second century AD) is suggested.

Sherds from MOU-7 such as the flat base Cat. No. 539, and also Cat. Nos 541, 542 and 551 are poorly preserved but they appear to be Roman (Fig. 2.6.32).

The existence of wheel-ridges in Cat. No. 546 also suggests a Roman date. The fabric and surface treatment of Cat. No. 536 are reminiscent of Middle Roman vessels.

A few sherds from MOU-7 appear to come from Roman amphorae. Cat. No. 535 has a similar decoration with the Late Roman I amphora from MOU-11

---

164   See for example Ntina 2010, 564 and pl. 3.

165   For a similar handle of the early seventh century AD, see Robinson 1959, 122, N7.

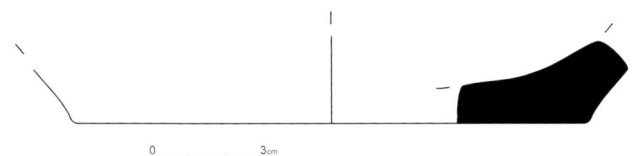

Fig. 2.6.32. Roman flat base sherd from MOU-7 (Cat. No. 539).

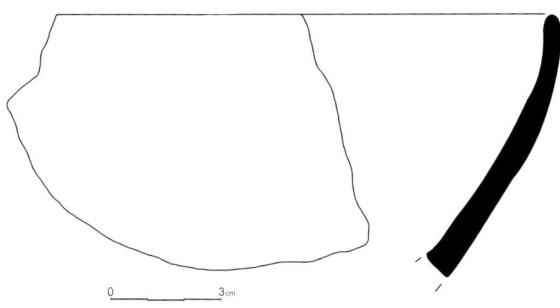

Fig. 2.6.33. Fragment of a possibly Hellenistic bowl with incurved rim (Cat. No. 576).

(see above) and could, therefore, be from another Late (or Middle) Roman amphora. Cat. No. 538 is earlier (Early or Middle Roman), but it is impossible to identify its shape and type. For the remaining two sherds of this cave (Cat. Nos 534 and 550), an Early Roman date is suggested. Judging by its fabric and shape, Cat. No. 534 is probably from a Dressel 1 amphora,[166] a popular vessel in the Mediterranean during the late second and first centuries BC. The main source of production of this type was Italy, and it was used for wine transportation. An approximate date for MOU-7's sherd is the first century BC. The fabric of Cat. No. 550 is reminiscent of Koan amphorae but its highly fragmentary condition makes its identification tentative.

Some amphora sherds from MOU-16 (Cat. Nos 563, 565 and 569-570) could be from the same vessel. Their fabric is again reminiscent of Early Roman

Dressel 1 amphorae with a date within the first century BC.

Another category of sherds with coarse fabric from MOU-16 suggests the presence of cooking vessels. More specifically, the mica and large schist and occasional quartz pieces (Cat. Nos 564, 567 and possibly 568) are reminiscent of Attic cooking wares of the Hellenistic period.[167] A date around the end of the second century and first century BC (Late Hellenistic–Early Roman) is suggested for these three sherds. Judging by fabric and surface treatment, Cat. No. 560 is slightly later (possibly first century AD), and falls within the chronological limits of Early Roman wares.

There are no securely datable features for the remaining sherds from MOU-16 (Cat. Nos 577-579, 582 and 584-587), but they all seem to fall in the period from the first century BC to the second century AD. The only exception is Cat. No. 576, which resembles typical Hellenistic bowls with incurved rim,[168] and could, therefore, be dated to the third century BC or slightly later (Fig. 2.6.33).

Lastly, an Early Roman date (end of first century BC to first century AD) is suggested for Cat. Nos 581 and 583. Cat. Nos 580 and 588 are undiagnostic and could belong to either Roman or post-Roman vessels.

---

166 Possibly from a Dressel 1B amphora, but it is difficult to be certain without having any clues about its rim. For Dressel I amphorae see Joncheray 1976, pl. IV, nos 41-45.

167 Rotroff 2006, 39-45. See also note 14 of this paper.
168 Rotroff 2006, 115. This type is the plain ware equivalent of the popular black painted echinus bowl, according to Rotroff.

2.7 Medieval and post-Medieval ceramic finds of the Pelion Cave Project

Joanita Vroom, Leiden University (NL)

Greek summary

Από τα 591 όστρακα κεραμικής που περισυνελέγησαν κατά τη διάρκεια της έρευνας, 244 είναι διαγνωστικά της Βυζαντινής και μετα-Βυζαντινής περιόδου. Αντιπροσωπεύεται μια ποικιλία σχημάτων που περιλαμβάνει κανάτες, κλειστά αγγεία, μαγειρικά σκεύη, ανοικτά αγγεία, αποθηκευτικά αγγεία ή πίθους, φιάλες/κούπες, λεκανίδες και κλειστά αγγεία με λαιμό. Η χρονολόγηση των οστράκων μπορεί χονδρικά να διαιρεθεί στην μετα-Βυζαντινή και Νεότερη περίοδο (περίπου 16ος-20ος αιώνας μ.Χ.) (91%), στην Ύστερη Βυζαντινή περίοδο (περίπου 13ος-15ος αιώνας μ.Χ.) (4%), στην Ύστερη Βυζαντινή/μετα-Βυζαντινή περίοδο (2%) και σε όστρακα άγνωστης χρονολόγησης (3%).

Χαρακτηριστικά παραδείγματα της Ύστερης Βυζαντινής περιόδου αποτελούν όστρακα της κεραμικής τύπου "Ζευξίππου", κλειστά σχήματα γραπτής δι'επιχρίσματος κεραμικής, μονόχρωμες κίτρινες εφυαλωμένες κούπες, καθώς και Πολύχρωμα αδρεγχάρακτα ("Sgraffito") αγγεία του ύστερου 15ουαιώνα με επίθετη διακόσμηση πράσινης και καφέ απόχρωσης.

Στα παραδείγματα μετα-Βυζαντινής κεραμικής περιλαμβάνεται μία λεκανίδα ή μεγάλη κούπα με λευκό επίχρισμα εξωτερικά και εφυάλωση εσωτερικά, καθώς και γραπτή διακόσμηση με πράσινες λωρίδες ή πινελιές. Ένα τμήμα προχοής ή στομίου προέρχεται από αγγείο του τέλους του 19ου-πρώιμου 20ου αιώνα μ.Χ. (ibrik). Χρησιμοποιήθηκε για μεταφορά ή σερβίρισμα υγρών καθώς επίσης και για λόγους υγιεινής όπως πλύσιμο χεριών ή μετάγγιση νερού σε λουτρό ("hamam"). Υπάρχουν, επίσης, θραύσματα από πινάκια ή λεκανίδες, γνωστές στην Ελλάδα ως "λεκάνες". Αυτά τα πολυλειτουργικά αγγεία χρησιμοποιούνταν για ζύμωμα ψωμιού, πλύσιμο ενδυμάτων ή και για λόγους προσωπικής υγιεινής.

Τα ευρύστομα εφυαλωμένα μαγειρικά σκεύη ("τσουκάλια ή τσικάλια") προέρχονται από το νησί της Σίφνου. Χρησιμοποιούνταν για το μαγείρεμα της τροφής και για την κατασκευή ή αποθήκευση τυριού και γιαουρτιού. Ορισμένες χαρακτηριστικές κανάτες (γνωστές στη Θεσσαλία ως "κανάτια") εμφανίζουν στο άνω τμήμα της εξωτερικής επιφάνειάς τους ένα λευκωπό επίχρισμα και κηλίδες ή πιτσιλιές πράσινης ή κίτρινης εφυάλωσης πάνω από αυτό. Μερικές φορές εμφανίζεται και διακόσμηση περιμετρικά, στον ώμο του αγγείου. Τα αγγεία αυτά χρησιμοποιούνταν, ως επί το πλείστον, για τη συλλογή νερού από πηγάδια ή πηγές, για μεταφορά υγρών με ζώα (γαϊδάρους, μουλάρια κ.ά.) ή απλώς για κατανάλωση υγρών. Φαίνεται ότι αποτελούσαν τοπικά προϊόντα του Βόλου όπου κατασκευάζονταν μεταξύ του 1940 και 1960 από πηλό που είχε συλλεχθεί σε όχθη χειμάρρου κοντά στο Διμήνι. Τέτοιου είδους κανάτια εντοπίζονται ακόμη και σήμερα σε κήπους ή στο εσωτερικό Πηλιορείτικων σπιτιών. Σε άλλες περιπτώσεις κτίζονταν σε τοίχους σπιτιών (πρόκειται για τα επονομαζόμενα "bacini").

Introduction

During the summer of 2010 I was invited to study the post-Roman ceramic finds of the Pelion Cave Project (PCP), carried out under the direction of the Danish Institute at Athens in collaboration with the Ephorate of Palaeoanthropology and Speleology of Northern Greece. The pottery finds had been collected during a diachronic, regional survey in and around caves, rockshelters and abandoned mines on the Pelion Mountain in southeastern Thessaly. The principal aim for me was to diagnose and to date the post-Roman ceramics from the field seasons 2006-2008. During my research, I have been looking at bags with Medieval and post-Medieval pottery finds from 29 selected sites, and I would like to thank the project director Niels H. Andreasen of the Pelion Cave Project for the chance to see this material.

One must keep in mind, of course, that most of these survey finds are abraded, fragmentary and unstratified. Furthermore, it is clear that some pottery types are, within the analysis of surface assemblages, chronologically more diagnostic and better defined than others.[169] Unfortunately, there are no large-scale excavations with published post-Roman ceramic finds in the direct neighbourhood of Pelion, and this lack of excavated comparisons makes the analysis, definition and dating of the later ceramic finds of the Pelion Cave Project problematic. For my research, I could rely only on parallel pottery finds from other regions in Greece.

Of the 591 sherds collected by the team of the Pelion Cave Project, I have recognised approximately 244 diagnostic pottery fragments of the Medieval and post-Medieval periods, and 41 roof tile fragments (mostly of unknown date). The diagnostic pieces include imported vessels from other parts of Greece as well as locally manufactured products. The local fabrics of the Medieval and post-Medieval fragments

found during the Pelion Cave Project are, in general, medium fine, red-bodied and quite micaceous (because of the existence of schist in the region).

For reasons of quantification, I have been looking at the differences in vessel parts, vessel shapes, types of wares and dated periods of the later ceramic finds (Tables 2.7.1-2.7.4). During this study, I have only counted the diagnostic 244 pottery fragments (and not the additional 41 roof tiles). The majority of the total vessel parts in the pottery assemblage consists of body fragments (85%), followed by rims (6%), bases (6%), handles (2%) and a limited amount of other vessel parts such as lids and spouts (1%; see Table 2.7.1).

Of the total vessel shapes, most are unknown (51%) due to the fragmentary character of the survey material, although the following shapes could be recognised: jugs (19%), closed vessels (10%), cooking pots (7%), open vessels (5%), storage jars or *pithoi* (3%), bowls (2%), basins (2%) and jars (1%; see Table 2.7.2). Most typical are Unglazed Wares (64%), followed by Monochrome Glazed Wares (21%), Decorated Glazed Wares (7%), Decorated Unglazed Wares (6%) and Transparent Glazed Wares (2%; see Table 2.7.3).

The dating of the ceramic finds can be roughly divided in the post-Medieval period (circa sixteenth–twentieth centuries; 91%), the Late Medieval period (circa thirteenth–fifteenth centuries; 4%), the Late Medieval/post-Medieval periods (2%) and pottery fragments of unknown date (3%; see Table 2.7.4).

Vessel shapes	%
Unknown	51
Open	5
Closed	10
Jug	19
Pithos	3
Jar	1
Basin	2
Cooking pot	7
Bowl	2

Table 2.7.1. Distribution of vessel shapes.

---

169   For problems regarding chronology and terminology of Medieval and post-Medieval surface finds, see Vroom 2003, 25-8, 45-6, 135-7.

Vessel parts	%
Other	1
Rim	6
Base	6
Handle	2
Body	85

Table 2.7.2. Distribution of vessel parts.

Wares	%
Monochrome Glazed	21
Decorated Glazed	7
Transparent Glazed	2
Decorated Unglazed	6
Unglazed	64

Table 2.7.3. Distribution of ware types.

Periods	%
Unknown	3
Late Medieval	4
Late Medieval / post-Medieval	2
post-Medieval	91

Table 2.7.4. Chronological distribution of sherds.

Medieval ceramics

Only a few caves yielded pottery finds of the Medieval period, and in particular of the Late Medieval period (circa thirteenth–fifteenth centuries). These include the rockshelters KAR-4, ART-9 and ART-10. In this last locality, seven fragments of Late Medieval times were recovered during the field season of 2006.

Among them is a body fragment of an open vessel (probably a bowl) of a so-called "Zeuxippus Ware Subtype" or "Zeuxippus Ware Variant" (Cat. No. 52). This is a type of incised glazed ware of the thirteenth–mid-fourteenth centuries, which was widely distributed throughout the eastern Mediterranean (and especially in northern and central Greece and on the western coast of Turkey).[170] The interior of the fragment is covered with a white slip (also known in

French as *engobe*) and a monochrome yellow lead glaze, and decorated with one gouged and four incised concentric circles in the centre of the vessel. This fragment can be dated to the Late Medieval/Late Byzantine period (c. thirteenth century).[171]

Of the same period is a body fragment of a closed vessel (probably a jug) of Slip-Painted Ware (Cat. No. 53). The sherd is decorated on the outside with circles of white slip, which were then covered with a green lead glaze, giving the white slip a light greenish colour. This type of decoration can also be seen on Late Medieval vessels from excavations in Arta (in northwestern Greece).[172]

The Late Medieval pottery assemblage from ART-10 further includes a body fragment of a Monochrome Yellow Glazed Ware bowl (Cat. No. 58), as well as four body fragments of Polychrome Sgraffito Ware with added colours in green and brown (Cat. Nos 54-57). These last pieces can be roughly dated to the (late) fifteenth century.[173]

Post-Medieval ceramics

As we have seen in Table 2.7.4, most of the later ceramic finds of the Pelion Cave Project (at least 91%) can be dated to the post-Medieval period (circa sixteenth–twentieth centuries). Within this group of ceramic finds, I will discuss some pottery types that deserve further attention.

The cave of VOL-4 yielded two base fragments of a basin or large bowl with a flat base of Painted Glazed Ware (Cat. Nos 130 and 131; Fig. 2.7.1).

170 Cf. Vroom 2005, 110-1.

171 Cf. Vroom 2003, 164-5, fig. 6.25: W16.1-5 with further literature; Papanikola-Bakirtzi & Zekos, 2010, 57, no. 44 (early thirteenth century).

172 Cf. Papadopoulou & Tsouris, 1993, 245-8, fig. 6, no. 11; Vroom 2003, 167, fig. 6.43: W21.1-3 with further literature; Vroom 2005, 124-5.

173 Cf. Vroom 2003, 171-2, figs 6.30-1 and 6.43-4: W26.1-16 and W26.ex with further literature; Vroom 2005, 144-5.

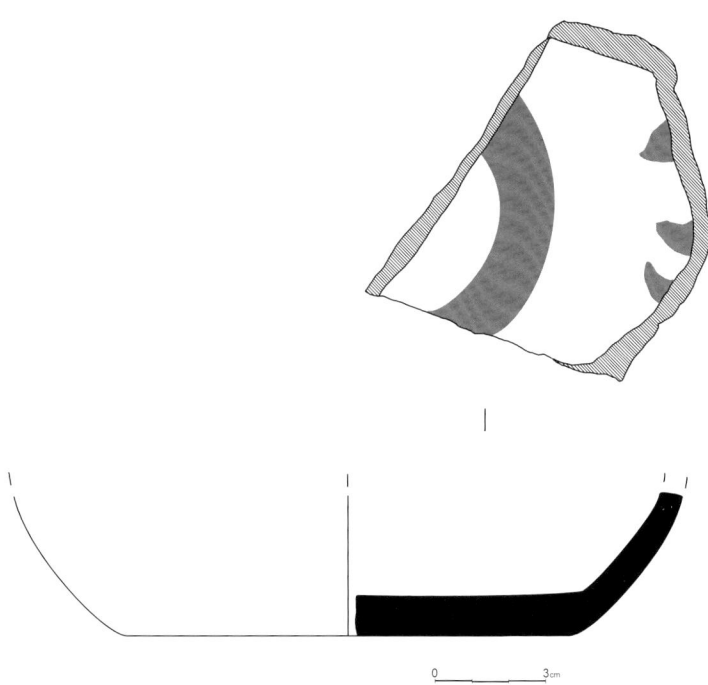

Fig. 2.7.1. Base fragment of a basin or large bowl with green strokes under yellow lead glaze (Cat. No. 130).

There is a white slip and yellow lead glaze on the inside of this vessel, enhanced with a painted decoration of green stripes or strokes under the glaze. If one compares these two base fragments with similar-looking examples from workshops at Lapithos on Cyprus, it is evident that these green-painted basins are quite late and can be dated around the late nineteenth to early twentieth centuries.[174]

At KAR-3, one spout fragment of a jug was found (Cat. No. 184; Fig. 2.7.2). The spout is glazed and decorated with a white slip-painted design. This design was painted directly on the vessel surface and then covered completely with a brown-tinted lead glaze. The colour of the glaze varies from light yellow where it covers the decoration to brown where it covers the un-slipped fabric.

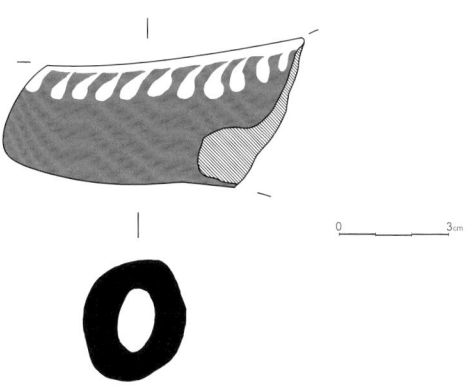

Fig. 2.7.2. Spout fragment of a jug with white, slip-painted design from KAR-3 (Cat. No. 184).

This type of jug with a spout (and often with one handle) is also known in Greece and Turkey as an *ibrik* (in Turkish), *briki* or *brikaki* (in Modern Greek).[175]

---

174 For examples of similar-looking painted vessels, see Korre-Zographou 2000, 177, fig. 174; Papademetriou 2005, 47 upper (early twentieth-century bowls from Lapithos on Cyprus).

175 See in general Vroom 2005, 176-7, figs TUR/VEN 16.2-3 for the shape of an *ibrik* with spout in an unglazed version; see Giannopoulou & Demesticha 1998, 66, no. 12 and 68, no. 26 and Vroom 2005, 194-5, fig. EMOD 7.5 for the slip-painted decoration on an unglazed jug or *ibrik* with spout.

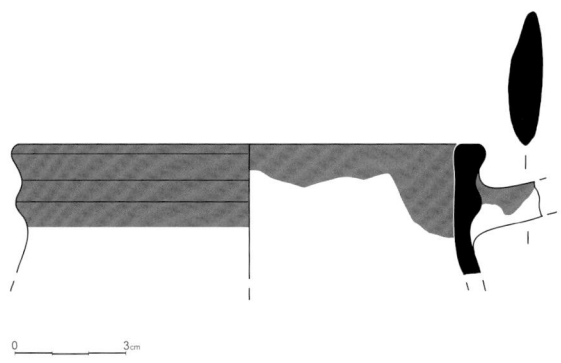

Fig. 2.7.3. Rim-handle fragment of a large jar of Monochrome Glazed Ware from KAR-6 (Cat. No. 201).

It was used for carrying and serving liquids (water, wine, raki, vinegar, olive oil), but also for sanitary purposes such as for pouring water in a basin during (hand) washing or for pouring water in a bathhouse (*hammam*).[176] This piece from Pelion can be dated around the late nineteenth to early twentieth centuries, as is shown by exactly analogous looking jugs with spouts decorated with slip-painted motifs from Bulgaria.[177]

Apart from slip-painted fragments, post-Medieval glazed wares are also present in the pottery assemblage. From KAR-6, for instance, comes a rim-handle fragment of a large jar of Monochrome Green Glazed Ware (Cat. No. 201; Fig. 2.7.3).[178] This piece has a white slip and green lead glaze on the interior and exterior upper part, and can be dated to the nineteenth to twentieth centuries. The glazed jar was probably used for the storage of goods or liquids (including olive oil).

At MIL-3, two rim fragments were recovered of wide-mouthed open vessels with glazed interiors (Cat. Nos 262 and 264; Figs 2.7.4-2.7.6). One sherd is covered with a white slip and green lead glaze on the inside, on top of the everted rim and on the exterior upper part. The other one has a white slip and yellowish lead glaze on the interior and on top of the rim but is also decorated on the interior with green painted stripes under the glaze.[179] Both fragments are probably parts of large dishes or basins, which are known in Greece as *lekani* (deriving from the ancient Greek words *lekos* or *lekanion* for 'dish'). These multi-functional vessels were not only used for kneading bread and washing clothes, but also for personal hygiene.[180]

In the same cave, I have distinguished three fragments of wide-mouthed glazed cooking pots or dishes (Cat. Nos 261, 265-266). They are covered on the inside with a vitreous transparent lead glaze. These fireproof cooking pots or dishes come from the island of Siphnos, and are known in Greece as *tsoukali* or *tsikali* (a diminutive of the Venetian word *zucca*, which means 'pumpkin').[181] They were used for cooking food on the stove, on the brazier or in the oven, as well as for the making or storing of pastoral products (such as cheese or yoghurt).

Finally, MIL-3 yielded three fragments of a type of jug (a so-called *stamna* in Modern Greek) with the exterior upper part of the vessel covered with a white slip and blotches or splashes of green or yellow lead glaze on top of the slip (Cat. No. 257; Fig. 2.7.7). Sometimes, there is also a roulette decoration on the shoulder (Cat. No. 259; Figs 2.7.8-2.7.9).

---

176   For the use of *ibrik* for sanitary purposes, see Vroom 2007, 3234, pl. 7c–f.

177   See Zanov 2000, 74 (right and left pictures below) and 133 (in colour).

178   For a similar-looking ware, see Papademetriou 2005, 19 below (nineteenth century), 21 below (early twentieth century), 37 below (early twentieth century).

179   I have noticed a similar painted technique on fragments collected at the Mallakastra Survey in Albania.

180   For a similar-looking *lekane*, see Vroom 1998, 145, fig. 10 with further literature; Vroom 2005, 190-1.

181   For examples of such cooking pots from Siphnos, see Vroom 1998, 145-6; Vroom 2003, 185, fig. 6.39: W43.1-2 with further literature; Vroom 2005, 192-3.

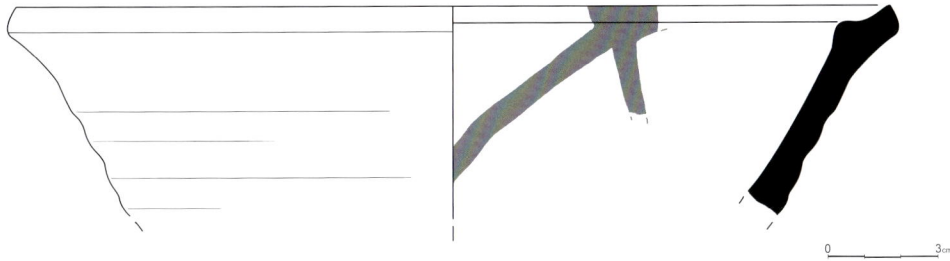

Figs 2.7.4. & 2.7.5. Rim fragment of a basin with yellow glazed interior from MIL-3 (Cat. No. 262).

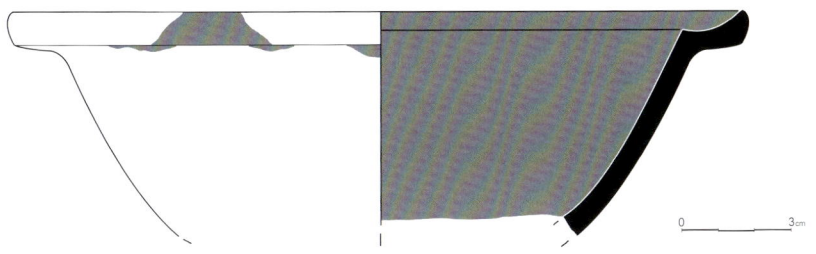

Fig. 2.7.6. Rim fragment with green lead glaze from MIL-3 (Cat. No. 264).

Fig. 2.7.7. Handle from *stamna* water jug with white slip from MIL-3 (Cat. No. 257).

These jugs are known in Thessaly as *kanati,* and can be dated in the twentieth century. They were mostly used for the collection of water from wells or springs, for the carrying of liquids on donkeys and mules, or simply for drinking. They appear to be local products from Volos in Thessaly, where they were made between 1940 and 1960 from clay gathered from a riverbed near Dimini (see below for more information about this production received through oral interviews with local potters). However, these water jugs (or *stamni*) were not exclusively made at

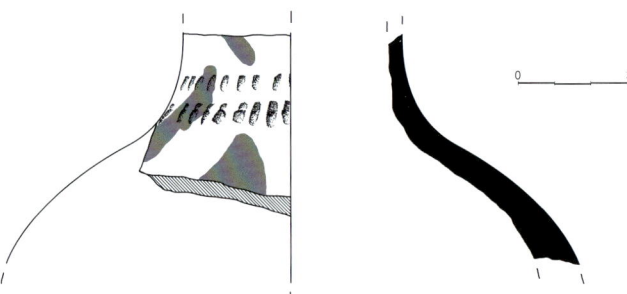

Figs 2.7.8. & 2.7.9. Fragment from *stamna* with white slip and splashes of green or yellow glaze from MIL-3 (Cat. No. 259).

Volos, but also at workshops in Karditsa (producing examples with a white slip and glaze splashes) and Tyrnavos (producing unglazed examples) in western Thessaly and in the coastal village of Agia (east of Larissa).[182]

It is interesting to know that MIL-3 (situated below the village of Milies) functioned until circa AD 1880 as a dwelling for a community of allegedly 150 Albanian Ghegs, who probably worked either on the construction of the Volos–Milies railway or as fieldworkers in the fruit orchards around Milies. The aforementioned basins, cooking pots/dishes and water jars found in this cave could have been used by the Ghegs for reasons of short-distance transport, storage and food preparation, as well as for purposes of beverage and food consumption.

A note on the production of Modern ceramics in Volos[183]

The following information was given to the Pelion Cave Project team by members of the Anetopoulos Family, and especially by Vaggelis Anetopoulos. This potter's family came to Volos from the mountainous village of Apeiranthos (or Aperathu) on the island of Naxos. This village had a rich history and folkloric tradition on Naxos.

The potter Basil Anetopoulos arrived in 1885 at Volos and established there the first and only workshop for the manufacture of ceramics (Fig. 2.7.10). After its incorporation into Greece from the Ottoman Empire in 1881, Volos had a population of only 4,900 inhabitants, but grew rapidly in the next four decades as merchants, businessmen, craftsmen and sailors gravitated toward the town from the surrounding area. Apparently, there was no ceramic production in the area of Volos before 1881.

After 1885, more potter's workshops appeared in Volos, in particular, those run by potters from the islands (such as Siphnos or Naxos). These islanders would already have been familiar with the local demand, as small boats from Chalkis and the Cycladic

---

182   See Papathoma 1999, 99, nos 17-8 for similar-looking jugs from Karditsa.

183   I would like to thank the Anetopoulos Family and Niels H. Andreasen for this information. See also *Δύο αιώνες ιστορίας στην κεραμική τέχνη.* Panthessalian News Portal ΤΑΧΥΔΡΟΜΟΣ. 07.09.2009.

Fig. 2.7.10. Complete example of locally produced water jug (*kanati*) with white slip and bright green-painted blotches on display in the Anetopoulos Pottery Museum.

Islands (e.g., from Lesbos, Skyros, the Sporades) regularly came to Pelion (especially the settlement of Kala Nera on the coast) to sell their wares at local markets. There was no ceramic production in the villages on Pelion, only in Thessaly. However, the Thessalian potters did not make large ceramic vessels (such as storage jars) – these were imported from other parts of Greece (for example from Koroni on the Peloponnese).

Clay for the manufacture of local pots was gathered from a riverbed near Dimini. The raw clay is best described as of a light brown colour. When it was fired in the potter's kiln, the clay would turn a 'vivid red' colour. Clays were also imported to Volos from the 1970s onwards. In fact, imported clays nowadays come mostly from Spain and Italy. Apparently, the Anetopoulos potters manufactured a variety of vessel shapes with names such as *tsanaki*, *kanati*, *vazo*, *karapha*, *stamnes* and *kardari*, of which most of them are related to the carrying and serving of liquids. It is interesting that some of these names appear in the 1964 pricelist of the Volos Association of Potters.

The following statement from K.A. Makris on the history of Magnesia gives us further information about this recent pottery production:

Pottery and ceramics were not highly developed crafts in the villages of Pelion, as the quality of earth and clay in the regions was not particularly suited to the art. Only in the neighbourhood of the villages Kanalia and Dimini there was any workable earth, but such was chiefly used by the artisans from the islands who moved to Volos following the liberation of Thessaly from the Ottoman yoke. The scruting or separation of the earth took place in the so-called *karoutes* (open shallow cisterns), into which the diluted clay was poured. There the sand was removed from the raw earth. The lower layer of precipitated impurities was not used by the potter and ceramists. When the purified clay had dried to the desired degree and consistency, it was kneaded with the feet prior to being worked on the potter's wheel [...]. The large jars used to store the basic agricultural produce of Pelion, olive oil, of which the region yielded considerable quantities, were distributed throughout the villages of Pelion by the major pottery centres of the coastal area of Ionia.[184]

According to Vaggelis Anetopoulos, the jugs (known in Modern Greek as *stamni*, and in Thessaly as *kanati*) with the exterior upper part covered with a white slip

---

184 Makris 1982, 167.

Fig. 2.7.11. *Kanati* water jugs in a garden on Pelion.

Fig. 2.7.12. Water jugs re-used as flower vase in Kalamaki village.

and bright green-painted blotches or splashes on top of the slip are of local production, from Volos and from other workshops in Thessaly. They were largely manufactured there between 1940 and 1960 from clay gathered at the Dimini riverbed. The jugs seem to have been used mostly for the carrying and drinking of water. A complete example can currently be seen in the Anetopoulos Pottery and Ceramics Museum at Ano Lechonia/Malaki on the western coast of the Pelion Mountain.

Fig. 2.7.13. Water jug used as *bacini*.

Various fragments of these water jugs, or *stamni/kanati,* from Volos (and perhaps from Karditsa or Agia) were found during the survey carried out by the Pelion Cave Project (see above).[185] These jugs from Thessaly can nowadays still be found in village gardens on the Pelion Mountain; members of the Pelion Cave Project noticed them spread over such gardens (Fig. 2.7.11). They were also sometimes spotted inside houses in the function of flower vases (Fig. 2.7.12). On other occasions, they were cemented into the walls of houses as so-called *bacini* (Fig. 2.7.13).[186]

Old photographs from Greece, taken in the 1930s, 1950s and 1960s, clearly show the water jugs (with the upper part covered white slip and green glaze blotches) for sale in markets and on bazaars in Larissa and Volos in Thessaly.[187] The jugs in these pictures sometimes have one handle, but more often two handles. They stand or lay as a group on the ground until customers buy them.

---

185 See note 14.

186 *Bacini* is the Italian word for decorated bowls which were embedded in façades of buildings (especially churches) from Medieval times onwards; cf. Vroom 2003, 44-5 for more information on *bacini*.

187 Papathoma 2001, 55, 57 and 59.

In the 1930s, these jugs became fashionable among the folklorist elite of Athens, where they were called *pissades* (from the Greek word *pissa*, which means 'tar') as the vessels were often covered with tar on the inside to make them impermeable. In Attica, they were used for the consumption of water and retsina, not only at home but also in local taverns. Apparently, the jugs disappeared in Attica (for use?) before World War II, but their production continued in Thessaly until the 1970s and in the coastal village of Agia until at least the late 1990s.[188]

2.8 Modern rock carvings and graffiti in the Pelion caves

Niels H. Andreasen

Greek summary

Ο όρος "γκράφιτι σπηλαίου", που χρησιμοποιείται κατά τη διάρκεια της έρευνας, αναφέρεται σε ένα ευρύ φάσμα απεικονίσεων σε βραχώδεις επιφάνειες μέσα στις σπηλιές ή στις εισόδους τους. Τα γκράφιτι συμβάλλουν στη διαφώτιση σχετικά με τη χρήση σπηλαίων στην σύγχρονη εποχή με διάφορους τρόπους: 1) με τα γκράφιτι γίνεται απτή σύνδεση μεταξύ των ανθρώπων και τόπων στο τοπίο, 2) χρησιμεύουν μερικές φορές ως μόνιμες μαρτυρίες του τόπου και της ταυτότητας, 3) σημαδεύουν τοποθεσίες και τιμούν τοπικά γεγονότα, 4) μπορούν να ρίξουν φως στις ζωές των ανθρώπων που χρησιμοποιούν σπηλιές και που δεν θα μπορούσαν με άλλο τρόπο να έχουν αποτελέσει το αντικείμενο των περισσότερων συμβατικών αφηγήσεων.

Τα χαρακτικά στους βράχους γύρω από τις σπηλιές έχουν τόσο θρησκευτικούς όσο και κοσμικούς συνειρμούς. Σταυροί χρησιμοποιούνται συχνά, ίσως για να δείξουν την ενάρετη κατάσταση του καλλιτέχνη. Μία επίσης κοινή χρήση των θεμάτων εξημέρωσης μπορεί να αντικατοπτρίζει μία πρακτική εξευμένισης για τις δυσκολίες της κτηνοτροφίας και της καλλιέργειας. Ενώ υπήρχε μία παράδοση γκράφιτι παράλληλα με την καθιερωμένη νομαδική κτηνοτροφία, γκράφιτι παρήγαγαν επίσης και άλλες κατηγορίες χρηστών σπηλαίων. Αυτές ήταν οι έμποροι, άνθρωποι που αναζητούσαν καταφύγιο, πεζοπόροι, ερημίτες ή περιστασιακοί επισκέπτες. Οι απεικονίσεις υποζυγίων και ιστιοφόρων πλοίων σε δύο σπηλιές πιθανόν συνδέονται με την άνθηση του εμπορίου στα χωριά του Πηλίου κατά τον 18ο και 19ο αιώνα, αλλά δεν μπορεί να αποκλειστεί μία μεσαιωνική χρονολογία για ορισμένες από αυτές τις εικόνες.

Εκτός από εικόνες, καταγράφηκαν 174 ημερομηνίες στους τοίχους 24ων σπηλαίων, όλες χρονολογούμενες μετά την προσάρτηση της Θεσσαλίας στο ελληνικό κράτος το 1881. Η δεκαετίες του 1930, του 1950 και του 1990 ξεχωρίζουν ως αυτές οι δεκαετίες όπου περισσότερα σπήλαια είναι χαραγμένα με έναν μεγάλο αριθμό ημερομηνιών. Εξηγήσεις για αυτό περιλαμβάνουν την αυξημένη ποιμαντική δραστηριότητα, την μετανάστευση Αλβανών βοσκών και μια αύξηση στον τουρισμό.

Το νόημα από τα σύγχρονα γκράφιτι σπηλαίων προκύπτει κυρίως σε σχέση με τις συγκεκριμένες αναφορές τους σε φυσικά πρόσωπα, με τη μορφή αρχικών και ονομάτων, που εκπροσωπούν διαφορετικές κοινωνικές και εθνοτικές ομάδες. Ως τέτοια αποτελούν σημαντικό στοιχείο για την ερμηνεία και τον αριθμό των ανθρώπων του μετα-οθωμανικού τοπίου του Πηλίου.

❋ ❋ ❋

---

188   I would like to thank Yorgos Kyriakopoulos for this information.

Introduction

The term 'cave graffiti' used during the survey refers to all images and inscriptions rendered on rock surfaces within the caves or at their entrances. Painted images are called 'pictographs' while pecked, carved or incised images are 'petroglyphs'. There are few systematic studies of graffiti and post-Medieval rock art in Greece, and the presence of cave graffiti on Pelion has largely been ignored by formal academic study and is even unnoticed by local residents in the region.[189] This unawareness is in contrast to the existence of a rich tradition in Greece of producing images and inscriptions on natural rock surfaces, throughout the post-Medieval period.

Cave users made cave graffiti for a variety of reasons. The majority are made by individuals who wanted to show that they had visited or used a particular cave – or as a pastime. It would be difficult to trace individual graffiti makers and without ethnographic evidence; most of the small, individual stories connected to such visits are inaccessible or permanently lost. Only in a few cases during the survey did we make a connection between carved initials and names and specific individuals. Yet despite the anonymity of most graffiti, a closer and more comprehensive look shows that it may provide insights on the different ways that people engaged with the rural landscape, something not easily recognisable in other ways. In particular, graffiti can shed light on cave use in several ways:

- Graffiti make tangible attachments between people and places in the landscape.
- Graffiti are sometimes bound up with issues of land use/ownership (this issue is further explored in Chapter 4), and thus serve as permanent statements of place and identity.
- Graffiti mark locations and commemorate local events. By doing so, they may hint at different intensities of use and value of places over time.

---

189 However, see Theocharis 1966.

- Graffiti can shed light on the lives of cave users who might not have been the subject of more conventional narratives.

Methodology

During the fieldwork, a wide range of dates, initials, images and marks were recorded at 42 caves and rockshelters. In some cases, graffiti were noted on open, horizontal and vertical rock surfaces not associated with caves. Although these potentially complement the wider picture, documentation was limited to cave graffiti. The presence of initials and names were always noted, but these were not reproduced in detail apart from in a few cases where the name had a special interest. Dates and images were documented in all instances. Field recording involved taking digital photographs of each graffiti group, enhanced by close-ups of individual graffiti text and images with a scale included. Digital images were later enhanced and re-drawn using image processing software. While also being extremely cost-effective, this approach provided us with an accurate record of both text and images on the rock walls.

Five case studies

In the following, five case studies are presented, which are representative of the range and types of cave graffiti found on the mountain.

KAR-10

Within this small crevice-like rockshelter, numerous initials, dates (ranging from 1884 to 1966) and small, incised designs are tightly clustered on a single panel on the right wall of the shelter. On the 1 m x 0.5 m large panel, one can see a great number of fine lines, small figures, initials and dates. The sharp line and fresher appearance of engraved graffiti dating to

Fig. 2.8.1. Abstract graffiti designs from KAR-10.

Fig. 2.8.3. "Man with Kite" from KAR-10.

after the 1930s (based on engraved dates) stand in contrast to numerous, underlying patinated lines. While it is impossible to estimate their age, these fine, crisscrossing lines may have a degree of antiquity. It is possible that the panel contains additional naturalistic images.

In the left-hand zone of one panel, four small, abstract designs were identified (Fig. 2.8.1). Two of them consist of incised, parallel lines, producing rectangular forms. In between the two designs are two lines. One has split ends and connects to a triangular shape; the other terminates in a kite/cross-shaped form.

Fig. 2.8.2. Budded Greek Cross (right) from KAR-10.

Fig. 2.8.4. "Angel" from KAR-10.

Fig. 2.8.5. "Hairy Monster" from KAR-10.

A depiction of a Greek cross accompanied by the initials and date "ΓΑΛ 1933" (Fig. 2.8.2) measures 4.5 x 4.5 cm and has four arms of similar length. Arm edges are straight but terminate in triangular forms with a bud or knob attached to each – a so-called Budded Cross.

Particularly intriguing is a depiction of a tall, human figure with a kite (Fig. 2.8.3). The figure is holding onto the kite itself rather than its flying line, which appears to be hanging from the kite to the left of the tail. A flat diamond shaped kite is outlined, the spine and spar visible. The kite has a long tail with many ribbons attached to it. To the left of the kite is a plain cross, which may support the idea that the scene refers to the celebration of "Clean Monday" (*Kathara Deftera*) before Easter when kite flying is a popular tradition in Greece. The depiction is lightly patinated in comparison to other incisions located on the same panel.

A depiction of an angel (Fig. 2.8.4) occupies the same area as the preceding figure, slightly further to the left. The figure is small, 3.5 x 4.0 cm; it has out-stretched arms and is flanked by three plain crosses positioned above and at the sides of the figure. The body is filled with cross-hatching and lines on the arms might be an attempt to outline the structure of the wings. Fine lines radiate from the head, possibly an indication of the angel's divine status.

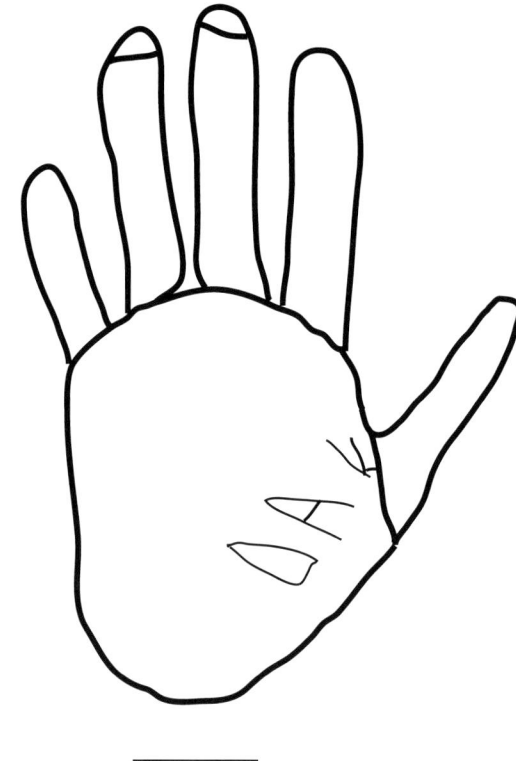

Fig. 2.8.6. Natural-scale depiction of a left hand from KAR-20.

Fig. 2.8.7. Two framed dates and initials from the beginning of the twentieth century.

Fig. 2.8.8. 1938 dates and Greek initials from MAK-23.

A small, enigmatic figure is clearly defined although the upper part is damaged (Fig. 2.8.5). While it may possibly portray a small figure or creature, the lack of definable upper limbs complicates its interpretation. Slightly curved lines of similar lengths extend from a sub-circular shape with two leg-like extensions.

KAR-20

KAR-20 is a sub-vertical cave, and a large area of horizontal bedrock around the entrance has been decorated with numerous initials, dates (many from WW2) and small drawings (Figs 2.8.6-2.8 7). A local landowner provided information that the cave was used as a hideout during World War II.

MAK-23 (Gidospilia)

MAK-23 is a small rockshelter above Makrinitsa. It is not near to any settlements or major pathways and is probably only known to individuals who have animals in the area. Both horizontal and vertical rock faces outside the shelter are decorated with initials, dates and images, and at least seven different Albanian names and four Vlach or Aromanian names are

present (Fig. 2.8.8). Among the images are several human representations.

A line portrait in profile depicts an apparently bald man with an emphasised jaw line (Fig. 2.8.9). Vertical and horizontal lines on the torso below the head give the impression that the person is wearing a sweater. Below the image is a date (April 4, 1966) and a statement that the artist is a forest guard (*dasofylakas*) from Makrinitsa.

Fig. 2.8.9. Human torso from MAK-23.

Fig. 2.8.10. A young man, MAK-23.

Fig. 2.8.11. Various designs, MAK-23.

A group of simple drawings – a heart, a tree (?) and a house – is probably the work of two artists, judging from the two sets of engraved initials (Fig. 2.8.11).

MAK-22 (Arkoudotrypa)

MAK-22 is a small cave situated on a steep slope of Sarakinos Mountain above a creek that runs west of Makrinitsa. Although the rocky outcrop faces east towards Makrinitsa, the opening is hidden from view by tall junipers and is not easily spotted from

Another line portrait shows a young man with long hair and a possible headband. He is wearing a shirt or coat with a folded-over collar (Fig. 2.8.10). The upper body is showed en face, but the head is drawn in semi-profile.

Fig. 2.8.12. Weathering processes result in the loss of delicate petroglyphs at MAK-22.

the footpath that leads into the village. A vertical panel at one side of the cave's entrance contains a wealth of small, delicate graffiti. Several are in a poor state of preservation due to erosion of the rock wall (Fig. 2.8.12). The images represent common activities of the community, such as trading (mule-transport) and fishing.

One category of images shows three scenes of horses or pack mules mounted with a rider and one scene of a human figure leading the animal by a rope or a tether (Fig. 2.8.13). One of the mounted riders seems to be holding two reins, one in each hand, of which one rein is tied to the tail of the animal. The rider is also holding a whip in his right hand. Three of the animals appear to be carrying pack loads tied onto their backs. The animal torsos are in all cases shown in profile view and sketched out with a single line. The human figures are always shown en face.

Another category depicts three figures in boats with crescent-shaped hulls (Fig. 2.8.14).[190] The boats are

probably rowing boats as there are no visible masts or rigging. In each case, there is a large human figure (with hands and feet) standing inside the boat, which in two cases is accompanied by two smaller figures / crew. The size of the figure suggests the artist is emphasising his presence by deliberate exaggeration. He clearly has a dominant and/or protective role, and in all three compositions, his arms are outstretched over the two smaller figures.

The large figure appears to be the helmsman, holding a steering oar from the boat's stern. The different positions and the different angles it subtends from the vertical are significant details and indicate that it is intended as a steering oar rather than a side rudder hanging down next to the stern. The "steering oar" may also be interpreted as a fishing or landing net, not least because of its patterned shape. The large figure is also holding something in its left hand that extends to the waist. In one of the images, the small figure in the stern appears to be holding a fishing line with an attached hook. Several details of the images are not preserved as the weathered surface of the rock wall is coming off in some places.

---

190 The term "crescent-shaped" is applied to a hull with curved extremities of equal height (Wedde 1990, note 30).

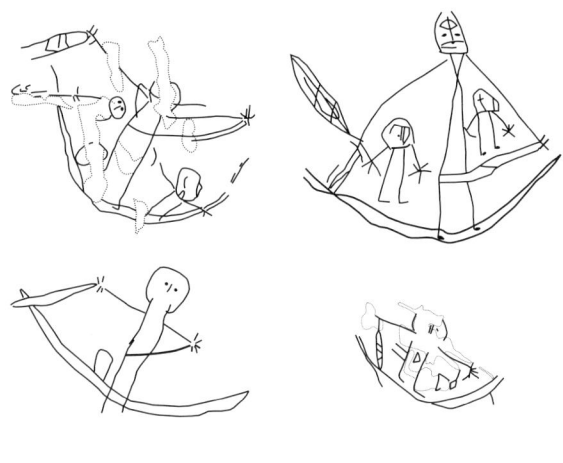

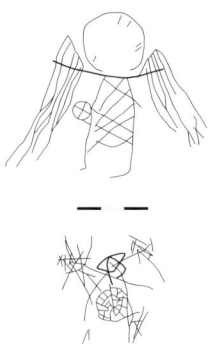

Fig. 2.8.15. "The Arkoudotrypa Angel" and an unidentified figurative image from MAK-22.

Fig. 2.8.14. Various representations of "Three Persons in a Boat" from MAK-22 near Makrinitsa.

There are stylistic variations that indicate the contribution of several individual artists, but all four images are nevertheless striking in their thematic similarity. This could signify that the images were made within a relatively short time span. Is the "Three Persons in a Boat" theme a religious portrayal of Jesus fishing with the two brothers Simon and Andrew? In the Bible, Jesus is depicted on several occasions as preaching from fishing boats and sailing in fishing boats.[191]

An angel with wings and a halo surrounding the head is holding a round object, possibly a holy apple (Fig. 2.8.15). No facial features are preserved. While another drawing can be tentatively identified as a human figure (Fig. 2.8.15), it is not possible to make out its individual parts with certainty. It consists of a semi-circular body filled out with short, semi-aligned lines, and a triangular head. Three (or four) concentrations of short, unconnected lines may represent hands and feet.

The cave has so far provided no direct evidence for the ages of the mixture of commercial and religious engravings. An interesting parallel of a human figure leading a pack animal has been found in a cave near Serres in Northern Greece, where its age was estimated to the fifth or sixth century AD.[192] The location of MAK-22 may provide some clues as it is close to one of the important regional trading routes just outside Makrinitsa. Transport to Pelion during the eighteenth and nineteenth centuries was carried out primarily by pack animals, singly or in caravans. Since Makrinitsa was perceived as the economic centre of West Pelion well into the beginning of the nineteenth century,[193] major roads emanated from it and most caravans passed through it. Different categories of travellers could have used the small cave as a temporary shelter before entering or leaving Makrinitsa. Scenes such as the ones depicted in Fig. 2.8.13 would have been common in the eighteenth and nineteenth centuries, when a steady stream of traders and muleteers with their heavily laden pack animals transported goods in and out of the region. Unlike many of the other villages on West Pelion, Makrinitsa did not benefit from construction of the railway at the end

---

191 "And passing along by the sea of Galilee, he [Jesus] saw two brethren, Simon who is called Peter, and Andrew his brother, casting a net into the sea; for they were fishers" (Mark 1:16).

---

192 See Mais *et al.* 1978, 75 (fig. 4).
193 Magnitos 1860, 55.

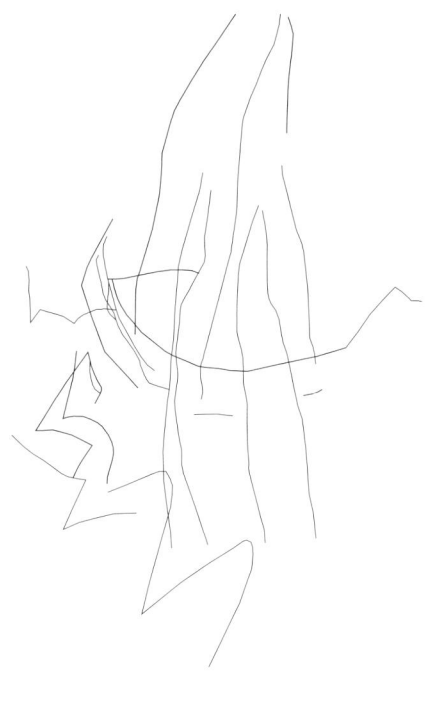

Fig. 2.8.17. Depiction of a sailing vessel at MOU-5.

is not curved and it appears to have several masts (Fig. 2.8.17). A straight vertical line is obvious on the port side, and while there is no sign of anything at the end of the vertical line, a mooring or an anchor is perhaps what was intended.

The discovery of ship graffiti on a mountain prompted questions about their meaning and significance. Although there is a risk of overemphasising the connection between the creators of ship graffiti and maritime activity, the ship graffiti does convey the relative importance of ships and shipping within the society of the artists.

Seafaring and shipping were important features of the Pelion economy for centuries and ships were not alien cultural objects.[194] Perhaps the drawings were the work of persons who belonged to a maritime society and saw sailing ships as important tools for trade.[195] On the other hand, the expedient character of the images suggests that the artist was either not familiar with ship construction or was not concerned about detail. They may have been carved simply as expressions of preoccupation and interest.

KAR-8 (Agios Athanasios)

A deep, horizontal cave was explored on the eastern edge of the Karla depression northwest of the Koryfi Hill. It has two chambers of which the deep-

Fig. 2.8.16. Depiction of a sailing vessel at MOU-5.

of the nineteenth century because of its geographical position high above Volos. Transfer of goods by animals would, therefore, have remained vital until the introduction of motorised transport.

MOU-5 (Koutra)

MOU-5 is a small cave on the northern side of the Milopotamo gorge on East Pelion. The cave consists of an outer and inner chamber, divided by parts of a drystone wall. Initials, two dates (1907 and 1938) and a few fragments from an unglazed ceramic jug dating to the nineteenth or twentieth century suggest that the cave was used as temporary shelter for local shepherds. Engravings in the outer part of the cave show side views of two ships. One hull has a curved outline and a clearly definable bow (Fig. 2.8.16). Unconnected scratched lines are probably parts of sails, masts or the rigging. Wavy lines beneath the ship could signify waves. The hull of the other ship

---

194   Makris 1982, 224.
195   Turner 2006.

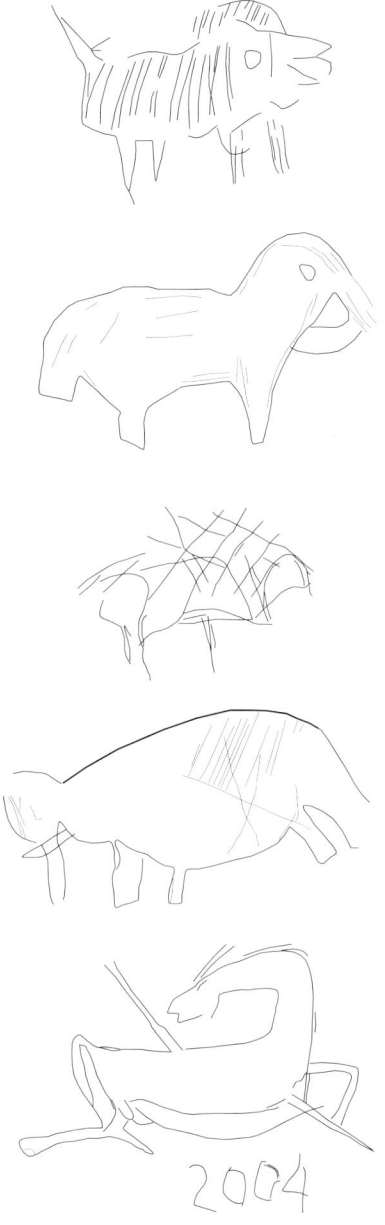

The images were first described in the 1960s by the Greek archaeologist D.P. Theocharis, who reported drawings of wild boar, ungulates, mammoths, a wounded cervid and human forms.[196] All the species are characteristic of Pleistocene megafauna. Erroneously believing the petroglyphs to be genuinely Palaeolithic, Theocharis soon after his discovery reported the findings to the archaeological community. The resemblance of the images to Upper Palaeolithic artwork is striking, and an effort has clearly been made to mimic such designs. An ochre-like colour has been added to four of the images to accentuate contrasts. All have a bright scratch and a fresh appearance.

Appraising Modern cave graffiti on Pelion

Graffiti images from Pelion are mainly figurative and include animals, human figures and specific objects. The less common abstract or geometric designs may have had specific significance to their makers, but we can no longer discover their meanings. With the exception of the peculiar "Ice Age drawings" from KAR-8, all images appear to represent domestication themes or elements. That the images have no obvious links to the "wild" is striking when one considers that Pelion and caves are often associated with the "untamed" (e.g. mountain, dragons, centaurs, severe weather, wild vegetation). One can speculate

Fig. 2.8.18. Modern reproductions of 'Palaeolithic'-style drawings from KAR-8 near Lake Karla. The 2004 date below the wounded cervid is a later addition.

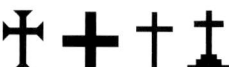

Fig. 2.8.19. Schematic representations of the types of crosses encountered during the survey. 1. Cross with straight arms that broaden into triangles, 2. Greek Cross, 3. Plain Cross, 4. Eastern Orthodox Cross.

est has spectacular dripstone formations. The outer chamber contains a possible Mycenaean burial that was recently plundered by treasure hunters. The latter chamber contains images of animals and a date (2004; Fig. 2.8.18).

196   Theocharis 1966a, 76-82, figs 4-7.

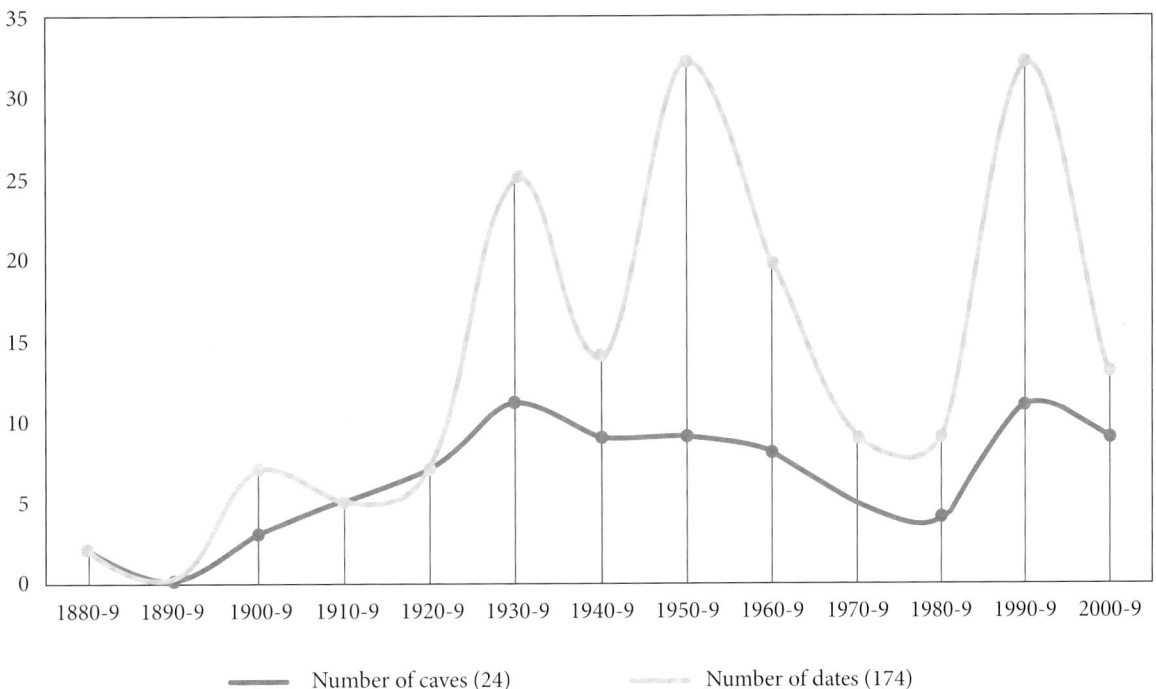

Number of caves (24)   Number of dates (174)

Fig. 2.8.20. Engraved dates from 1880-2009 documented on cave walls in 24 caves during the survey. Note that most caves contain dates from several decades. The large drop in documented dates in the previous decade (2000-2009) at least partly reflects the fact that fieldwork was carried out in 2006-08 and consequently did not have the opportunity to document dates made after the fieldwork.

whether the domestication themes reflect a propitiating practice for the hardships of animal husbandry and cultivation.

The most common graffiti design is the cross, found in various types (Fig. 2.8.19). The presence of crosses and angels demonstrate that rock engravings around caves sometimes have religious connotations. Crosses often occur with initials, showing the artists' desire to connect his identity to expressions of faith or moral purity. It is interesting in this aspect that we did not encounter sexual images or writing indicating sexual themes, which are otherwise common in some "sheepherder art" and in urban graffiti.

Graffiti artists were not only literate in visual recording methods but could as a minimum write the initials of their own name. Modern graffiti mainly consist of identifiable writing, often dates, initials or a combination of both, reflecting the time of visita-

tion. Thus, "ΓΚΑΚΑΣ 1940" (MAK-3), would most likely refer to Gakas' 1940 visit to the particular cave. Personal identification represented by names or initials were in a few cases accompanied by a reference to a village or the outline of a torso, presumably portraying the visitor. There is a clear orientation towards personal rather than group identification and social references such as sports teams, etc. were surprisingly absent. A single case of "political graffiti" was discovered at the small hermit's cave of Askitario (KER-5) where the slogan *Kokini Epanastasi* (Red Revolution) was signed by "Dimitris, May 1 1944". Most likely, Dimitris belonged to the group of EAM/ELAS partisans who had their headquarters and illegal printing facility at the nearby Sourvies monastery.[197]

_____

197   Monk at the Sourvies Monastery (pers. comm. 2009).

Time-block	Culture-historical developments	Graffiti characteristics	
Pre-1920s		Very few dates prior to 1900	Delicate abstract or figurative drawings with both religious and secular themes
1920-1940	Immigration of Asia Minor refugees	Many dates from 1930s	Initials and designs often carved elaborately with a knife
1940-1949	WW2 & Civil War	No dates from the end of this decade	
1950-1970	Migration to lowland and urban centres	Many dates from 1950s	
1970-1990	Tourism increases	Few dates from 1970s and 1980s	
1990-	Influx of Albanian shepherds and field hands in the countryside	Dates increasingly associated with Albanian, Vlach and other foreign names Use of the first name more common	Graffiti from the 1980s onwards tend to list many foreign (non-Greek) names
	Increase in domestic and foreign tourism	Dates tend to be written out with day and month A city of origin often stated that is not on Pelion	Graffiti superficial, often not very deeply cut
		A drop in dates from the last decade	Pen or paint is occasionally used

Table 2.8.1. Characteristics and temporal developments in cave graffiti on Pelion.

In total, 174 dates were documented on the walls of 24 caves (Fig. 2.8.20). The oldest observed date was 1884 and the most recent 2008. The number of dates (and caves) is larger if dates are included that are no longer readable, were overgrown by moss or were simply overlooked by the survey team. Older dates may be biased against because they fade, erode or become overgrown.

Caves with dates divide into sites with five dates or fewer (n=15), and sites with eleven dates or more (n=9). Some caves are engraved with a large number of dates covering a long time span. Examples of this are MAK-3 with 24 dates (1916-1991). and KAR-20 with 19 dates (1900-1994).

The 1930s, the 1950s and the 1990s stand out as these decades saw more caves being engraved with a high number of dates (Fig. 2.8.20). While it is difficult to establish the significance of dates, this apparent clustering may suggest increased pastoral activity in and around the caves during these periods. The near-absence of dates from the Greek Civil War (1946-1949) is understandable, as the mountains were perceived as unsafe territory. Partisans frequently used caves as hideouts and were renowned for requisitioning livestock from villagers.[198] This would likely have put limitations on normal shepherding practices.

The 1990s saw both an increase in tourism and an influx of Albanian immigration, and there is little doubt that both groups had an impact on the number of dates being engraved, especially since the latter group often found employment as shepherds upon their arrival in Greece. One may also ask why a drop in graffiti dates is seen in the most recent

198   Mazower 1993, 131.

Category	Item
Human	figure, bust, hand
Animal	donkey/horse, mammoth, wild boar, deer
Objects	tree, sword/knife, house, boats, anchor line, pack loads, fishing tackle, spear, kite, tether
Symbolic	Christian crosses (various types), stars, hearts
Abstract	parallel and crossing lines, unconnected lines, frames (for dates, initials)
Letters and numbers	year and/or date of visit, age, profession, home village, declaration of love, political slogan

Table 2.8.2. Categories and types of graffiti on Pelion.

decades. A possibility is that increasing availability and use of digital cameras has influenced how people document their visits to places in the landscape.

The generally 'simple' graffiti documented during the survey is dissimilar from the elaborate panels of graffiti known from earlier periods in Greece. [199] Attempting to understand the meaning of the graffiti in the caves of Pelion requires us to place them in their proper social, economic and chronological contexts. Early twentieth-century pastoral culture provides the lens through which many of the engravings should be understood, and it is clear that there existed a graffiti tradition alongside established pastoralism. This is certainly not unique to Pelion but is a phenomenon that can easily be observed by visiting caves throughout the Mediterranean. Most graffiti are located in areas where herders would pass regularly, daily or on a seasonal basis. In some of these cases, graffiti was made to commemorate a stay at a locality rather than marking the event of arrival. However, other categories of cave users also produced graffiti, such as traders, people seeking refuge, hikers, hermits or casual visitors, and people who wanted to deceive archaeologists.

Although images and initials are impossible to date accurately, it seems reasonable, based on the asso-ciated dates and the "freshness" of most graffiti, to argue that the majority of the graffiti was made during the first part and the middle of the twentieth century. There are some possible exceptions to this. The depictions of pack animals, boats and sailing ships from MAK-22 and MOU-5 are likely connected to the flourishing trade in the Pelion villages during the eighteenth and nineteenth centuries. Three graffiti dates from these two caves (1907, 1933 and 1938) appear to be too late to be connected to these activities. It is surprising that so few pre-1900 dates were found, especially in light of the intense travel, trade and building activity on the mountain in the Ottoman period. While the deterioration of dates due to natural erosion of the cave walls may provide some of the explanation, it cannot fully account for their near-absence. It seems unlikely that caves were used less in these centuries and it is more probable that a graffiti tradition only slowly developed after the end of the Ottoman period in Greece. The reasons for this are unknown, but one explanation could be that the rate of literacy was on the rise throughout the nineteenth century. Although Pelion as such had a high rate of literacy (around 40%) at the beginning of the twentieth century,[200] the well-educated "village urbanites" on Pelion contrasted with the transhumant pastoralists and local shepherds who used many of the caves around Pelion.

---

199   See Chatzilazaridis 2000 for an interesting overview of Prehistoric rock art from Northern Greece.

200   Sivignon 2009, 460.

Most graffiti were either carved or incised on a smooth surface of the rock with a sharp and fine-tipped implement, presumably a knife or other metal object, or broken glass. A characteristic of the use of metal tools for carving is the resulting symmetry in width and evenness in the depth of the carved lines (as opposed to a stone edge, which creates lines that are irregular). Pecking (indirect percussion with a metal tool) is more rarely used and mostly for initials and dates. While hikers spend little time in the wilderness and even less producing elaborate graffiti, goat/sheep herders undoubtedly learned rock-carving techniques from each other and had time to perfect their skills. While most graffiti are expedient and naïve, in several cases, the visitor took great care to form elaborate letters and numbers or to accentuate his initials by circling, framing or underlining, especially on panels where other names were already present. This is particularly in evidence for dates of the 1930s (e.g. MAK-1, MAK-3, MIL-1, KAR-20).

The challenge in documenting pastoral graffiti at caves on Pelion is to identify and interpret their meaning, attempting to answer such questions as: Why were they drawn? Why were they drawn in caves? Who were the artists? Without the support of documentary, ethnographic, or other evidence, it is difficult to make more than informed assumptions about possible motives for creating these drawings and inscriptions. Meaning is mainly derived from Modern cave graffiti in relation to their specific references to individuals (of different social and ethnic groupings) in the form of initials and names. Graffiti may indicate that caves were visited frequently; such locations would provide an obvious focal point for people and animals moving about the landscape. The regular incidence of pastoral graffiti in the caves on Pelion indicates that the cultural intentions for the graffiti were much more than doodling by idle hands and minds. Carefully engraved initials and dates are an appropriation of place that conveys information to other herders and users that the particular cave is being used. In some cases, an engraved date is "updated" by the same shepherd in subsequent years, which indicates that the purpose of the graffiti goes beyond a "visiting card", namely to show that the use is recent. Thus, graffiti may point to how people are not just passive consumers of places, but active participants, whose actions subsequently influence the perceptions and actions of others.[201]

Our research has demonstrated that the tradition of creating pastoral cave graffiti on Pelion lasted at least 110 years. The tradition possibly dates further back, but older graffiti may not have survived in the archaeological record. Some caves were situated along major regional transport routes between the wealthy Peliote villages and the Thessalian lowland; others were close to paths regularly used by local and transhumant shepherds. Graffiti caves are predominantly associated with shepherds and are archaeological evidence of such pastoral activity. The site-specific nature of these drawings furthermore suggests that these graffiti and any related activity was gender-based, male activity. Shepherd graffiti is not unique to Pelion and is known throughout mainland Greece and the islands. While the practice of writing names and dates in caves continues to this day, the introduction of pastoral open-air structures rendered part of this long-standing tradition redundant, as people no longer spent significant time in and around caves. The archaeological record of ordinary people inscribed on cave walls can provide a new if somewhat unorthodox means of exploring the history of places with seemingly uncontroversial pasts. Not only does graffiti sometimes point to past transformations of the local community, but it also introduces a dynamic perspective on caves, which are often assumed to be static places.

---

201   Oliver & Neal 2010, 20.

2.9 Cave use on Pelion during World War II and the Civil War

Niels H. Andreasen

Greek summary

Κατά τη διάρκεια της Γερμανο-Ιταλικής Κατοχής και του Ελληνικού Εμφυλίου Πολέμου, στο Πήλιο υπήρχε εκτεταμένη δραστηριότητα στις σπηλιές, καθώς το βουνό έγινε ένα από τα κύρια Θέατρα των επιχειρήσεων των ανταρτών κατά τη διάρκεια και των δύο συγκρούσεων. Πολλές σπηλιές πρωταγωνιστούν σε μύθους και ιστορίες για αμάχους που βρήκαν καταφύγιο σε αυτές από χερσαίες και εναέριες επιδρομές, και έκρυψαν προσωπικά αντικείμενα σε σπηλιές. Αργότερα, η σκληρή μεταχείριση αμάχων, για τους οποίους υπήρχαν υποψίες ότι ήταν κομμουνιστές ή ότι συνδέονταν με τον Δημοκρατικό Στρατό, από τις μονάδες του Εθνικού Ελληνικού Στρατού (Ε.Ε.Σ.), ανάγκασε επίσης πολλά άτομα και οικογένειες να αναζητήσουν καταφύγιο σε σπηλιές. Κρίνοντας από τις ιστορίες που συνδέονται με την σύγκρουση, η χρήση σπηλαίων μπορεί να ήταν και εντονότερη.

Οι αντάρτες χρησιμοποιούσαν σπηλιές για να κρύψουν πολεμοφόδια καθώς και στρατιώτες, και επίσης λειτουργούσαν σαν βολικά σημεία συνάντησης και σαν στρατηγικές θέσεις για παρακολούθηση και επίθεση σε αντίπαλα στρατεύματα. Ένα ελάχιστα γνωστό αντάρτικο νοσοκομείο-σπηλιά βρισκόταν πάνω από το χωριό Μούρεσι. Είχε μετατραπεί σε ένα προσωρινό υπαίθριο νοσοκομείο για τον Δη-μοκρατικό Στρατό κατά τη διάρκεια του Εμφυλίου Πολέμου, εφόσον είχε χρησιμοποιηθεί σαν κρυψώνα για επτά οικογένειες κατά τη διάρκεια της γερμα-νο-ιταλικής κατοχής.

Μία σπηλιά νότια του Βενέτου έπαιξε σημαντικό ρόλο σαν αποθήκη πυρομαχικών για τους αντάρτες του Βορείου Πηλίου. Ο Ελληνικός στρατός εκκένωσε τη σπηλιά από πυρομαχικά και εκρηκτικά κατά την δεκαετία του '50, αλλά άφησε πίσω τουλάχιστον 40 κράνη μάχης και μία μοναδική συλλογή στρατιωτικών μεταλλικών πλαισίων για το φόρτωμα μουλαριών. Τα ευρήματα από την σπηλιά υποστηρίζουν τις υποψίες ότι υπήρχαν καλά οπλισμένοι αντάρτες στο Πήλιο, αλλά μαρτυρούν επίσης την έλλειψη διατηρημένου υλικού πολιτισμού σχετιζόμενου με αυτήν τη διαμάχη, κάτι που μπορεί να μας δώσει πληροφορίες και να στηρίξει τις μαρτυρίες των ανθρώπων που συμμετείχαν σε αυτή ή τη βίωσαν. Παρόλο που η εθνογραφική ομάδα αναγνώρισε τουλάχιστον 30 σπηλιές με πληροφορίες που αφορούν τις διαμάχες κατά τη διάρκεια του 2ου Παγκοσμίου Πολέμου και/ή του Εμφυλίου Πολέμου, αντικείμενα που σχετίζονται με τις εν λόγω περιόδους συνελέγησαν μόνο από τέσσερα σημεία.

Historical background

There is little concrete information about Pelion caves being used for military purposes or for refuge earlier than World War II, although further back, several undoubtedly served as hideouts for the infamous Thessalian *kleftes* and brigands, and as refuges from pirate attacks for the villagers. During World War II and the Greek Civil War, Pelion saw extensive activity in caves all over the mountain, as it became one of the main theatres of operation for the partisans during both conflicts.[202] The background for the use of caves on Pelion during the German-Italian occupation and the Civil War is complex, and an important source for the events played out in Magnesia during the 1940s is Voyiatzis' book from 2008, which includes local witness statements, newspaper cuttings and archive studies.[203]

The Italian occupation forces in Thessaly included the 24th Pinerolo Infantry Division of the Third Italian Army Corps, stationed in Larissa, which consisted of five infantry regiments, a mortar battalion and several support units. In the same region, there were also units from the 36th Forli Infantry Division and various smaller units – some 10,000-14,000 men in total.[204] On Pelion, smaller forces of Italian soldiers were stationed at Chorefto, Tsagarada and Kato Lechonia.

The Greek People's Liberation Army or ELAS was the military arm of the left-wing National Liberation Front (EAM) during the period of the Greek resistance until February 1945. By the beginning of 1943, the top priority for the growing ELAS partisan force on Pelion was to arm itself. After a successful strike on an Italian military post at Tsagaradha, weapons started to percolate into the newly formed EAM/ELAS regiments on North Pelion. The attack would furthermore have the effect of discouraging the presence of Italian patrols and military outposts on Pelion, where reinforcements could not be employed quickly from the Italian garrison in Volos.[205]

Patrols of the occupation forces controlled the major roads, but early on ELAS took control of the

mountain, and bands of partisans continuously roamed the forests.

The Italian forces capitulated in Volos on September 9, 1943, and immediately after the capitulation, negotiations were begun between ELAS and isolated Italian garrisons. Most of the Pinerolo Division under its pro-British commander, General Infante, joined forces with ELAS together with volunteers from various Italian garrisons in Northern Greece.[206] Many of the war-weary Italian troops, however, soon became disenchanted with fighting a war that they felt was no longer theirs.

The capitulation of the Italian forces in Thessaly meant that ELAS suddenly gained access to considerable amounts of weapon and equipment. In Volos, EAM/ELAS units managed to transfer the contents of a military arms depot to an elementary school on the outskirts of the town. After stocktaking, the weapons and ammunition were transported on mules to Pelion.[207] Such achievements enabled ELAS to form and equip a new unit in Thessaly, the 16th Division. At Pelion, it was the 54th Regiment of this newly formed division that formed the main body of resistance against the German forces from the beginning of 1943 until the regiment was dissolved in March 1945.[208] The 54th Regiment was created in Kerasia (Pelion) in September 1943 and consisted mainly of partisans from Mavrovouni, Kissavos and Magnesia, totalling a force of 3,435 men.[209] It was well equipped with Italian, German, English, American and Greek weapons and included a battery of heavy machine guns, a mortar unit, light artillery and a saboteur unit. For transportation purposes, it furthermore had 10 trucks, 5 cars, 7 motorcycles and a large number of horses and mules. Forty-four were mules of special burden or "gun-mules", for transporting arms and ammunition on the mountain.

---

202   For overviews and general discussion of these conflicts, see Sarafis 1980; Iatrides 1981; Baerentzen *et al.* 1987; Close 1993; Mazower 1993; Margaritis 2000; Alexandrou 2014.

203   I am grateful to Giorgos Papamichelakis for his comments on this chapter and for identifying various military artefacts collected during the survey.

204   Lazaros 1977, 65-6.

205   Voyiatzis 2008, 32.

206   Sarafis 1980, 181.

207   Voyiatzis 2008, 44.

208   Arseniou 1977, 184.

209   Voyiatzis 2008, 80.

Civilian/Military	Type of use	WWII	Civil War	WWII/Civil War
Civilian	Refuge			16
	Bomb shelter	1		
	Hiding place for personal objects	1		
Military	Partisan hideout			7
	Munitions / equipment storage	1	1	
	Field hospital		1	

Table 2.9.1. Cave uses associated with World War II and the Civil War identified at Pelion. Some caves were used for several purposes during the conflicts.

German repressive action certainly escalated horrifically after the Italian troops left, and led to massacres in 1943 in Zagora, Drakeia, Portaria and Milies. It was partly in response to these atrocities that the 54th Regiment of ELAS stepped up their activities on the mountain. However, while these tragic events meant an escalation in the confrontations between partisans and occupation forces, it also in some cases divided the local Greek population. Many villagers would blame the partisans and local supporters of the resistance movement for the massacres, and this produced intense mutual hostility and internal violence during the ensuing Civil War.[210]

During the Civil War, villagers were sometimes caught in the struggle between the National Greek Army (anti-communist paramilitary units supported by the government) and the Democratic Army of Greece (DSE), founded by the Communist Party of Greece. The latter mainly grew out of ELAS and was responsible for organising a highly effective guerrilla insurgency against the right-wing army in the mountains throughout Greece.[211]

The survey data

The 1940s were destined to leave a legacy of folklore surrounding caves on Pelion. Many caves are at the centre of legends and tales about men, women and families who took shelter in them to avoid detection by the occupation forces, and partisans who used the caves for hideouts and for storage of munitions and other supplies. Despite the increased role of caves during this decade, there is a dearth of preserved material culture at most caves, which might otherwise have supported the testimonies of people who used them during wartime.

Around 30 caves with information or material evidence relating to conflicts during World War II and/or the Civil War were identified during the survey (Table 2.9.1). The caves are distributed evenly on both sides of the mountain. Most information is derived from the ethnographic fieldwork, but in three cases we recovered military artefacts that with some certainty can be related to the conflicts in the 1940s. An impact-deformed projectile from a World War II military rifle was found at MAK-12 and perhaps is evidence of a skirmish below Koukourava and Makrinitsa. An unfired Mauser cartridge for a German Karabiner 98k rifle was found inside a rockshelter (MOU-24), which supports the fact that ELAS possessed a number of German-produced rifles.[212]

210   Van Boeschoten 2006.
211   See e.g. Baerentzen *et al.* 1987; Close 1993; Margaritis 2000.

212   Voyiatzis 2008, 82.

Most striking was the discovery of a cave with a large quantity of military equipment, mainly helmets and mule pack-frames. Finally, a confirmed littoral refuge cave (AFE-7) contained civilian artefacts that are consistent with a date in the 1940s. These include a colourless chemical bottle with some onset of glass sickness and some poorly preserved fragments of a straight-sided metal vessel.

Fourteen graffiti dating from the 1940s were found in nine caves – two of them confirmed refuge caves, but none containing conflict-related artefacts. Half of the dates are from rockshelters around Lake Karla, and two of these (1948 and 1949) are the only engraved dates from the Civil War years that we found on the mountain. With the exception of the Askitario hermitage, most of the caves containing dates from the 1940s appear to be long-established pastoral sites that were used by shepherds both before and after the wars.

Military uses of caves

The nature of the mountainous region of Pelion, particularly the irregular topography and the presence of caves for hideouts, afforded strategic advantages to mobile partisans familiar with the terrain. ELAS forces used caves on Pelion to conceal ammunition as well as troops, and caves above the villages functioned as convenient rendezvous points and strategic positions for surveillance and attack on enemy troops. Caves on the coast that were difficult to access were used by partisans as hideouts and for maintaining communication with the 4th Naval Squadron of the 54th Regiment. Several measures were taken to keep these locations hidden. A partisan unit above Makrirachi would cover the entrance of their rockshelters with nets interwoven with plant material; fireplaces were maintained only in places that were not visible from the roads and villages. Sometimes locals moving around in the landscape would spot the partisans. An informant recounted

how he as a child during the Civil War came across a group of partisans resting in front of a small cave on East Pelion. Unnoticed, he quickly ran back to the village and alerted the local force of the National Greek Army. They made a sweep of the area, but the partisans had by then managed to escape.

Theodoros Kasidis was a notorious partisan leader who would hide with his men in an area with rockshelters above Kissos. He came to Pelion at the beginning of World War II, reportedly to recruit young men to fight for EAM/ELAS. He stayed in the area to raid (to the displeasure of some Pelion residents), making his headquarters in caves, moving frequently from cave to cave to avoid the Germans, Greek collaborationist military groups and later the anti-communist forces of the National Greek Army. Undoubtedly, Kasidis eluded capture largely due to his network of cave hideouts.

The need for medical facilities to treat injured soldiers prompted the establishment of field hospitals in caves. Cave hospitals are known from other parts of Greece, e.g. Kotili Cave at Grammos and a cave near Vrontero at the Prespa Lakes.[213] At Mavrovouni, to the north of Pelion, there was a mobile field hospital and an underground shelter in which the non-mobile injured would be placed during the operations of the army. The shelter was near the sea and a small fountain. Its entrance was covered so well that one could stand right above it without noticing it.[214] On Pelion, the only cave hospital on the mountain was situated above Mouresi village (MOU-12). During the German-Italian occupation, it was used as a hideout for seven families, but during the Civil War, it was converted into a provisional field hospital for the Democratic Army. The main chamber of the cave is relatively large, with 35 m² of floor space and a 5-m high ceiling. The chamber is accessed vertically from a sloping entrance chamber

---

213   Voyiatzis 2008, 13; Standring 2009, 23.
214   Voyiatzis 2008, 229.

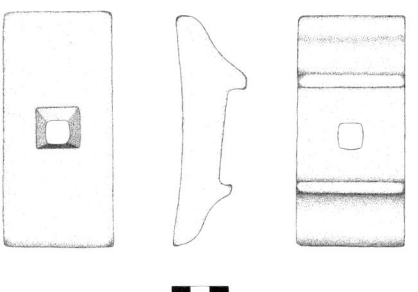

Fig. 2.9.1. Unidentified weapons part of cast iron from KER-8, possibly for mountain artillery.

Fig. 2.9.2. Metal ring with small loops.

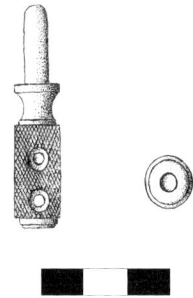

Fig. 2.9.3. Unidentified weapons part.

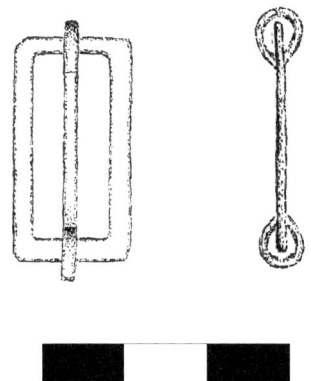

Fig. 2.9.4. Buckle from helmet chinstrap.

and is difficult to enter without a ladder or rope. A small fireplace was maintained near the entrance in such a way that it could not be spotted from Kissos below. In Mouresi, we had the chance to meet the daughter of a man who was injured and hospitalised for a short period in the cave. In unfortunate cases, injured partisans received help too late. According to local legend, the cave of Kremmida (MIL-28-e) above Agios Georgios Nilias contained a dead partisan holding his rifle.

Apart from concealing smaller combat units and their use as field hospitals, caves played additional roles in military logistics planning. Occupation forces in Volos used the cave of Dios Milichiou (POR-5) as a bombproof storage facility for ammunition and equipment. Secure storage of ammunition and equipment was of even greater concern for mobile partisan groups in constant danger of bombing raids or sweeps by enemy troops. The area above Ano Kerasia further to the north was an important base of operations for the 54th Regiment of ELAS during World War II, and a concealed cave south of Veneto, *Spilia me Kranoi* (Cave with Helmets; KER-8), played an important role as munitions storage. The Greek army emptied the cave for ammunition and explosives in the 1950s and a newspaper article covering the event shows a picture of the cave packed with ammunition boxes.[215] However, the army left behind at least 40 Italian M16 and M33 combat helmets not considered salvageable. Other remaining artefacts include chinstrap- and pack buckles, as well as corrugated aluminium spacers and leather padding that has separated from inside the helmets (Figs. 2.9.1-2.9.4).

A piece of preserved Italian newspaper dated to December 6, 1941, was used as padding inside one

---

215 A member of the survey team spotted the reproduced photo in a book, but unfortunately, we have been unable to find the book again.

helmet.[216] A unique assemblage of metal mule pack frames found near the entrance of the cave illustrates how the rocky terrain made the military deployment of pack animals necessary. A quick glance at the landscape of North Pelion is all that is required to realise the importance of mules, which could negotiate the treacherous mountain trails that led to partisan hideouts and gun emplacements. Mules fitted with heavy pack frames could individually carry about 100 kg of payload. In this way, supplies, ammunition, guns and storage batteries for the wireless telegraph could be moved around in the landscape.

After the Occupation and the disarmament of all armed forces, large amounts of weapons and ammunition were concealed in caves by the partisans. Voyiatzis cites a testimony by Thanasis Oikonomou from Karditsa who, as a 17-year-old, was tortured in 1947 at the local police station. The police wanted information regarding the weapons and ammunition that his father had hidden in various caves around Karditsa following the Agreement of Varkiza.[217]

Civilian uses of caves

Civilian cave use on Pelion during World War II involved villagers seeking refuge from land- and air raids, and hiding personal possessions in caves, a situation that can be closely paralleled in late antiquity by the occupation of cave sites during unrest or epidemics.[218] Personal valuables would be buried or hidden in keeping with long-held practices during war and unrest. An informant remembers how she hid her dowry in *Kotroni* cave (MIL-25-e) near Agia Triada to keep it safe.

While written sources rarely refer directly to refuge caves, there are plenty of oral testimonies, which indicate their use for this purpose. We met several retired shepherds and farmers who had recollections of spending time in caves during the 1940s. According to informants, families, relatives and friends found refuge in caves that were adjacent to their fields and most caves used for hideouts during the struggles were well known by members of the local communities, as many of them were used for pastoral purposes. People took refuge in caves for periods ranging from a few days to a couple of weeks. The longest stay we have information on is the case of *Sapospilia* (ART-3), below the village of Agios Lavrentios, where a Jewish family from Volos took refuge when the Germans started patrolling intensively in the area. Here the young mother, Nina Atoun, gave birth to a boy in December 1943.[219] The story of this episode has been covered by local media and is well known.[220] Persecution of ethnicities other than Jews did not take place in Greece during the occupation, but an informant described to us the story of a man from Agios Lavrentios, who would hide in a cave above the village. Because of his dark skin, he was afraid of being persecuted; the cave in which he hid was later named *Gyftospilia*. Whether he was in fact of Gypsy origin or nicknamed "Gypsy" because of his skin colour is not clear, but the story illustrates how the mere possibility of persecution would prompt people to seek shelter outside the villages.

In reprisal for partisan actions, German troops massacred villagers and disposed of the bodies in a vertical cave at Alikopetra between Portaria and

---

216 It is possibly "Corriere della Sera". The newspaper is in Italian and may have been published as an Athens edition as it contains ads of potential interest to Italians stationed in Athens, e.g. Italian-speaking dentists living in the city.

217 Voyiatzis 2008, 378. The Agreement of Varkiza on 02.12.1945 entailed that all armed forces were to disarm and a new National Army was to be formed.

218 One such example is the famous case of Andritsa cave in the Peloponnesus, where a group of people took refuge sometime in the late sixth century – and eventually died there.

219 It is noteworthy that during this period caves, otherwise exclusively male spaces, would have been co-occupied by women.

220 Papadopoulos 2011.

Fig. 2.9.5. Military helmets and pack frames for mules scattered outside the entrance of KER-8.

Chania. Such measures make it easy to understand the panic in the local villages during the fighting on North Pelion when villagers would hide from German retaliation in ravines and caves.[221] When villagers on Pelion asked the local ELAS *kapetanios* (commander) to cease attacking the Germans, in order to save their villages from reprisals, they were told that ELAS would continue to fight as long as the Germans continued to establish garrisons in the rural area.[222] In other cases, bands of partisans looking for supplies would sometimes alienate the villagers due to their methods of requisitioning food. A villager from Kissos described how the partisans would wait for

the villagers to harvest their crops, after which they would steal them.

The situation did not become easier for the civilian population during the ensuing Civil War. The harsh treatment by units of the National Greek Army of civilians suspected of being communists or having connections to the Democratic Army forced many people to flee to the mountain, where they would stay either alone or with their families in caves or other hideouts.

Memories of atrocities and of the conflicts, in general, are part of the reason for the secrecy surrounding World War II and Civil War caves that persist today. An elderly woman from Makrinitsa would not tell us the location of a cave where she had hid during the war for fear that "she might get into

221 Arseniou 1977, 188.
222 Mazower 1995, 178.

trouble". Her motivation for not helping us appeared to be that she did not want to be accused of handing out information to strangers. Her behaviour is reminiscent of mistrust towards outsiders that must have been widespread during the 1940s and possibly in the years that followed.

Chapter 3

An ethnography of cave use on Pelion

Nota Pantzou & Dimitris Papadopoulos

Greek summary

Στα πλαίσια του ερευνητικού προγράμματος για τα σπήλαια του Πηλίου εφαρμόστηκαν τόσο η εθνογραφική όσο και η αρχαιολογική έρευνα. Η χρήση εθνογραφικών μεθόδων, εκ παραλλήλου με την επαρκή αντίληψη των αρχαιολογικών ερευνητικών ερωτημάτων, προσέφερε μία μοναδική ευκαιρία εμπλουτισμού και διεύρυνσης τόσο της έρευνας πεδίου όσο και της καταγραφής, δημιουργώντας έναν δίαυλο επικοινωνίας με την τοπική κοινωνία.

Η εθνογραφική έρευνα συνδύασε μια σειρά μεθόδων και τεχνικών όπως συνεντεύξεις, ανεπίσημες συζητήσεις, πεζοπορία στο πεδίο και έρευνα των αρχείων. Κατά τις περιόδους έρευνας 2007-2008, η ερευνητική ομάδα επισκέφθηκε περίπου 32 μικρά και μεγαλύτερα χωριά, και πήρε συνεντεύξεις από ντόπιους σε έκταση περίπου 362 τετραγωνικών χιλιομέτρων, με εξαίρεση τις δύσβατες και ακατοίκητες περιοχές του Πηλίου. Συνελέγησαν εθνογραφικά δεδομένα για 225 θέσεις, εκ των οποίων η αρχαιολογική ερευνητική ομάδα εντόπισε τις 160.

Η πλειοψηφία των ερωτηθέντων ήταν κυρίως άρρενες άνω των 50 ετών απασχολούμενοι με την γεωργία, την κτηνοτροφία ή την υλοτομία. Οι σημαντικότερες πληροφορίες προήλθαν από κύριους χρήστες των σπηλαίων, κυρίως άντρες ηλικίας μεταξύ 70 και 80 ετών.

Από αρχαιολογικής άποψης τα σπήλαια είναι εν γένει χώροι παραγωγής και κατοικίας, αλλά τα σπήλαια του Πηλίου ειδικότερα είναι έντονα συνδεδεμένα με την πνευματικότητα, την εξερεύνηση, την αναζήτηση, την περιέργεια και τη διασκέδαση. Τα αποτελέσματα της έρευνας καταδεικνύουν ότι τα σπήλαια κατέχουν μια περιθωριακή και ταυτόχρονα δυναμική θέση στο τοπικό φαντασιακό και ζωή.

Διακρίθηκαν έξι διαφορετικές κατηγορίες τοπωνυμίων βάσει καταγεγραμμένων ονομασιών. Αν και η ονομασία μιας περιοχής καθρεφτίζει μια συγκεκριμένη περίοδο στην βιογραφία ενός σπηλαίου, η ανάλυση των τοπωνυμίων προσφέρει μια ενδιαφέρουσα προοπτική της τοπικής ιστορίας και του τοπίου του Πηλίου.

Η επίγνωση των θέσεων των σπηλαίων δεν είναι προφανής ούτε αποκαλύπτεται αμέσως από τους κατοίκους των χωριών του Πηλίου. Η υπόδειξη σπηλαίων και βραχοσκεπών απαιτούσε μία βαθύτερη γνώση του βουνού και την ικανότητα ανάκλησης ονομασιών της περιοχής, αναγνώρισης χαρακτηριστικών στον ορεινό όγκο ή υπόδειξη τοποθεσιών στον χάρτη, γνώσεις και ικανότητες βασισμένες στην εμπειρία. Κυρίως ηλικιωμένοι απέδειξαν πως κατέχουν αυτό το είδος γνώσης. Η ικανότητα του εντοπισμού και της περιγραφής σπηλαίων απαιτούσε μια βαθύτερη γνώση της αλλαγής του τοπίου σε βάθος χρόνου, όπως έχει αποτυπωθεί μέσω των διηγήσεων από γενιά σε γενιά. Υπό αυτήν την έννοια, η γνώση των σπηλαίων του Πηλίου, των χρήσεων και συσχετίσεών τους, δεν έμεινε ανεπηρέαστη από τον χρόνο ούτε είναι ανεξάντλητη. Σταδιακά μεταμορφώνεται, ξεθωριάζει ή ακόμη και εξαφανίζεται μαζί με την "γνώση του βουνού" των προηγούμενων γενεών.

Αν και συχνά εκλαμβάνονται ως περιθωριακοί τόποι εκτός των ορίων κατοίκησης, τα σπήλαια λειτούργησαν ως σημεία παραγωγής πλήρως ενταγμένα στην τοπική οικονομία και κυρίως βασιζόμενα στην γεωργία και την κτηνοτροφία. Τα σπήλαια και οι βραχοσκεπές απέκτησαν αρκετές, συχνά ανάμεικτες ή συγχρονικές λειτουργίες και σκοπούς. Χρησίμευαν ως αποθήκες, ως καταφύγια και ως χώροι φύλαξης ζώων. Είχαν ιδιοκτήτες και κατοικούνταν, ή δέχονταν ενίοτε επισκέπτες για ανάπαυση. Η ανθρώπινη δραστηριότητα, παρουσία και ιδιοκτησία ενσαρκώνονταν με υλικά μέσα (π.χ. κτιστές κατασκευές, στέγες, τοίχοι από ξερολιθιά, εστίες, γκραφίτι).

Ωστόσο, ο τρόπος αντίληψης των σπηλαίων είχε επίσης σημαντικούς, ασαφείς συσχετισμούς, που επιβίωσαν μέσα από την προφορική ιστορία, τις προσωπικές αναφορές και τη συλλογική μνήμη. Δεμένα με μύθους και τοπικούς θρύλους, ή συνδεδεμένα με προσωπικές ιστορίες και ιστορικά γεγονότα (συμβάντα του 2ου Παγκοσμίου Πολέμου) τα σπήλαια έχουν αποκτήσει μια μυθοπλαστική συμβολική αξία, η οποία έχει δημιουργήσει και ακόμη θρέφει μία σειρά από αφηγήσεις, που αντανακλώνται στην πολιτισμική κληρονομιά και τις προσπάθειες για τουριστική αξιοποίηση.

3.1 Situating ethnography: Subjects, sites and practices

Within the scope and context of the Pelion Cave Project, ethnography went hand in hand with the archaeological survey. Research practices and methodologies blended and overlapped throughout the planning and fieldwork stages of the project, since the two research teams, working in parallel in the field, engaged in a continuous exchange of information, ideas, thoughts and concerns. The blending of archaeology and ethnography was further affected by the fact that the ethnographic fieldwork was carried out by two Greek researchers with a background in archaeology and heritage studies.

In this respect, motivation and inspiration derived from the emergence of new trends in archaeological thought and practice known as "Ethnographic Archaeologies"[1] or "Archaeological Ethnography",[2] which place ethnography at the core of archaeology as a means of readdressing its aims and re-establishing the latter's social and political role in a constantly changing world. It is in this context that Meskell in describing her conception of Archaeological Ethnography suggests that "there is a significant difference when archaeologists conduct their own ethnographic work".[3] Her claim is based on the observation that "archaeologists have requisite insider expertise".[4] Thus, employing ethnographic methodologies while also having a good grasp of the archaeological research questions and objectives offers a unique opportunity for enriching and widening archaeological fieldwork and documentation by creating a communication platform with the local community, in our case with the village communities of Pelion.

However, opting for and applying this dual, parallel research pathway was challenging. We had to

---

1   Castaneda & Matthews 2008, 1.
2   Meskell 2005, 81; 2007, 384; Hamilakis & Anagnostopoulos 2009, 65.

3   Meskell 2007, 384.
4   Ibid.

adapt our ethnographic work to the geographic coverage and spatial context, the time constraints and practicalities of the project in only two main field seasons. Moreover, as it progressed, fieldwork also acquired its own dynamics through interaction with the Pelion village communities. Thus, in order for our research to be successful, we had to develop our own "particular research style".[5]

Our fieldwork covered several sites in various contexts and periods between 2007-2008. We attempted to trace cave sites and their biographies by endorsing the virtues of Marcus' (1995) conception of multi-sited ethnography, adjusted to the context of this research.[6] In temporal terms, the ethnographic fieldwork did not entail a continuous and long-term process. Rather, it combined, often within limited time frames (ranging from 3 days up to 2-3 weeks), a series of methods, techniques and tools,[7] such as semistructured interviews, informal discussions and field walking, as well as archival research.[8] This lead to a hybrid practice that we could probably call a "focused and rapid archaeological ethnography", to borrow Knoblauch's definition of the term "focused ethnography".[9] Such a characterisation is also dictated not only by the duration and frequency of the field visits, but also by the intensity of data collection and analysis.

An added parameter is the conscious decision to have two researchers carrying out the ethnographic fieldwork. Adopting such a strategy had certain benefits. First, it provided the opportunity to crosscheck collected evidence and resources before passing it over to the archaeological survey team. Second, it allowed for more intense and efficient data collection. For instance, the simultaneous presence of two Greek-speaking researchers proved to be of great value in several cases and various contexts where on-the-spot discussions were initiated with multiple discussants, mostly at *kafeneia* and *tavernas*, leading to a double, parallel process of note-keeping and local knowledge input by different sources and informants. The two researchers' nationality of origin also proved to be crucial, granting them an insider status and helping to establish grounds for communication with the informants. Finally, the fact that the researchers conducting fieldwork were a man and a woman meant together they reached a wider base of informants, and in some cases further facilitated an increased "insider's" familiarity effect.

The ethnoarchaeological research on Pelion by the Danish Institute at Athens in collaboration with the Ephorate of Speleology and Paleoanthropology of Northern Greece was officially launched in September 2007. The previous year, a small group had "scanned" the field, laying the ground for the opening season. Preparation for the ethnographic fieldwork was initiated soon after. One of the first steps, apart from deciding upon the methodological tools and familiarising ourselves with the region, was to work towards a research design that could facilitate the constant flow of information between the two field teams. To this end, a documentation form was set up with various fields and information categories in order to record essential data. Unquestionably, the first field week helped reset priorities and clarify objectives that would establish a flexible and dynamic cooperation between the two groups. This required day-to-day planning of fieldwork and joint outings, either in search of new locations or informants or to confirmed and already recorded cave sites. The latter practice served to provide context regarding landscape, structures and finds, indispensable when speaking with local informants.

---

5 Forbes 2007, 102.
6 Marcus 1995, 96-7.
7 Voice recorders and cameras were used for interviews and fieldwork recording, combined with note keeping.
8 Considering archival/historical research as an indispensable component of ethnographic work in the context of Archaeological Ethnography, the present analysis often combines both ethnographic and archival/historical information.
9 Knoblauch 2005, 5.

Municipality	Villages	Hamlets	Field season
Mouresiou	Kalamaki, Lambinou, Xourichti, Mouresi, Kisos, Anilio	Taxiarches, Agia Kiriaki, Mavroutsa, Agia Paraskevi	2008
Zagoras	Makriachi, Zagora, Pouri		2008
Mileon	Milies, Pinakates, Vyzitsa, Agios Georgios Nilias	Kato Gatzea, Ano Gatzea, Aghia Triada	2007
Artemida	Agios Vlasios (Karamba), Ano Lechonia, Agios Lavrendios		2007
Agria	Drakia		2007
Portaria	Portaria, Katichori	Stagiates, Alli Meria	2007
Makrinitsa	Fytoko, Makrinitsa		2007
Karla	Kato Kerasia, Kanalia	Ano Kerasia	2007

Table 3.1. Range of fieldwork, 2007-08. Twenty villages were covered during the 2007 field season and 13 villages and hamlets in the 2008 field season.

The second field season started in September 2008 and ended in late October of the same year. During 2007-2008, the project team visited and identified sites, and interviewed local informants across an area of approximately 362 km² (excluding the difficult to access and non-inhabited parts of Mount Pelion). For research purposes and despite the time constraints, we visited about 33 villages and hamlets in total (see Table 3.1). It is essential to note that the process of identifying informants was not limited to villages and residential areas but expanded to mountain tracks, dirt roads, apple orchards and cultivation fields.

Regarding to the fieldwork's spatiotemporal context, it is essential to note that following the defined scope and geographical limits of the project, we had the small villages of Kato Gatzea (West Pelion) in 2007 and Aghios Dimitrios (East Pelion) in 2008 as our operation bases, moving southeast to the northwest. Having these two locations as starting points has greatly affected our own perception of Pelion's landscape features, such as driving distances, village limits or elevation. More concretely, we moved around Pelion's landscape and engaged with sites in various locations by:

- Driving from village to village
- Walking with and talking to locals at village squares, *kafeneia* and tracks
- Field walking around village limits (tracks, cultivation fields), usually following directions given from locals to identify cave sites
- Field walking with the archaeological survey team
- Exploring sites with the locals
- Returning to villages and informants to ask for more information and discuss collected evidence
- Revisiting cave sites based on new information or collected evidence

Soon after the first field season had started, we realised that we were carrying our own preconceptions and connotations of Pelion based on childhood summer/winter holidays, touristy imagery, coffee table books and centaur mythology. Often transfixed by the deep shade of mountain trees or the view over the Pagasetic Gulf, we understood that one had to escape this visual-centric approach to locate less visible or seemingly less attractive sites such as caves and nearby goat pens.

One evening in late September 2007, an old man sitting by the main cobbled street of the traditional settlement of Makrinitsa drew our attention, since he was the only one gazing up towards the mountain, ignoring the view reaching up to the city and port of Volos. When we asked him why he was not enjoying the view he responded perfectly naturally that he was "tired of looking down, everyone was there for the view" – he had had enough of it and the mountain was of more interest to him. It did not take us long to realise that the old man in Makrinitsa was not just reacting to the touristy clichés dominating Pelion: he was simply turning his gaze to what has been a familiar landscape to Pelion villagers for hundreds of years. The mountain, at a higher elevation than that reached by most visitors, has been for most Peliorites an almost natural extension of their fields and villages, their grazing lands or charcoal-making sites.

This realisation of the entire mountain as a familiar landscape was a promising one since we expected to locate most cave sites within that zone. We did indeed manage to locate most cave sites, with the guidance and feedback of the locals. Identifying the sites, however, was far from trouble-free. To sum up some of the main problems that limited our own landscape perception compared to local views, we shall refer in brief to issues of accuracy, orientation/visibility and fragmentation:

Accuracy: Even in cases where we managed to get detailed descriptions from locals, tips and directions could not be as accurate as maps, particularly when referring to higher altitude sites. This difficulty raised an issue of incompatibility of the disciplinary requirement for accurate and valid site documentation with local perceptions of distances, altitude and geomorphology. Although in most cases such problems were resolved by combining means and information (e.g. maps, GPS, local descriptions) and through several site revisits, the epistemological issue of the ways in which local, experiential dimension of lived places could be "translated" onto the archaeologi-

cal survey map or in GPS coordinates remained a concern throughout the project, triggering a lot of debate.

Orientation/visibility: In several instances, caves were hard to see and locate, even when one was almost standing at the entrance. Thick vegetation, overgrown mountain tracks and steep elevation greatly hindered orientation and visibility.

Fragmentation: With site identification during short visits being our main objective, we had a focused but often fragmented perception of Pelion's landscape. Caves and mountain sites were parts of a spatial continuum of interwoven placenames, markers, boundaries and networks in which they were traceable and identifiable. Moreover, mountain sites have changed a lot during the last few decades. The landscape we were looking for and the locals were referring to did not exist anymore. The "living mountain"[10] of the early twentieth century, which was full of the sounds of people working in the fields or crossing each other at the creeks and mountain tracks, had to be reenacted and re-animated through local narratives and archives in order to achieve a wider perspective. And it was often an incomplete one, since for several informants it wad been a long time since their last visit to cave sites. However, vivid recollections of the not so remote past by an old woman from the village of Kissos merit our attention. Contemplating the mountainscape as a living and dynamic space, she characteristically emphasised the sounds, the voices and the flickering lights as Peliorites navigated mountain trails during nighttime.

Given these problems of orientating ourselves in the field, we should stress that the ideal way of locating sites was when locals (shepherds and villagers) would take us to locations themselves. This proved to be valuable not just for cave site identification but

---

10   This is an expression often used by informants when talking about Pelion in the past.

also in terms of unraveling placenames, stories, land uses and landscape features on our way to each site.

During the three field seasons, the number of sites for which ethnographic evidence was collected reaches 225, though only 160 sites were actually located by the archaeological survey team. [11] This discrepancy is a direct corollary of an array of factors. A large number of sites are located at a very high altitude and in heavily wooded areas, thereby hindering access and impeding site identification. Yet it was not merely a matter of remote location and accessibility. Like many archaeological projects, PCP took place during autumn, taking into consideration, among other things, weather conditions and availability of potential informants. In terms of weather, the months of September and October in Pelion are ideal for such projects given the average temperature and limited precipitation. With respect to locating informants, Pelion thrives in bustling communities that remain socially and economically active all year round. In the summer season, Peliorites are mostly preoccupied with tourist business activities, during the autumn period with chestnut- and apple-picking and from November onwards to early spring with gathering olives and other agricultural activities. Therefore, reaching informants and retrieving detailed information on caves proved to be a rather demanding endeavour.

The main challenge was not simply identifying interviewees or discussants, but rather finding the "right" people to talk to, meaning Peliorites who had a rich experience and a deeper knowledge of the mountain (we defined these informants as "key informants"). Such individuals stood out from the majority of our discussants who had a focused, selective or fragmented perception of the landscape based on the routine of their daily activities. From

the outset, we also acknowledged the apparent difference between the kind of people the two teams thought of as key informants and those whom fellow villagers pointed out during conversations as possible sources. On numerous occasions, we were directed towards people who had great knowledge of local history but not necessarily of caves and agropastoralism. These realisations led to the chosen strategy of identifying key informants that employed a combination of approaches, as explained earlier. All collected information was assessed and treated as equally significant and of potential importance for the research.

From the very first days out in the field, we realised that caves were heard of and often known as placenames but where not readily open and identifiable to us. In some instances, a typical first answer to the question "Do you know about caves on Pelion?" was "There are no caves". Hence, their recollection required a certain ability to re-assemble various landscape features (e.g. placenames, morphology) that some key informants seemed to know about. In order for us to locate and explore them, villagers had also to reconstruct them and the landscape surrounding them through placenames, features and narratives. Where this was not enough, they had to take us there themselves. This process of locating and "reconstructing" the seemingly marginal sites was the most challenging, critical and at the same time fascinating part of the ethnographic fieldwork. Yet routine and field activities often prevented locals from joining any of the fieldwork teams in locating sites. Reluctance in a few cases was also linked to looting practices. One informant with good knowledge of the mountain in one of the Western Pelion villages abstained from disclosing information on caves with treasure-hunting interest (albeit he kindly accompanied the field team in order to help explore other local caves). It is noteworthy that on a few occasions informants expressed their willingness, to no avail, to help us locate sites for a small fee. These two incidents reminded us of the entangled

---

11   It is essential to note here that the cited number reflects information gathered by both teams during both the preliminary 2006 research and the 2007 and 2008 field seasons.

Fig. 3.1. Local informant below the dripline of KAR-33 above Kanalia village.

character of archaeological research and the various stakeholders involved in the valuation and use of a past that archaeologists treat as heritage.

While walking in the fields or driving on local roads, sipping coffee and *tsipouro* in *kafeneia* and exploring the Pelion villages, we encountered fieldworkers, shop owners, goat herders, retired schoolteachers, housewives, priests, nature-lovers, landholders/land-owners, etc.[12] However, the majority of our informants as highlighted in Chapter 1 were male, over 50 years of age and occupied in agriculture, animal

husbandry or logging.[13] The explanation for this is to be found in the considerable changes that both the landscape and villages of Pelion have undergone in the last 50 years.[14] Among the most important key informants, primary users of the caves were mostly men aged between 70 and 80 (Fig. 3.1).

> PCP: Do locals have knowledge of caves?
> A: *Everybody knows they exist, apart from the younger people.*[15]

Males dominate animal husbandry and land cultivation. Yet, there are also examples of women involved in goat herding and agricultural activities. During the apple-picking period in the Zagora area, we encountered several women working in the fields. In Vyzitsa on West Pelion, we met a middle-aged widow who herded a small flock. The shepherdess explained to us that few goat herders remain in the area, while in the past each household kept its own

---

12 In recent years universities, research institutions and organisations have developed rules and procedures in order to protect research subjects in ethnographic research. PCP has taken into consideration ethical issues that may occur in the process of the research. We carefully considered the effects of our involvement with individuals and the consequences of our work on groups and individuals. Since confidentiality and anonymity are basic principles of participatory research, we maintain all participants' anonymity by using pseudonyms, even if some of the participants may have specifically wished to be named and acknowledged.

---

13 On key informants, see Bernard 2000, 346.
14 Liapis 2002, 36.
15 Field notes, Milies.

flock. Herding has now largely been taken over by immigrants of Albanian origin. On the whole we can speak of three types of cave users or informants according to their degree of familiarisation with cave sites, referring to the frequency of their visits and use, as well as to their direct/indirect knowledge of caves: 1) primary users – informants who have regularly used cave sites themselves, 2) secondary informants, or cave site visitors and lastly c) indirect informants. This last term refers to those who have heard of cave sites, but never physically accessed them.

3.2 The living mountain: Cave use and landscape perception on Pelion

Throughout the ethnographic fieldwork in both the 2007 and 2008 field seasons, Pelion villages were the spatial epicentre or the starting points of our research.[16] The village square, the cobbled alleys around it, the local *kafeneia* or perhaps one of the last households at the end of the village, were our usual "entry points" into Pelion's landscape. Villagers of East and West Pelion communities were both our cognitive and "on the ground" practical guides to the mountain. This was an intense social experience of daily encounters and interactions, as we came to perceive the mountain landscape with people and through the discussions and walks we had with them.

People were to be found in their villages in their daily routines or – often – at weekend social activities. They were also to be found outdoors, in their fields, apple picking in the orchards or crossing the road with their sheep and goats.[17]

Even so, we soon realised that on both West and East Pelion, we were always moving within extended village boundaries covering large agricultural zones, or through networks connecting the villages and their territories. We used to initiate discussions with the locals by asking them about their immediate surroundings and the spatial range of their life-paths: village population and history, village boundaries and agricultural territories, grazing lands, mountain landmarks, movement and seasonal travelling based on their occupations and life stories. Our intention was to become familiar with the ways in which people perceived, through daily practices, the mountain landscape and were moving from place to place within it. Or, in other words, to get a grasp of the ways in which the locals had built up, throughout the years, a "sense of place".[18]

The *Kentavrou* cave,[19] close to Milies, was one of the first sites we visited and as such was our introduction to landscape perception in Pelion:

> Following the path that starts right by the railway track ascending towards the mountain (which is also used, as we later found out, to access the Taxiarches chapel), we reached the Kentavrou cave, as it is mostly known. It is quite a deep cave that has been turned into a goat pen with a drystone walling structure at its entrance. Beyond its obvious animal-keeping use, there were also several interesting features such as re-used objects and rock surface graffiti.
> […]
> We left the archaeological survey team working on a ground plan of the cave and headed towards the village (Milies). After some wandering around, we found a group of old men at the *kafeneio* near the square and

---

16   In a way, this was the reverse approach of the one applied by the archaeological survey team that focused on site identification as its main task. Identifying cave sites was also one of our main objectives, but to do so, we had to achieve a level of landscape familiarity and this could only be done by asking the locals to be the mediators. They were the ones to introduce us to the "unchartered" territory of the mountain.

17   Depending on the time of year. We were there only in September and October.
18   See Basu 1997, 25.
19   Also known by the locals as "Taxiarchon" (MIL-1).

we had the chance to talk with them. The discussion quickly took the form of a semi-structured interview mainly with one of our discussants, Vasilis, who was born in 1937 and raised in Milies, and is taxi driver and restaurant owner.[20]

Vasilis proved to be one of our "key informants", having a deeply rooted knowledge and a still vivid mapping of the mountain landscape. He introduced us to the village boundaries and landmarks as perceived by the locals in Milies.[21] This is part of the discussion we had with him:

PCP: What are the village limits here in Milies?

Vasilis: *The village limits reach up to that bridge called Kakia Skala. There is a creek, where you went to see the Taxiarches cave. The village boundary follows that creek up to Kakia Skala Bridge and all the way up to the mountain. These are our limits; and then downwards from the railway track to the seashore at Kala Nera. There is a bridge close to Kala Nera with a sign saying "Kala Nera Limits". This is where our village boundaries end. They go all the way down to the sea.*[22]

Boufa is a hamlet belonging to Milies. Moving from Boufa towards Afissos, there is a creek. In the upper part of the creek are our borders with Neochori and Kalamaki. And moving up towards the mountain top towards Xourichti up to our village limits and then on the mountain… it is vast up there. From this side our

borders are with Vyzitsa and Kala Nera. That side of the creek (showing right) belongs to Vyzitsa. This creek is called Pelegrino and the other one is called Platanorema. We call it Bufa creek, Batsi creek, or Platanorema. This is how we call it…

PCP: How many churches are there in the village?

Vasilis: *Here, within the village are Taxiarches, Agios Georgios, Agia Marina, Agios Konstantinos, Agia Paraskevi, Holy Mary at the cemetery. All these are in the village. Then there is Agios Nikolaos, a big church, Agios Dimitrios and then Agios Dimitrios again, the new one, then at the village limits we have Agios Haralampos, Profitis Elias, just right above the village. Agios Dimitrios, Agios Haralmpros and Profitis Elias are on the way to Vyzitsa. Then there is Agios Athanasios.*[23]

As eloquently described by Vasilis as well as by several of our interviewees, certain landscape features or landmarks were perceived as spatial markers of the village boundaries.

Sacred sites, such as churches, chapels or monasteries, not only delineate the internal structure of the village but also relate to the extension of its outer limits, defining its cultivation zone. Churches and cemeteries have been traditionally used to demarcate the inner and outer limits of the village.[24] The ability of Vasilis to recall all churches and chapels within the territory of Milies reflects the significance these places had to the structuring of landscape perception and the limits of village life.

Sacred, religious sites also often shape the perception of surrounding features. The *Kentavrou* cave, mentioned above (MIL-1), is mostly known by locals in Milies as the *Taxiarchon* or *Taxiarches* cave, sharing the same name with the chapel on the top of the same rock cliff.

---

20   Field notes, 11.09.07.

21   In almost every village on Pelion we encountered certain individuals, mostly in their 60s or 70s, who had acquired a key status regarding their knowledge of placenames, landmarks, mountain morphological features and past activities both within and beyond the village. Although we use the term "key informants" to describe the involvement of these locals in our research, the term "key participants" perhaps best reflects their significant role and interaction with the ethnographic team.

22   Interview, 11.09.2007.

23   Interview, 11.09.2007

24   See Nixon 2006.

Creeks, often marking the boundaries between villages on a vertical level from mountain top to seashore, were not only perceived as demarcating features but also combined with springs and the use of water resources (e.g. irrigation, water mills), showing the economic and cultural significance of the hydrographic network in Pelion.

These markers also define the intersection of certain land use zones perceived through a series of factors such as primary use, proximity to village, accessibility or visibility. Three main land use zones could be identified based on ethnographic data, namely a habitation zone (inner village limits, hamlets), a production zone (fields and cultivations, agro-pastoral activities) and a transitional/marginal zone (only limited cultivation or production activities on higher altitudes).

Creeks, chapels and mountain tracks are commonly perceived as spatial markers in Pelion. The boundaries defined by these markers, however, are not always clear or undisputed. The perspective of villagers in Vyzitsa, 5 km from Milies, was interestingly different in certain aspects. No caves were mentioned at the immediate vicinity of Vyzitsa by informants, but two caves, *Kentavrou* (MIL-1) and *Laios* cave (MIL-4) on the other side of the creek, were mentioned to us as sites belonging to the territory of Vyzitsa. People in Vyzitsa were also aware of the so-called Taxiarchis' horseshoe mark imprinted on a rock step just outside the entrance of *Kentavron*. We realised that there were several ways not only to access a cave, depending on the starting point, but also to associate a site with village limits.

Land use and cultivations have dramatically changed in Pelion within the last 100 or 150 years (see Chapter 1). The rural landscape that we were exploring was significantly different from the one described and narrated by our discussants. As explained by Vasilis:

All mountain tracks have now closed since no one uses them. There is a group of volunteers that clears the tracks every now and then. In the old times they had cultivations on the mountain, they had wheat and corn there, they were threshing; now, of course, it's all wooded. The cobbled path is still there, though. At the fields, you can see terraces all around – all these are drystone constructions. We were digging, planting the olive trees and then another terrace (pezoula). The drystone terraces were mostly made by the Ghegs at those times. That was still in the Ottoman times.

PCP: Were these terraces also used as field boundaries?
Vasilis: *Yes, I had a family property. And today I know we have 100 trees and the field boundaries are east and south and so on…*

Vyzitsa had 1,200 people, and olive groves all the way to the sea. On higher altitudes, above the village, they had more fields cultivating apple, cherry and chestnut, using wells and cisterns for irrigation.

In the afternoon, we went to Pinakates to try to get more information on caves around the village. A man at the local convenience store responded that there are no caves near Pinakates. If they were any caves, we (the locals) would know, he said.[25]

This is a response pattern that we often encountered on Pelion when asking about the existence of any cave sites in the area. Cave "awareness" seemed to relate to a deep knowledge of a "bygone" landscape, experienced through traditional practices that have now vanished, when distances were covered on foot or with animals. The caves were mostly known to those working on the mountain, i.e. field workers, shepherds, charcoal-makers:

The day ended with dinner at Agia Triada where we had a chance to talk for a while with Theodoros at his tavern. He mentioned five caves that he knew of

---

25   Field notes, Pinakates.

around the village of Agios Georgios. Theodoros used to lodge, carry wood and produce charcoal on the mountain. He had walked to many of the locations and had been to some of the caves he mentioned.[26]

In trying to define what should be termed a proper cave (by depth, size, visibility) or for us to explain what kind of sites were of interest to us or what a rockshelter is, a captivating typology of local names for various cave formations was unfolded by the locals. Some of the most often used terms were *charavlo* (crevice), *stefani* (wreath), *tympano* (drum), *fournoi* (ovens), *thalassospilia* (sea cave) and *spilia* (cave).

Village caves

There were cases of cave sites that were fully integrated into the cultivation zone of the villages and the daily agrarian practices. One of the most striking examples that drew our attention while still in the early days of ethnographic research was the Argyraki cave (MIL-2). The Argyraki cave made us review our own perception of the caves as marginal sites that were not inhabited, or used or associated with everyday activities. Based on the account of Vasilis from Milies and other interlocutors we were able to outline a profile of the site's use:

> We managed to get information on three caves. *Kentavrou*, which is known as a goat pen but mostly associated with Agioi Taxiarches chapel, one cave close to the Pelegrino creek and the *Argyraki* cave (named after an individual called Argyrakis).[27] It is just a few metres away from the bridge, on the track leading from the lower part of Milies to Kala Nera. A large

group of Gheg field laborers[28] lived inside this cave.[29] Vasilis said the cave is big and charred by the fires that the Ghegs used to build inside.[30] He did not mention, however, any houses, buildings or structures outside the cave.[31]

Argyraki provided a case with distinctive features: it is within the extended, outer limits of Milies, integrated into the production zone, or the agricultural landscape of the village; it is closely associated with a family's land property and cultivation fields (olive grove); and it was used for seasonal dwelling. The proximity to the village via the track (making it easier to carry goods), its association with cultivation and storage and its use for dwelling were clear indicators that this site was fully integrated into the agricultural territory of Milies.

Shepherd caves

Laios cave (MIL-4), at the boundaries between Milies and Vyzitsa, was used by Laios, a shepherd from Vyzitsa who, according to local oral history accounts, was carried away by the flooded creek with his flock, probably in 1912 or 1913. According to the story, he used to sleep in the cave with his flock. We met Apostolia, a widow of around 60 who keep her sheep and goats just outside the village of Vyzitsa. She took us along a path to a point, on the

---

26   Field notes, Agia Triada, 12.09.2007.

27   We later found out that the property on which the cave is located belongs to the Argyrakis family. This is the case for many terraced cultivation properties ranging below Milies and Vyzitsa, almost all the way to the shoreline. All the area ranging below Milies and Vyzitsa towards the coastal lowland is called Argyreika.

28   The Ghegs, a north Albanian ethnic community, were often mentioned on this side of Pelion as immigrant seasonal workers either working in the fields or at the construction of Pelion's railway in the late nineteenth century. It is unclear when Ghegs abandoned the area, but it was probably by the annexation of Thessaly to Greece.

29   Vasilis mentioned that 150 people were living in the cave. Although this is perhaps an exaggeration, the cave could certainly provide shelter to a few dozens of people.

30   We had the chance to validate this when visiting the cave the same day.

31   Field notes, Milies.

freshly paved road, from where she showed us where the Laios cave is. She called it the *Katsikadamou* cave. Katsikadamou was a shepherd family, she said. From that point, we could also have visual contact with Taxiarchis and Stavros (Holy Cross) chapels. Apostolia said that there were several more herders back in the early days.[32]

Laios cave is a case of a purely pastoralist site, an "animal cave" as we used to call these sites while on Pelion. Even in this case, however, their use as short-term shelter or for overnight stays for shepherds, particularly during bad weather, were not unusual. Laios cave was a typical site used by a shepherd. His tragic story, however, integrated the site into local storytelling, thus making its recollection available to more people through this narrative.

Caves as temporary "homes"

Caves on Pelion have occasionally served as habitation sites. Habitation caves (MIL-3, MIL-22, ART-1) represent a rather interesting category among the other types. This is due to the contrasting nature of these natural enclosures. Their particularity is based on the fact that, to quote Richard Buxton, "a cave is both like and not like a house: unlike, because natural; like because sheltering", "also open, yet impenetrable".[33] In the case of Pelion, MIL-3 and ART-1 represent two interesting examples that are somehow related. The first cave was used by Albanian seasonal workers towards the end of the nineteenth century, whereas the latter provided shelter to modern Albanian immigrants. MIL-22 (Marouko's Cave), on the other hand, was linked to the life of a single woman who chose to spend a significant part of her adult life – isolated from family and social life – in a cave that was finally named after her.

However, as was pointed out by informants, caves function mostly as temporary shelters or ref-

uge when the weather is bad, since they are "cold" and conditions do not favour the long-term stabling of flocks, and most importantly human habitation. Hence, for the purposes of this research, in addition to habitation caves we introduced the category of short-term shelter caves.

Mixed uses

The case of *Damari* (Vrochia) in Kato Lechonia is an example of a locality with multiple uses. One of our informants came to the area from South Albania in 1937 when he was still a young boy. He stayed with his family in stone-built huts next to the cave and he recalls how the place resembled a hamlet with at least 100 people working at the olive groves of Agios Georgios. The cave was originally a small quarry that was exploited until the 1940s, after which it became an animal shelter and a storage space for firewood.[34]

3.3 Naming places, recalling stories: Cave myths, narratives and biographies

Caves as production sites and dwelling spaces dominate the archaeological record. Yet a set of other aspects and uses of caves have emerged in the course of the ethnographic and archival research, reflecting a largely human inclination to satisfy secondary needs. Hence, Pelion caves are also linked to spirituality, exploration, research, curiosity and enjoyment. In other words, they have been frequented by hikers, hermits, archaeologists, geologists, tourists, nature lovers, hunters, worshippers, couples and young children. A middle-aged man recounted to us how he as a youngster used to chase birds with other kids in the cave of *Vrochia* near Kato Lechonia. Similarly, a 60-year-old informant explained to us how he used

---

32   Field notes, Vyzitsa, 12.09.2007.
33   Buxton 1994, 107-8.

34   Field notes, 15.09.07.

to chase toads in the cave of Kalamaki (AFE-1) as a child. He remembers how at the age of 12 the cave served as a changing booth when they went swimming at the Kalamaki beach. Later on, it became a "dating spot".[35]

Several locals visit caves out of curiosity, or because they happen to be passing by.

> PCP: Do you often visit the local caves?
> A: *I have been there occasionally, Well… during a leisurely walk; while going towards the fields.*

Quite a few informants have visited caves while hunting. It is known that since the Paleolithic period, caves have served as hunting camps.[36] In modern times, hunting practices are no longer related to survival; rather, they can be classified along with other leisure activities.

Archival data also demonstrates how caves satisfy the desire for exploration and research. Several travellers, occultists and archaeologists give accounts of Pelion caves in their writings. For example, the speleologist Anna Petrocheilou produced a detailed report on *Kentavrou* (MIL-1), substantiating the hypothesis that this was the legendary Chiron's cave:

> These observations allow us to make the bold thought that if Centaur Chiron had a personified existence, it could be that this was his workshop where he crafted metals and weapons. He also taught his craft to – according to mythology – many Homeric heroes, Jason and Achilles among others.

Chiron's myth is unquestionably a dominant one in the region and worthy of our attention, given its locus in popular beliefs. Caves residing within the mythical and sacred realm seems to be commonplace. One encounters spiritual associations of caves in diverse geographical, cultural and historical contexts. In Mesoamerica, according to Earle, "caves are the place from where people all emerge, and to which all return at the end, following the path of the ancestors".[37] In Greek mythology, caves operated as sanctuaries for mythical figures such as Pan and the Nymphs and "were associated with activities perceived as outside the norm".[38] On Pelion, the myths of the Centaur Chiron pervade caves in particular, and the entirety of the mountainscape in general. Chiron is one of the most well-known centaurs in Greek mythology. A hybrid deity, he was a therapist and a mentor of mythical and historical figures such as Achilles, Jason and Asclepious.[39] Moreover, he mastered astronomy, hunting and botany. He resided in a cave on Mount Pelion whose identification has attracted the interest of past travellers, modern tourists and researchers and fuels the imagination and hopes of locals over tourist development. The English traveller Martin Leake, in his fourth volume of travels in northern Greece in a brief account on the area, describes a cavern – Achilles' Cave – above Drakeia on the Pliassidi peak that is thought of as Chiron's school. What really made an impression upon Leake was that the location of the cave corresponded to Dicaearchus'[40] description from the fourth century BC.[41]

> The north western summit, called Plesshidi, rises immediately above Portaria; to the southward of which, one hour and a half above Drakeia, which lies between the two tops, there is a fine cavern, commonly known by the name of the cave of Achilles: it is supposed to have been the place where Achilles was instructed by the Centaur Chiron; and in fact the situation accords

---

35   Field notes, Kalamaki, 10.09.2008.

36   E.g. Marin Arroyo & Gonzales Morales 2007, 63.

37   Earle 2008, 81-2.

38   Richard Buxton 1994, 106.

39   Ibid., 105, 108.

40   A student of Aristotle, Dicaearchus was a philosopher from Messina, Sicily.

41   Laurent 1830, 109; he refers to the cave as Chironium Cave.

exactly with the data of Homer and Dicaearchus, the latter of whom states, that in the same place there was a temple of Jupiter Actaeus, to which it was the custom for many of the sons of the principal citizens selected by the priest to ascend at the rising of the dog-star, clothed with skins on account of the cold.[42]

On Modern Pelion, local myth has it that Chiron lived in MIL-1 at the creek of Kakoskali, some 12 km southeast of the so-called Achilles' cave of Leake. Claims over the original Centavros' cave are not limited to the Milies community. At Mouresi on the other side of the mountain, another Centaur cave (MOU-11) comes to confirm the diachronic allure that mythological narratives about this legendary half-man half-beast exert on people. A couple of years ago, two new theories regarding the location/identification of the cave were proposed by Sofias, a surveying engineer, and Galoukas, a paleontologist-geologist.[43]

Yet Chiron is not the sole mythical creature that resides on Mount Pelion. Alongside his myth, legends about dragons and other mountain spirits also surface in local folklore. For instance, where the territory of Makrinitsa meets the territory of Zagora,[44] there is a complex of caves known as Drakospita. According to a local study of Makrinitsa's placenames, tobacco smugglers fabricated stories about dragons in order to discourage locals from going to their den and getting in their way. Not surprisingly, informants on both sides of Pelion when in conversation with us brought up several placenames with the prefix 'drako', such as *Drakospilia, Drakou Skamni, Drakopigado*, etc.

Various myths and tales are embedded in the biography of the *Malaki* Cave (ART-8), near a seaside

hamlet of the same name. Throughout the fieldwork, there were recurrent mentions of this particular cave. We encountered few informants on West Pelion who did not refer to Malaki. There were even informants from East Pelion who mentioned the cave as a site of interest. According to popular belief, a cave/tunnel linked Malaki with the medieval site of Palaikastro, above Ano Lechonia. This was used as an escape route when the site was under attack by the Venetians.

Transcending myths and folklore, we also recorded information about caves lying on another spectrum of spirituality, in that they operated as religious spaces, e.g. chapels and hermitages. For instance, in the Fakistra area, *Kryfo Scholeio* (MOU-10) above the steep cliffs has served as both a chapel and a hermitage. Caves as secluded spaces proved ideal for those wanting to live in spiritual and social seclusion. This is also the case, according to local narratives, for a cave overlooking the village of Drakeia (AGR-2).

This multifariousness and complexity of cave uses and associations reflect in a unique way how these rather dark, cold, damp subterranean geologic features hold a marginal and at the same time dynamic place in local imagination and living. Most importantly, memories, narratives, tales, myths and legends indicate the heterogeneous practices performed in and near cave sites, and shed light on diverse patterns of human activity.

Back in time: Personal recollections, cave biographies

Collecting information on caves required that informants embark on a trip back in time, an often painstaking endeavour of recalling stories and locations and naming places, given the time distance and the apparent marginality of caves in Modern Pelion life. It thankfully turned out to be a fruitful process that enriched our understanding of cave

---

42   Leake 1835, 384-5.
43   http://www.hellinon.net/NeesSelides/NEOTERES/
     SpiliaKentavrou.htm; http://www.ethnos.gr/article.asp
     ?catid=22733&subid=2&pubid=2642800; http://www.
     sourlas.gr/index.php/writings/14-book3
44   Nanou-Skotinioti 1988, 128.

uses, personal stories and local history. It soon became apparent to us that when caves were related to family sagas, historical events or events of local significance, they seem to have left the strongest imprint upon local memory. For example, it was easier in these cases for informants to recall details about location, adjoining structures and space arrangement.

> We walked along the train track in order to visit Mrs. V., a possible informant. On our way to Mrs V., we met Mr A. X. He told us that he knows the Karvounari cave. His family was using it. He said, "That is where we were born, we lived, we slept".[45]

A few years ago, the daily newspaper *Ta Nea* published a moving story of a young Greek Jewish couple during the German Occupation.[46] A cave below the village of Agios Lavrentios was their home for almost two years and the place where Nina Atun gave birth to their son, while hiding to avoid deportation. It is of interest, as Nina Atun explains to the journalist, that today she still cannot banish from her memory "the dark room" where she delivered her baby. Almost 65 years later (June 2011), the news of a local hermit and beggar's body recovered by police in a difficult to access cave not far from the neighbouring village of Agios Vlassios comes to enrich further the Pelion cave folklore.[47] Death and life intermingle in Pelion caves, confirming what Dietrich noted with respect to caves in antiquity, that these sites "were a place where one got born, lived and died".[48]

Even if it was sometimes hard for informants to recall cave names, the ones recorded due to their help construe useful and complementary sources with respect to cave biographies. For instance, in Agios Georgios Nileias, in the outskirts of the village, lies a rockshelter (MIL-5) that overlooks the Pagasetic Gulf, known as *Tis Grias to Pidima* (Old Lady's Jump).[49] Pertaining to this quite peculiar name, a couple of locals recounted the story of an old woman jumping from the end of the hill. This placename is also common in the Makrinitsa area and according to archival data derives from similar stories dating back to the Ottoman Period.

Based on the recorded placenames, we can distinguish six categories. First, there are those toponyms that derive from prominent features of the local landscape or caves' particular features (location, geomorphology, size), such as *Vigla* (Vantage Point), *Tsouka* (Mountain Top), *Tholos* (Dome), *Timpano* (Drum), *Frameni Spilia* (Blocked Cave), etc. Second, we repeatedly came across cave names that consisted of two words, one of them being either *trypa*/hole or *spilia*/cave and the other referring to its human or animal use, e.g. *Hionotrypa* (Snow Hole), *Mandrinia* (Enclosures), *Damari* (Quarry), *Alepotrypa* (Fox Hole), *Arkoudotrypa* (Bear Hole), *Gourounotrypa* (Pig Hole). At the same time, several caves' names signify ownership (when situated within someone's property), proximity to someone's property or the cave's last inhabitant/dominant user, e.g. Maroukos' Cave at Agia Triada (Marouko dwelled in the cave). Fourthly, we encountered mythological placenames like *Kentavrou* (Centaur), *Tou Drakou to Skamni* (The Dragon's Stool), *Drakotrypa* (Dragon Hole), *Drako-*

---

45   Field notes, Agia Triada – Ogla, 07.11.2007.
46   Papadopoulos 2010.
47   See http://skopelos-news.blogspot.com/2011/06/blog-post_4243.html.
48   Dietrich 1974, 77. Again, these examples bring to mind the special place that caves also hold in Christian traditions, with the Nativity in Bethlehem and the Church of the Holy Sepulchre in Jerusalem.

49   Nanou-Skotinioti 1988, 128.

pigado (Dragon Well).[50] The fifth category contains toponyms that evoke past events such as *Tis Grias to Pidima* (Old Lady's Jump), *Gyfto Spilia* (Gypsy Cave), etc. In the last category fall placenames for which the aforementioned classifications do not apply, like *Chamorigani* (Oregano) or *Syrtades* (Sliding Doors). Although placenames mirror a specific period in the biography of these natural landscape features, a toponym's analysis undoubtedly offers an interesting perspective on local history and Pelion's landscape. As Nanou-Skotinioti expresses it: "Place names were and are like today's postal addresses, house roads and numbers, helping rural people orientate in the landscape".[51] To such an interesting remark, we can add that caves were not solely landmarks but a type of monument, a palimpsest of memories and information with respect to the natural environment and cultural heritage of Pelion.

On cave use ethnography: Some observations

Doing ethnography of cave use on Pelion was a revealing, challenging and at times intense experience. It involved multiple sites on West (2007 field season) and East (2008 field season) Pelion, several actors and constant movement across the mountainous landscape, often within limited time constraints. This often painstaking process exceeded the limits of merely collecting ethnohistoric evidence on Pelion cave sites and land use and resulted, through multiple field re-visits and established links with the locals, in a deeper, situated awareness and engagement, not only in terms of spatial perception but also with respect to local knowledge construction and narrative building.

Seen as such an engagement, the ethnographic fieldwork enabled us to obtain a "sense of place" that, combined with archival resources, would provide a broader interpretational context not just for archaeological analysis but also for local traditional economy and landscape history. Exploring the several and often mixed modes in which Peliorites have used, owned, dwelled in or altered cave sites and the ways in which they perceive and associate caves within the mountain landscape unlocked a whole set of "entry points" into landscape history and new research possibilities. At the same time, it has also delineated the disciplinary and practical constraints of the fieldwork.

Knowing about caves

Cave site awareness is not evident or readily revealed among Pelion villagers, several of whom initially responded that they did not know of any caves in their village's proximity. Naming caves and rockshelters required a deeper knowledge of the mountain landscape and the skill, forged through experience, of recalling placenames, identifying features on the mountain range or pinpointing locations on the map. It was mostly elderly individuals among our discussant ho proved to hold this kind of knowledge.

This led us to the realisation that although conventionally seen as fixed, natural sites comprising geological formations, caves and rockshelters, are not perceived as static but have a rather temporally dynamic character. Vegetation, earthquakes, animal activities, human interventions, road construction and settlement expansion works are just some of the factors that significantly affect the locating of cave sites and alter their main features (e.g. entrance, size,

---

50   There seems to be no clear relationship between cave toponyms and proximity to the villages. For instance, it would be reasonable to expect that cave names that describe mythical beings or wild animals (e.g. *Drakotrypa, Lykotrypa, Alepotrypa*) predominate outside the village's agricultural territory, while caves closer to "home" bear names that refer to ownership or domesticated themes (e.g. *Tou Vatsiou, Gidospilia, Mandrinia, Polychroneika*); this was not the case.

51   Nanou-Skotinioti 1988, 121.

visibility). The ability to locate and describe caves required a deeper knowledge of landscape change over time as recorded and narrated through generations. In this sense, knowledge about Pelion caves, their uses and associations are not unaffected by time, or inexhaustible. It is slowly transforming, fading or even vanishing along with the "mountain knowledge" of past generations.

Using caves

Although often perceived as marginal places outside habitation limits, caves functioned as production sites that were fully integrated into the local economy, mostly based on agriculture and pastoralism. Caves were fully lived and worked features of the mountain, perceived within a continuous production zone of combined uses, resources or "taskscapes".[52]

Cave sites and rockshelters acquired several, often mixed or synchronic functions and purposes. They were used for storage, refuge or animal keeping. They were owned and dwelled in, or occasionally visited for leisure. Human use, presence and ownership were articulated through material production (e.g. built constructions, roofs and drystone walls, hearths, graffiti).

However, cave perception also had significant intangible associations surviving through oral history, personal accounts and collective memory. Tied to myths and local legends, or linked to personal stories and historical events (such as World War II incidents), caves have acquired a myth-making, symbolic value that has produced and still nourishes a series of narratives – something that has recently been reflected in heritage and tourism promotion.

Reflecting on cave use ethnography

The ethnographic fieldwork also has some interesting implications in terms of fieldwork practices, disciplinary knowledge production and insider–outsider relations. Seen in a self-reflexive manner, the ethnography of cave use problematised archaeological surveying practices that solely rely on the truthfulness of the archaeological record. It seemed critical to maintain throughout the duration of the project a lively, constant debate between the archaeological survey and the ethnographic fieldwork, not only to interrogate or validate archaeological interpretation but also to inform ethnographic focus areas and cross-fertilise approaches and field "strategies". Adopting such a dual, conversational approach was particularly important for PCP as a project dealing with the Modern and contemporary past.

The cave use ethnography was at the same time an intervention in (albeit only seasonal) and an interaction with the local communities. This became evident to us in several ways, including the interest that the PCP attracted to potential heritage sites and touristic destinations, and some concerns raised about cave looting. It was clear throughout our presence on Pelion that, although in most cases we were welcomed, we were an added factor in the insider–outsider dynamics on Pelion. Although Pelion, due to its highly touristic profile, is quite familiar with visitors and "outsiders", our seasonal but repeated visits, along with our particular interest in cave sites, was in some cases the source of reticence among locals.

It has been our concern to create a virtual and dynamic bridge between the material record, and landscape as perceived and experienced through the biographies of local villagers. Based on the project's objectives and methodological approach we tried not to narrow our attention merely to sites or finds, but to embrace the plurality of different landscape narratives and interpretations of the past. In our days, it has become a truism that archaeologists are not the

---

52 Ingold 1993, 158.

sole negotiators and stewards of cultural heritage. Multiple agents and stakeholders ascribe meaning to cultural resources. Our mediators with the past were the locals from Pelion, whose subjectivity and personal understanding of heritage often challenged our academic perspective/perceptions.

In any case, this interplay of insider–outsider relations constituted one of the main factors that not only to some extent shaped our experience on the field, but also created a post-fieldwork sense of responsibility in terms of recording – if possible – this experience on paper.

Chapter 4

Identifying finds and cave uses: Dialogues between the archaeological record and the ethnohistorical context

Niels H. Andreasen, Nota Pantzou & Dimitris Papadopoulos

Greek summary

Οι παρατηρήσεις της έρευνας πεδίου δείχνουν ότι συγκεκριμένοι παράγοντες καθορίζουν αν ένα σπήλαιο χρησιμοποιείται ή όχι. Το φυσικό περιβάλλον ασκεί ισχυρή επιρροή σε αυτούς τους παράγοντες: εξοικείωση, φυσική πρόσβαση, απόσταση και υψόμετρο, γοητεία της σπηλιάς.

Οι περισσότερες δραστηριότητες που έλαβαν χώρα σε σπήλαια τα τελευταία χρόνια θα χαρακτηρίζονταν, κάπως γενικευτικά, ως οικονομικές. Η πιο συχνή χρήση τους είναι ως βραχύβιο καταφύγιο σχετιζόμενο με ποιμενικές δραστηριότητες. Παρόλα αυτά, η έρευνα έδειξε ότι η αρχαιολογία μπορεί καλύτερα να αναγνωρίσει οικονομικές χρήσεις, ενώ τα εθνογραφικά ερωτήματα αναγνωρίζουν συχνότερα οικιστικές και άλλες μη οικονομικές χρήσεις των σπηλαίων. Αυτή είναι μια σημαντική παρατήρηση, με πιθανές επιπτώσεις σε αρχαιολογικές έρευνες, οι οποίες μπορεί να αποτύχουν να εντοπίσουν αποδεικτικά στοιχεία που να αποδίδουν στα σπήλαια χρήση καταλύματος ή καταφυγίου.

Περισσότερα από χίλια ευρήματα συνελέγησαν από 50 σπήλαια και βραχοσκεπές. Η ταξινόμηση και η ανάλυση των αρχαιολογικών αντικειμένων οδήγησαν σε ορισμένες γενικές παρατηρήσεις:

- Αντικείμενα που σχετίζονται με την γεωργία δεν εντοπίζονται συχνά.
- Σπήλαια που χρησιμοποιούνται για διανυκτερεύσεις και ως προσωρινό κατάλυμα ή σπήλαια που χρησιμοποιούνται συχνά για ανάπαυση περιέχουν ένα ευρύτερο φάσμα ευρημάτων. Πρόκειται για διάφορα δοχεία, υφάσματα/είδη ένδυσης και άλλα αντικείμενα που σχετίζονται με προσωπική υγιεινή και ψυχαγωγικές δραστηριότητες.
- Μόνο οι θρησκευτικές θέσεις σχετίζονται με θρησκευτικά αντικείμενα και ένα περιορισμένο εύρος οικιακών αντικειμένων. Τα τελευταία χρησίμευαν κυρίως στις θρησκευτικές δραστηριότητες.
- Τα υλικά κατασκευής είναι κυρίως για την ανέγερση και την συντήρηση των περίκλειστων χώρων διαβίωσης ζώων παρά για την κατασκευή καταφυγίων για ανθρώπους.
- Οικιακά αντικείμενα εμφανίζονται συχνά σε μη οικιακά πλαίσια.

Συχνά τα αντικείμενα από τα σπήλαια του Πηλίου σχετίζονται με οικιακές δραστηριότητες. Παραδείγματα αποτελούν αντικείμενα που σχετίζονται με προσωπική χρήση, για διασκέδαση και για κατανάλωση και προσωρινή αποθήκευση. Τα οικιακά αντικείμενα συχνά χρησιμοποιούνται σε μεταβατικές καταστάσεις, απλά για να διευκολύνουν μικρής διάρκειας παραμονή στα σπήλαια.

Η βραχυπρόθεσμη κατοίκηση συνήθως ακολουθεί ένα εκ των δύο μοτίβων: 1) σύντομες ή παρατεταμένες απλές διαμονές σε σπήλαια κατά την διάρκεια ασυνήθιστων γεγονότων όπως πόλεμος, σύγκρουση και επιδημία, ή 2) συχνή χρήση των σπηλαίων για οικονομικές δραστηριότητες, έχοντας ως βάση κάποιο χωριό. Και στις δύο περιπτώσεις η διάρκεια της παραμονής δεν υπερβαίνει μία εποχή του χρόνου (καλοκαίρι ή το θερμότερο μέρος του έτους) και μόνο λίγες από τις χρήσεις μιας τυπικής μόνιμης κατοικίας έχουν μεταφερθεί στο σπήλαιο. Τυχόν παρούσες κατασκευές συνδέονται συνήθως με ζώα. Διακριτές διαιρέσεις χώρων είναι σπάνιες. Δεν εντοπίστηκαν μόνιμες εστίες ή διακοσμητικά αντικείμενα και η πλειοψηφία των υπολειμμάτων τροφίμων που βρέθηκαν ήταν από 'πρόχειρα γεύματα'.

Σε διάφορες δραστηριότητες των σπηλαίων του Πηλίου μπορεί να αποδοθεί θρησκευτικό περιεχόμενο, και συγκεκριμένα η χρήση των σπηλαίων για αρχαίες ή χριστιανικές ιεροτελεστίες και η πιθανή παρουσία αναθηματικών αποθέσεων, καθώς και η ταφική χρήση των σπηλαίων. Τα ανθρώπινα οστά που συνελέγησαν από τέσσερα σπήλαια είναι πιθανότατα μέλη διαταραγμένων ταφών των ιστορικών χρόνων. Σε όλες τις περιπτώσεις συνδέονταν με κεραμεική της Υστεροελλαδικής και της Ελληνιστικής-Ρωμαϊκής περιόδου, που μπορούν να ερμηνευ-

τούν ως υπολείμματα μικρών ταφικών προσφορών που περιείχαν κρέας, σιτηρά, κρασί ή λάδι.

Ο σπηλαιοτουρισμός αποτελεί μία ξεχωριστή χρήση σπηλαίου και μπορεί να συμπίπτει με πιο επαγγελματικές χρήσεις των σπηλαίων, συμπεριλαμβανομένης της φωτογραφίας, της χαρτογράφησης και διαφόρων άλλων επιστημονικών ειδικοτήτων σχετιζόμενων με σπήλαια. Οι λιγότερο σχεδιασμένες επισκέψεις σε σπηλιές είναι συνήθεις, αλλά δεν υπάρχει τρόπος να εκτιμηθεί πόσο συχνές είναι. Οι παράλιες σπηλιές του Ανατολικού Πηλίου προσελκύουν ένα μεγάλο αριθμό επισκεπτών το καλοκαίρι, ακριβώς επειδή βρίσκονται στην ακτή. Άλλα σπήλαια που είναι σηματοδοτημένα και χαίρουν τοπικής αξιοποίησης ή σπήλαια ευδιάκριτα στο τοπίο, επίσης δέχονται ένα μερίδιο επισκεπτών, κυρίως ξένων ή Ελλήνων αστών εκδρομέων του σαββατοκύριακου. Κάποια από τα γκραφίτι που βρέθηκαν χρονολογούνται από την δεκαετία του 1990 έως σήμερα και προφανώς έχουν γίνει από επισκέπτες τέτοιου τύπου.

❉ ❉ ❉

This chapter is presented in three sections. First, a consideration of physical factors that encourage or discourage cave use is followed by a classification of the uses we encountered on Pelion and the artefacts that are associated with them. The second part looks in more detail at the main categories of cave uses that we have encountered on Pelion. These can broadly be divided into production sites, dwelling sites, spiritual sites and sites used for other non-economic purposes. In the final section, we attempt to weave together the various strands of data with our understanding of Modern cave use to give a diachronic overview of socio-economic change on Pelion – from a cave use perspective.

4.1 Factors affecting the use of caves: introduction

While there may be some legitimacy to the claim that the caves most in use are in the most favoured locations in the landscape, the term "favoured" does little to emphasise that the decision to use a cave for a specific purpose derives from a consideration of several factors. Identifying these factors is of obvious importance because they bring us closer to human decision-making. Observations from the fieldwork on Pelion indicate that one set of factors influence whether a cave is used, and another (related) set of factors influence the character of this use (Fig. 4.1). All are natural factors or factors strongly influenced by the natural environment. While a stone wall is a man-made feature, one can pre-exist and be experienced by the user as part of the physicality of the cave and facilitate later uses.

The use to which a cave can be put depends on a number of factors, not least its size and shape. In terms of their suitability for use by humans or domestic animals, it is possible to classify caves according to whether they are day-lit caves that in some cases are indistinguishable from rockshelters or deep caves. The former can be, and frequently are, used for economic activities, whereas this is much less common in deep caves. A third category consists of cavities formed within rock falls (talus caves), although these have more in common with rockshelters than caves.

Whether a cave is used is influenced by a number of factors that can be listed in a plausible order of importance:

- *Familiarity*. Pelion is well endowed with caves, but they can be difficult to find. Caves must, therefore, be known to exist by any potential users. Knowledge of the location and character of a cave can be spread by word of mouth or by visits to caves along with previous or current users. The naming and identification of caves are crucial for the establishment and maintenance of their identity. During the ethnographic fieldwork, it became clear that at least three different levels of awareness of caves exist that are strongly related to their use (see Chapter 3).

- *Physical access*. For use, a cave must be accessible. Physical factors include vertical entrances, which reduce regular use to near zero. Other caves are difficult to find, overgrown or require long hikes from the nearest road or path to reach the entrance. Factors affecting general access might include access control by the landowner (ART-2) or cultural authorities (KAR-8), by completely closing off the entrance to the cave, or by building pastoral structures that serve as a barrier around the entrance. Physical access may change over time. New ownership may lead to a change in access control, and changes in land use patterns and path maintenance can affect accessibility. During the survey, we observed that thorny bushes and other vegetation in some cases completely covered the entrances of caves that had seen regular

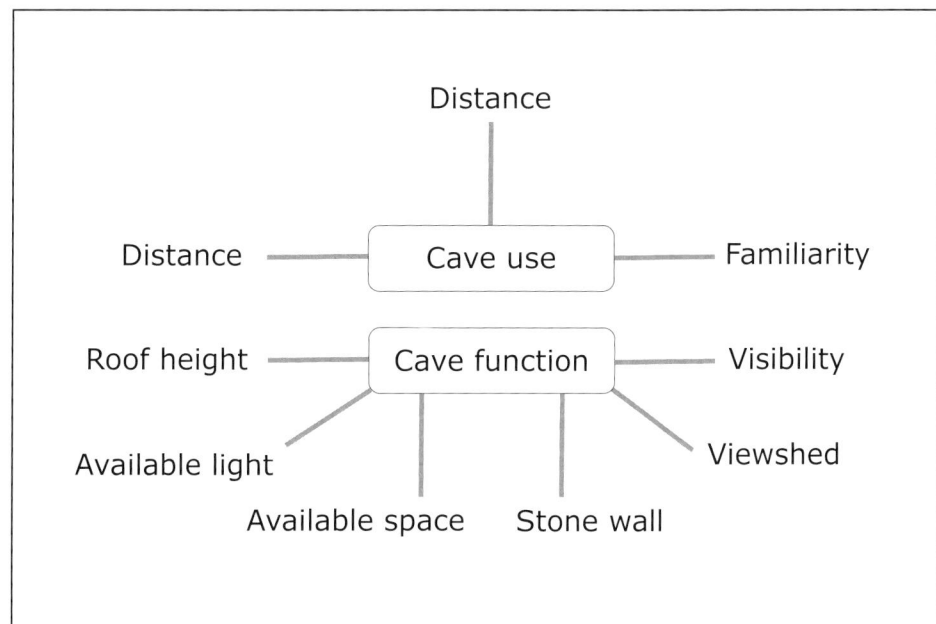

Fig. 4.1. Factors influencing the use of caves and rockshelters. Based on data from Pelion.

use in the recent past (e.g. ART-5 & ZAG-5). Such obstacles discourage exploration or other use by potential visitors.

- *Distance and elevation*. It is reasonable to assume that a higher percentage of caves were used near villages or in areas that are relatively easy to access. However, this again depends on the function of the cave. For instance, caves used for refuge or hideouts during times of conflict are usually located outside the agricultural territory of the village (Table 4.1). In addition, they do not have stone walls or other structures associated with them, and graffiti is rare or absent. Neither are they connected to well-defined paths. Hermitages are another example of a cave function that is not influenced by distance, as hermits usually preferred caves that were isolated or in remote locations. Distance can be perceived in terms of the actual walking distance from a village, but most villagers on the mountain measure distance according to how long it will take to reach a destination. Locations high on the mountain or in less travelled terrain would require more time to reach.

   On Pelion, where habitation is confined to villages and hamlets rather than individual farmsteads, one would expect to find a relationship between cave use and its proximity to a village and its fields and paths (Table 4.1). For instance, caves for cool storage of cheese would be found in the immediate surrounds of the village while caves used as stables by professional pastoralists tending to large flocks would be in pastures located outside the orchards and crop fields of the villages.[1]

- *Cave appeal/attractiveness*. Caves have one or more features that appeal to the potential user, such as size, roof height, a level floor, view shed, temperature, availability and abundance of water resources, or perhaps unique or unusual geologic features. Other features might have the opposite effect, such as bats, insects or humidity. The attractiveness of a cave to users often depends on the degree to which potential users can fulfil certain tasks or carry out certain practices. Residential activity, for instance, requires adequate space and light. Size is not always a suitable indicator of use as illustrated by the large, tunnel-like cave of Roumbos (AFE-3) on East Pelion. While dry most of the time, the cave is hazardous to humans and livestock when rainwater is discharged in large quantities through the chambers during violent downpours.

Deep caves without natural light, difficult to access, damp and often poorly ventilated, have rarely been the scene of economic activities. They were instead occasionally used in extreme circumstances of social unrest when they could offer a measure of security.

4.2 Cave use classification scheme

The range of different activities undertaken in caves by humans in the Modern and contemporary periods fall into several categories; most of these may be described, rather loosely, as economic. The most frequently encountered use is as a short-term shelter in connection with pastoral activities.

   The identification of cave uses relied on cave morphology (ground plan), the nature of existing structures, the behavioural indicators indicated by the associated artefacts and the ethnographic interview data. The cave use categories employed by the project and discussed in the following did not function as a fieldwork tool in the sense that we attempted to provide evidence on particular categories

---

1 Smaller domestic flocks kept near the village would usually be allowed to graze and eat fallen fruits within the orchards.

	FOCUS OF MOST PRODUCTION ACTIVITIES – LIMIT OF MODIFIED ENVIRONMENT		PERIPHERAL RESOURCE SPACE*
CAVE USE	Within built village	Within village's agricultural territory**	Outside village's agricultural territory
Dwelling	0	✓✓	0
Short-term shelter	0	✓	✓✓
Pastoral	0	✓	✓✓
Storage	✓	✓✓	✓
Refuge	0	✓	✓✓
Quarantine	0	✓	0
Natural resource use	0	✓✓	0
Spiritual	0	0	✓✓
Recreational	0	✓✓	✓✓
Research	0	0	✓

* "Peripheral" is based on what was experienced as marginal at the time of the survey. This is a somewhat subjective term and furthermore is dependent on the period in question. In the case of ethnography, this subjective experience of both our team and the locals (mostly the latter) was rather sought as a means to acquire a "sense of place". In any case, it would be difficult to avoid the arbitrariness of any attempt at strictly defining limits and territories.

** Agricultural territory: here loosely defined as an area of land with crops belonging to a village that is involved in agriculture.

Table 4.1. Relationship on Pelion between a cave's use and its proximity to a village and its fields and paths. 0: no caves; ✓: one cave; ✓✓: two or more caves.

of use. Instead, the use categories were employed as an interpretative means in order to identify broad patterns of use across the mountain. In this regard, caves are associated with a certain type of use rather than being characterised by it. It should be stressed that the categories define the primary cave use as illustrated by the evidence (Table 4.1). However, the identification of "primary" use can be problematic: mixed practices would mean in some cases that different uses could occur simultaneously in a cave (e.g. dwelling/storage), or be replaced by another within a short time interval (Table 4.2). This can produce a sort of horizontal stratigraphy in which several processes in modern times have produced their own particular patterns of material remains.

Even a particular type of cave use can include a confusing array of data. For instance, graffiti dates and initials indicate that different shepherds used particular caves consecutively on a "time-share basis". What objects are discarded will depend on the activities and scheduling of the individual shep-

	No.	%
No specified use	92	*41.1*
Single use	80	*35.7*
Dual use	39	*17.4*
Multiple uses (3 or more)	13	*5.8*

Table 4.2. Distribution of the number of "primary" uses identified in individual caves on Pelion.

		Archaeologically identified cases		Ethnographically identified cases	
		No.	%	No.	%
Economic	Pastoral	47	37.3	28	23.9
	Storage	6	4.8	7	6
	Natural resource use	16	12.7	8	6.8
Habitation	Dwelling	2	1.6	7	6
	Short-term shelter	36	28.6	47	40.2
Religious	Religious	8	6.3	4	3.4
Non-economic	Recreational	6	4.8	11	9.4
	Research	1	0.8	5	4.3
	Looting	4	3.8	0	0
		126		**117**	

Table 4.3. Identification of cave uses on Pelion (several uses are often identified at the same site).

herd, and over an extended period, this will lead to the deposition of a mixture of artefact types in the cave record.

Fourteen caves (of which four are vertical) and 14 rockshelters included in the archaeological survey did not reveal any trace of human activity. Virtually all these were shallow and exposed rockshelters or narrow, uncomfortable caves or caves that are difficult to access. While these characteristics may have made them unattractive options for providing shelter or other uses, such attributes should not be used as criteria for excluding them during a survey. Many shallow, narrow and difficult to access caves have revealed evidence of use. We therefore do not suggest that these caves were not visited, but rather that their use was so limited and incidental that little or no evidence is preserved. Subsurface testing would undoubtedly throw up sporadic cultural remains in at least some of these sites.

The results of the fieldwork show that archaeology may be better at identifying economic uses, while habitation and non-economic uses are more frequently identified by ethnographic queries (Table 4.3). This is an important observation with potential implications for archaeological investigations, which may fail to detect evidence of dwelling and sheltering.

4.3 Cave uses and cave site artefacts

A variety of artefacts from the caves provides information about cave use in the post-Ottoman period (nineteenth and twentieth centuries). Unfortunately, Greek material culture from the last 100 years is difficult to date with the necessary accuracy. For instance, most of the finds are ceramic sherds, but few of these can be dated more accurately than to the nineteenth and twentieth centuries.[2] Such a chronological resolution is insufficient to reveal changes in patterns of deposition during the last 200 years. Most needed are regional typologies for ceramics and glassware frequently found in surveys and excavations that will allow archaeologists to run Greek-produced and common imported utilitarian bot-

---

2   Vroom, this volume (Chapter 2.7).

Group	Collected	%	Surveyed caves
Ceramics	606	59.2	44 (27.3%)
Animal bones	241	23.6	22 (13.7%)
Human bones	24	2.3	4 (2.5%)
Invertebrates	33	3.2	5 (3.1%)
Glass	45	4.4	12 (7.5%)
Stone	6	0.6	5 (3.1%)
Metal	31	3	11 (6.8%)
Plastic	17	1.7	9 (5.6%)
Miscellaneous	17	1.7	7 (4.4%)
Smoking-related	3	0.3	2 (1.2%)
Total	**1023**		

Table 4.4. Number and percentage of collected artefacts by category group and by their occurrence in the surveyed caves.

tles or significantly sized bottle fragments through a series of questions based primarily on diagnostic physical, manufacturing-related characteristics or features, to determine the approximate manufacturing age range of the item. Furthermore, most bottle shapes were closely associated and identified with a certain product or products, as "form follows function" to a large degree in bottle shapes and styles.

A total of 1023 artefacts were collected in 50 caves and rockshelters for further identification and analysis. The finds comprise mainly pottery sherds and animal bones, with other artefact categories being significantly less common (Table 4.4). Fragments (typically ceramic containers or glass bottles broken in situ) were counted individually and the actual number of artefacts is therefore somewhat lower.[3]

As the finds are from a large number of caves and rockshelters, examination and analysis of the artefacts concentrated on identification of use, provenance and dating. The nature of the survey was such that the results were not suited to a form of analysis where the collective content of each cave – artefacts and ecofacts – could be exploited to give a cultural-historical interpretation of each locality. What the artefacts were expressions of, and the everyday life of which they were a part, could not always readily be ascertained from the finds alone. This deficiency was addressed through a close dialogue with the ethnographic survey team, which in several cases could confirm or elaborate on activities in particular caves.

The idea of regarding and classifying plastic bottles, batteries and toothbrushes as archaeological artefacts can initially seem peculiar, but to document contemporary activities in caves, it was necessary to document Modern artefacts as well as ecofacts that could aid interpretation. For instance, unmodified bones of wild animals in caves are not *per se* cultural objects and of immediate archaeological or ethnographic interest, but they are culturally significant non-artefactual materials.[4] The remains of both domestic and wild animals can be useful in shedding light on the fauna around the caves, and furthermore serve to illustrate both animal-related cave names (e.g. wolf, bear, fox, goat, pig, crab) and ethnohistorical narratives of wildlife (wolf, fox and owl) in connection with caves and shepherd stories.

Artefacts that belong to the familiar recent past (contemporary) did not require explicit study as we already think we know what they were. Nevertheless, although the original functions attributed to most artefacts are identifiable, they are often arbitrary. For instance, batteries can provide power to any of a variety of appliances, plastic bottles and ceramic vessels can hold different substances, a toothbrush may be used to clean a hunting rifle, etc. The emphasis on the analysis was to identify the objects and

---

3   For instance, 62 ceramics collected from MOU-11 all belong to one vessel, as is the case with 25 sherds from the rockshelter of MIL-4, 50 sherds from ART-1, 12 fragments of a metal container from AFE-7 and 19 fragments of bottle glass from MAK-12.

4   Binford 1964, 332; Shackley 1981.

determine their datable range to see whether they could be associated with activities in the caves.

All finds were divided into five groups that were intended to organise the wide array of objects and aid interpretation of the activities in the caves (Fig. 4.2). These groups were further divided into categories and types.

While some artefact types are restricted to particular activities (e.g. vaccination bottles for pastoral use, an icon for religious use), other activities involve a mix of domestic objects and other artefact categories due to their multifaceted character.[5]

The mixture of different types of objects includes materials necessary to build and repair/maintain the structures inside and around caves, or materials relating to activities in the cave and in the immediate vicinity. The latter would include resources associated with subsistence/consumption, and the resulting waste products. The Modern period finds include no evidence of cooking or food storage, although these activities undoubtedly took place. Finds related to consumption were in all cases "expedient" foods, consisting of conserves, snacks, nuts and fruits. Edible invertebrates, the remains of which were found in several caves, must be presumed to have been collected on the coast, but most likely not in the Modern period (see Chapter 2, this volume).

A wide range of beverages was consumed, including coffee, juice, water, beer, water, wine, liquor and soft drinks. In the past, herders were required to make trips to and from springs and/or wells to collect water for themselves and their flocks. This included the use of water jars, a number of which were attached to the saddle of a donkey or mule. Over time, many of these water jars broke and were left in and around the caves where they had been used. In addition, water was also kept for drinking

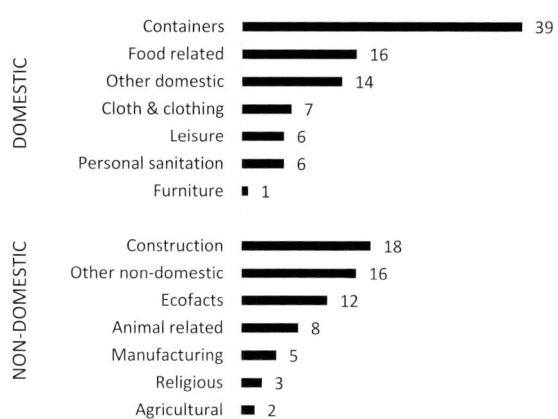

Fig. 4.2. Frequencies of domestic and non-domestic artefact categories in 66 caves. Values indicate the number of cases in which each category has been encountered.

by the shepherd himself. The many nineteenth- and twentieth-century water jar fragments around the caves substantiate this use of such vessels. Some of these two-handled glazed water jars would have been suspended from a rope for use at wells and for transport. Other types of ceramic vessels were used to bring food (usually hot food for the midday meal) to workers in the field or to store pastoral products.

In recent decades, plastic and glass bottles, or single-use containers such as juice cartons and soft drink cans, have replaced re-usable ceramic containers. Large 1.5-litre plastic bottles were found more frequently than the smaller 0.5-litre bottles, the latter so familiar in urban settings. The reason is undoubtedly that the cave users anticipate relatively protracted stays away from immediate sources of water.

Classification of the archaeological objects led to a number of general observations:
- Objects associated with agriculture are uncommon.
- A picture frame was the only secular decorative object found (MAK-5). This suggests that caves only in rare cases were perceived as 'homes'.

---

5   The inability of archaeology to unravel such complexity may lead to flawed interpretations. See Murray & Chang's (1981) investigation of a pastoralist site in the Didyma sinkhole for an example of this.

- Caves used for overnight stays and as temporary accommodation or caves used frequently for resting contain a wider range of artefacts. Such assemblages are dominated by various containers, cloth/clothing and artefacts associated with personal sanitation and leisure. Leisure activities include listening to music, reading newspapers and, in particular, smoking. Both the amount and variety of smoking debris indicate that tobacco was commonly used and different brands of store-bought cigarettes in the same cave indicate use by several individuals. Caves used for these activities are usually not associated with animal-related objects, although animal-related activities may take place outside them.
- Religious sites are only associated with religious objects and a limited range of domestic objects. The latter serve to support the religious function.
- Construction materials are for building and maintaining animal enclosures rather than for building human shelters.
- Domestic objects regularly occur in non-domestic contexts.
- Caves without domestic objects are in most cases not suitable for domestic use, or activities other than domestic ones appear to be dominant.

4.4 Caves as production sites on Pelion

Traditional rural communities have often incorporated caves in their local economies. The role of caves as supplementary or supporting features in the production of agro-pastoral goods is, therefore, a familiar notion in the Mediterranean, with the ubiquitous stonewalled caves and rockshelters being a familiar sight in the landscape.[6] However, caves have mostly been regarded as low-status sites as they are not critical to a sedentary agro-pastoral economy.

This is perhaps best reflected in the associated expedient and low-investment structures that underline their main practical function as auxiliary resources.

Key aspects of the Pelian economy cannot be directly related to cave use, as caves are rarely used for non-pastoral economic activities. Only a few can be connected to traditional economic activities on the mountain, such as mining and quarrying, logging and charcoal burning.[7] Nor do we have examples of caves being used as workshops. The only cases of non-pastoral uses (except tourism) were storage of field equipment (ART-2 and AGR-8) and shelter for field hands during the olive harvest (MIL-3). Most caves would have been unsuitable for storage of agricultural products.

Therefore, when we refer to caves as production sites, it is mainly to their pastoral use as "animal caves". It is in this capacity that they most demonstrably are part of the village economies. However, while pastoral use is the most frequently encountered cave use category (identified at 61 localities), we only found firm evidence of pastoral activities in less than half (45%) of the total number of caves with an ethnographically and archaeologically identified use. The identified number of pastoral caves would undoubtedly be higher if the survey had been supplemented by geo-archaeological testing of surface sediments to detect indicators of dung deposits,[8] or if we had been in contact with a wider range of people with experience in husbandry.

Herding and penning practices on Pelion

In contrast to other highland communities in Greece, the general shortage of mountain pastures

6   Faure 1964, 217-20; Brochier et al. 1992.

7   An informant told us that people used to burn charcoal in a cave 4-5 km east of Kerasia, near Panagia Lestiani (not located).

8   Brochier et al. 1992; Canti 1997.

on a large part of the wooded mountain meant that large-scale breeding of sheep and goats could not be sustained. In the middle of the nineteenth century, only Makrinitsa and the villages of Veneto, Kerasia and Keramidi on North Pelion had larger areas of grazing land available to support flocks.[9] Today, livestock herding is still practised mainly in the most open landscapes on North Pelion around the villages of Veneto, Pouri, Kanalia and Kerasia.

In contrast, livestock breeding was common on the Thessalian plain, where wool became a profitable product in response to intense local demand from the developing textile industry in Zagora and other villages during the 1700s and 1800s. The edges of and foothills around Lake Karla would have provided excellent conditions for the winter pasturing of flocks. From the draining of the lake in 1962 and until a few years back, specialised pastoralists would also graze their flocks in the stubble fields of the dried-out lake.

The well-watered lowlands were suitable for grazing large numbers of ovicaprids, and the development of massive sheep flocks to satisfy the demand for wool became a dominant feature of livestock farming in Thessaly. In 1911, there were still lowland municipalities in Thessaly where the number of livestock reached 30,000.[10] In comparison, the same tally from 1911 estimated that municipalities on Pelion each had 1,000 to 5,000 sheep and a larger number of goats. The dominance of goats was less pronounced in the lowland municipalities of Nilia, Orminion, Iolkos and Makrinitsa around Volos, Agria and Lechonia. Other species of livestock were insignificant and only around Lake Karla were some cattle found. According to observations by the mid-nineteenth-century traveller Magnitos, oxen may earlier have been common in Veneto and Kerasia,[11] and we know from informant interviews that cows

were occasionally sheltered in Alafoklisi cave above Fytoko.

While goats and sheep dominated among the livestock, other species were also important to individual households, such as pig, chicken and turkey. Their bones are rarely encountered in remote caves because they were raised within or close to the villages. Each family had their own small herd of horses, donkeys or mules. These animals were valuable for traction since many villagers were dependent on transport of agricultural and manufactured products to markets in other villages and in Volos. Horses and donkeys were also crucial to shepherds as they transported necessities when moving between grazing areas, as well as carrying their dairy products from the mountain to the markets in Volos. Donkeys and mules were also used for carrying salt and olives, and they are still used today to transport building materials on the stone-paved paths and into those parts of the villages where trucks cannot go.

The dominance of sheep and goats is reflected in the fact that 69.9% of the total faunal assemblage can be identified as these two species.[12] The bones collected at the caves show that the meat of sheep, goats, cattle, pigs and probably equids, too, was consumed. Cattle, sheep and goats were the most important in terms of meat, but many age categories in the faunal sample, from newborn to adult animals, indicate mixed strategies of animal exploitation for milk (and cheese), meat and probably wool. Intentional culling of animals at a young age, shortly after birth, takes place when goat/sheep herders are interested in eliminating competitors for milk. The pig was regarded exclusively as a meat-producing species and Pelion would have provided a good quantity of acorns, chestnuts and other foodstuffs for pigs in the autumn.

As with pigs, much of the sustenance of other domesticates is also wild. That poses a challenge to mountain-based flock owners during the win-

---

9   Magnitos 1860, 57, 98-100.
10   Sivignon 2009, 463.
11   Magnitos 1860, 98-9.

12   Panagiotidou, this volume (Chapter 2.3).

Fig. 4.3. Bundled leaf-fodder, goat faeces and degrading roof structure elements at MIL-1.

ter. The Little Ice Age, a period of unstable climate with a trend towards lower temperatures between the sixteenth and eighteenth centuries, saw one of the coldest intervals during the Holocene: the Maunder Minimum (1645-1715 AD).[13] While these cold snaps were noticeable even in the lowland zones, the effects would have been magnified at high-altitude pastures. As the weather turned significantly colder, the summer gap between the snows shortened and perhaps disappeared altogether. In some periods, then, the summer pastures that are taken for granted today would not have existed; transhumance, if it was viable at all, would have been practised within a more constricted environmental range.

When winters were particularly severe on the mountain, the snow cover was regularly so deep and long lasting that livestock could not exploit any winter grazing that might be present. In addition, flocks needed sufficient shelter, particularly during autumn and early winter; roofed animal shelters were a necessity in the winter months. The only option was to keep the animals indoors during the winter and

bring the fodder to them. Because livestock flocks on the mountain were generally small, it meant that it was easier to provide year-round supplies of fodder in one place. Arboreal fodder for animals commonly consisted of leaf-bearing branches that could be bundled and hung from the roof of the cave pen, as documented at MIL-1.

Microscopic analysis of macrofossils from goat faeces collected on the cave floor of MIL-1 shows that these consist of chewed, fragmented stem and leaf fragments, possibly of the same species as the bundled leaf-bearing branches found in the cave (Fig. 4.3).[14] The analysed samples did not contain seeds or seed coats and the faeces have a high content of well-preserved pollen representing a wide variety of plants and flowers. This means that the faeces were deposited in the cave in spring/early summer. Based on the available evidence, it cannot be claimed with any certainty that the faeces samples and the fodder bundles are contemporaneous (from the same season). If this were the case, however, then the fodder would have been collected

---

13   Lamb 1977.

14   Pers. comm. Morten Mortensen 14.11.2011.

Type	Flock size		Flock size	Cave examples
Localised village herding (i.e. stationary pastoralism)	Usually unspecialised in terms of its products and practised alongside other forms of husbandry; flocks are herded on a daily basis on a radius of up to some kilometres round the village in relatively easy terrain. Back at the village at night. *Flock is for everyday needs of meat and dairy products.*		Small	MIL-1 MIL-2 MIL-4 MIL-5
Village-associated occupational shepherding	Field-station outside villages, but frequent contact with the village community. Flock is capital.		Up to 500 animals? Must be large enough to sustain herder	MOU-7 KER-9 KAR-43-e
Transhumant shepherding	Family-based "occupation", live in temporary structures outside the villages. Come e.g. from Macedonia and northern Thessaly and spend the winter e.g. in Makrinitsa. Flock is capital.		Large	MAK-3 KER-13-e

Table 4.5. Types of pastoralism found on present-day Pelion.

in spring or early summer before the trees develop seeds. In contrast, if the fodder had been used during winter,[15] then it would have been collected late in summer or in autumn, and thus be associated with seeds. As this is not the case, the leaf fodder (if contemporaneous with the faeces) was more likely cut opportunistically, playing only a supplementary role in the animals' diet in the spring/early summer.[16] A comprehensive analysis could possibly reveal more details on grazing habitats around Milies, but goats are omnivorous and this is reflected in their wide range of food choices when grazing in the landscape.

Stationary pastoralism can still be found on Pelion, albeit rarely,[17] and exists alongside village-associated occupational shepherding and transhumant shep-

herding (Table 4.5). Village-associated occupational shepherding is the commonest type of shepherding practice on Pelion. These shepherds tend medium-sized flocks owned either by themselves or by several owners in the village. While the functions of the same cave may change from season to season,[18] these three types of pastoralism possibly make use of caves in different ways. Stationary pastoralism is generally not expected to involve the use of caves, as the animals are often stabled in a pen inside or near the village. Exceptions are caves situated in convenient proximity to the village.

Transhumance is the seasonal movement of people with their livestock, typically to higher pastures in the summer and lower valleys in the winter. Herders have a permanent home, typically in valleys. The term "transhumance" is also occasionally used for semi-nomadic pastoralism – migration of people and livestock over longer distances.

Usually, only men and shepherds accompany the flocks on their journeys between the summer and winter pastures. Occasionally entire families or small pastoral communities move to the sum-

---

15 Haas *et al.* 1998; Halstead & Tierney 1998.

16 We would like to thank Mette Marie Hald and Morten Mortensen from the Natural Science Unit of the National Museum in Denmark (NNU) for looking through the samples.

17 While women are by no means excluded from these activities, we only came across one female herder practising stationary pastoralism, outside Vyzitsa. For a discussion of women's role in pastoralism, see Nixon & Price 2001.

18 Nandris 1999, 115.

mer pastures, but this type of (semi-)nomadism is in decline; yet traditional herding practices are sporadically surviving. In the autumn of 2007, we encountered a 30-35-year-old transhumant shepherd and his extended family above Fitoko. Perpetuating a long-lived family tradition, they move a large herd of goats on a seasonal basis between Trikeri and Kanalia and rent summer pasture from the municipality in the mountain range above Makrinitsa and Fitoko. Magnitos reports in the mid-nineteenth century how Vlach shepherds would buy the right to use the tracts of uncultivated land above Makrinitsa and that they brought their flocks from northern Thessaly and Macedonia to spend the winter there.[19] Specialised pastoralists are found in larger numbers in the nearby Óthris mountain range, which marks the border between the Mediterranean climate of the south and the continental climate of the Thessalian plains. Modern pastoralists of the surrounding plains and those of the villages on the slopes of the mountains, including villagers as well as (semi-)nomadic Sarakatsanéi, Aromanians (Southern Vlachs), take their flocks to the pastures of the Óthris in the summertime.

Shepherds used both their animals and their products as a medium of exchange for acquiring agricultural goods produced by their lowland neighbours. Family-based pastoralism would have been an effective way of dealing with the logistics involved in taking care of the flock and producing, transporting and selling dairy and meat products. In addition, it is demanding for a single person to herd larger flocks in the mountainous, wooded terrain and despite the presence of sheepdogs, there is an increased risk of losing animals. If it were not possible to join up with other shepherds on the upland pastures, a viable solution would have been to involve one's own family members. An informant on Pelion told us that when he was young, his family went to the mountain for six months every year to graze the animals and to make cheese. His father used their only horse to go to Volos, to sell their cheese and buy necessities for the family.

Sometimes the brigands would not only steal the animals but would actually kidnap the shepherds themselves in the hope of extracting a ransom from their families. A song from Makrinitsa ("Kidnapping of a Shepherd") describes how a group of shepherds decided to share the animals of a kidnapped fellow herder between them.[20]

Caves in the pastoral economy

Seasonal exploitation by pastoralists of different parts of the landscape had obvious implications for the use of caves and rockshelters. Archaeologically, seasonal use can in some cases be discerned from neonatal bones of ovicaprids found in the caves and from dates engraved on the cave walls, but seasonality is otherwise difficult to demonstrate without detailed analysis of sediments from within the cave pens or ethnographic data.

The timing of rotation in the landscape was determined by a combination of seasonal weather patterns and religious celebrations in the annual calendar. For instance, St George's Day in the spring (23 April) and St Dimitrios' Day (26 October) in the autumn often marked key points in the year for business and legal arrangements, and when transhumant groups moved between lowland and upland pastures.[21] However, within this overall pattern were (and still are) significant variations in seasonal moves, depending on topography and tradition.

A shepherd from Veneto on North Pelion would alternate seasonally between three different locations on the mountain, none of them too far from each other. Until the early 1990s he would stay with his goats near a series of shallow rockshelters (KER-9)

---

19   Magnitos 1860, 57.

20   Liapi 2006, 249.
21   Wace & Thompson 1914, 48, 77.

until Christmas, when the goats would give birth. Graffiti in the rockshelter confirms this with two engraved dates indicating use in late December and early January. The shelters would be used when the weather was bad. During spring, he would take the goats to "Soudies" and in the winter he would stay where his open-air *mandri* is now situated. It is worth noting that such micro-rotation in the landscape is partly facilitated by the presence of rockshelters, which provided protection to both herders and the flock during autumn and winter. In other cases, the use of rockshelters was more coincidental. A shepherd from Kanalia, who in the past moved his flock on a seasonal basis between the mountain in the summer and Lake Karla in the winter, would use rockshelters whenever their presence corresponded with the need to rest for the night.

There exist several excellent ethnoarchaeological studies of Modern pastoralism around the Mediterranean, which include descriptions of pastoral structures.[22] Some of these also touch upon the dominant issue of how to define pastoral activities from the simple description of semi-deserted structures and the presence of specific objects. None of these studies, however, specifically discusses the pastoral use of caves, and whether caves were used for the same purposes as open-air sites is unclear. As such, data from the Pelion Cave Project provides a useful counterbalance to case studies from open-air sites in Greece.

A fundamental difference between open-air sites and caves is that the former could generally be constructed where they were needed, while caves were used only if they were located in a convenient position. In this way, pastoral cave use was determined by the cave's relationship to other sites and to its position in the cultural landscape.

Animal pens in caves can be viewed as specialised pastoral sites, and their use comprises a variety of functions of which only a few are archaeologically detectable, including feeding, watering, milking, birthing, penning, isolation of sick animals and cheese-making operations.[23] In fact, the animal shelters themselves can be so inconspicuous that they may be overlooked during a survey.[24]

The spatial organisation in open-air sites can rarely be reproduced in caves and rockshelters and it is therefore of interest to know how the limited space in a cave was utilised. Certainly, cave size would influence the way that activities were organised within them. Most caves are small and could only be used as short-term shelters for a shepherd and a small flock. That size was an important factor is underlined by the fact that cave characteristics are commonly given by locals by referring to the number of livestock the cave could contain (e.g. a cave could hold "300 goats and 4 mules"). Thus, a herder would take shelter only in caves that he knew in advance could fit his flock, his support animals and himself.

The construction of fences and walls around and beyond the opening of a cave would increase its usability and flexibility, and roof constructions would increase the number of animals it could shelter. It is therefore relevant to look at how structures were used in organising pastoral and other activities in and immediately around caves and how, vice-versa, the shape and location of the cave affected spatial arrangements and artefact discard patterns (Table 4.6). Unfortunately, too few intact pastoral sites are still in operation on Pelion for strict distinctions to be drawn between them. In most cases, we can only say that the surviving structures, artefacts and dung deposits can be clearly associated with the penning of goats or sheep.

---

22 Murray & Chang 1981; Murray & Kardulias 1986; Chang 1992, 1993, 1999; Chang & Tourtellotte 1993; Efstratiou 1999; Mientjes 2004. We do not aim at discerning the full range of pastoral activities and structures in the Modern Greek landscape as this topic has already been covered by a number of researchers and survey projects.

---

23 Chang 1981, 68.
24 Whitelaw 1991, 421.

SITE TYPE	DESCRIPTION	ACTITIVITIES	DOMINANT ARTEFACT PATTERN
Corral with rockshelter	Open-air, enclosure wall built in connection with a rockshelter or small cave, which is often inconspicuous; animals have ample space to roam inside the enclosure.	Used for congregating and corralling animals.	Domestic; artefacts rare Graffiti rare
Herder/animal shelter	Rockshelter or small cave without open-air enclosure wall; a small wall or a few stones may be present to provide shelter, but actual enclosures are absent.	Used primarily by shepherds as a temporary resting place or shelter. May be used by shepherds to shelter a few animals or for resting while monitoring a grazing flock.	Domestic Personal
Cave pen / Animal keep	Well-built enclosure immediately in front of a cave or larger rockshelter; may be roofed and have divisions. Some cave pens may have formal hearths to support a human presence.	Normal pen activities include "herding activities such as feeding, watering, milking, birthing, tethering, and housing animals" (Chang 1981, 68). Pastoral "centre of operations". Apart from pen activities, domestic activities such as food preparation and eating may take place.	Animal related Structural Domestic Graffiti frequent

Table 4.6. Hypothesised pastoral site types on Pelion.

Artefacts within the animal shelters are not plentiful, but some are directly animal related, including feed troughs, shepherd's staffs, a goat collar, a vaccine syringe and vaccine packaging. Other artefacts from animal shelters can be classified as domestic or structural, including such items as building materials and water jars. The largest diversity in artefacts is found around the more elaborate and work-intensive structures, such as animal folds. It is also here that activities that are more diverse and longer lasting take place. Shepherds commonly burn their trash before abandoning a shelter. Such "cleansing fires" can be mistaken for expedient hearths. Shepherds also commonly use plastic bottles to light their fires during wet weather, when dry firewood is difficult to find. Both practices entail that even non-perishable materials, such as plastic, may not be preserved.

Domestic animals often die in the caves, accidentally or intentionally. However, many (if not most) of the animal bones present in caves are introduced naturally, either as wild prey transported by hunting or scavenging animals or as dying and sick domestic animals seeking refuge before death. An informant from Pinakates once lost a goat that she later found had taken refuge in the *Katsikadamou* rockshelter (MIL-4). Apart from disarticulated bones, we found goat carcasses in various stages of decay at MOU-7 and KER-4, and a horse carcass at KAR-20. Such finds can be used to support a claim that herds of domestic animals frequent the area, but any complete carcass left in a cave may challenge the suggestion that humans currently use it as shelter. The fact that carcasses are rarely left in regularly used caves may explain the relatively low quantity of bone remains and the low minimum number of individuals.[25] At the same time, it shows that the domestic use of caves was occasional or limited to a few shepherds,

_____

25   Panagiotidou, this volume (Chapter 2.3).

who did not make use of their herd as a meat resource.

Throughout their use as pastoral facilities, dung accumulated in caves and rockshelters as the result of the congregation of goats and sheep. While dung accumulations in caves and rockshelters can be the result of the spontaneous congregation of free-ranging goats, thicker accumulations can generally be used to infer deliberate livestock concentrations such as penning or stabling. These deposits typically consist of the characteristic intact droppings and of loose debris derived from dried, crushed, and pulverised dung. Deposits of goat droppings are unconsolidated because the droppings are naturally dry. Over decades and centuries, a process of mineralisation, the loss of degradable organic matter through oxidation, slowly transforms dung into a layer of phytoliths, calcareous spherulites and detritic dust. These mineral residues of manure accumulated in caves and rockshelters are potentially important indicators of herding activities at both Prehistoric and Modern sites.[26]

In MIL-1, we encountered a dried surface comprised of a trampled accumulation of sheep dung. These had broken into plate-like shapes and deep desiccation cracks had formed in the top layer. Such surfaces indicate penning of sheep.[27] Other caves have been partly levelled in recent times by the activities of modern goatherds and are capped by a layer of compacted goat dung. A characteristic of Tsounaga (MOU-1) and Kentaurou (MIL-1) is that the origin of the surface sediment is strongly linked to human activities, as it derives from herbivore dung. These data show that the cave was used by shepherds who used it as a stable for their flock.

Manure had economic value before the introduction of artificial fertilisers and its presence on the pasturelands could be used to reduce the level of rent payable, especially since one sheep produces 500 kg of manure each year.[28] A more work-intensive method was to extract manure from deposits in caves and distribute it on agricultural land. This would also be a way of getting rid of unwanted organic waste in caves used for stabling. Dung extraction would have caused the removal of the top part of the deposits at many cave and rockshelter sites (at prehistoric sites an important cause of floor lowering is the loss of organics and volume reduction of manure). Dung extraction to be used as fertiliser in nearby fields would have been practised in the past, but getting insight into the frequency of this practice is difficult since archival information on the topic hardly exists. Dung deposits are rich in nutrients like organic matter, phosphorus and nitrogen and the inherent alkalinity of ash from burnt bedding or fodder and fresh or burnt animal dung would have been beneficial to gardens and orchards around the villages on Pelion.

Due to the logistical challenges of transporting the dung, it was probably only extracted from caves with easy access. This practice has now disappeared from caves due to the use of industrial fertilisers. However, one informant told us that he would give animal dung from his open-air pen away to the villagers for no payment, as people still use animal manure in their gardens and orchards. The presence of disturbances in the top levels of MOU-7 and MIL-3 may signal its removal for use as fertiliser. In Italy, small-scale extraction of cave manure for local orchards was still occasionally practised a few decades ago.[29]

The burning of manure is another way of getting rid of dung deposits from animal pens, but for obvious reasons it is an activity that is exclusively associated with caves. Burning may have been used in situations where it was not economical or practical to

---

26 Balme & Beck 1992; Brochier *et al.* 1992; Canti 1997, 1998, 1999; Mlekuz 2009.
27 The local traditional name of MIL-1 is *Stani* – Sheep Pen.
28 Nixon & Price 2001, 16.
29 Brochier *et al.* 1992, 64.

utilise the dung, or where the shepherd intended to reduce the risk of disease among the animals by getting rid of pests, as is known from the ethnographic record.[30] To allow the dung some time to dry up, this would have taken place sometime after the animals had abandoned the cave. Modern veterinary care and medicines have rendered this practice unnecessary and indeed, we did not find any evidence of it on the present-day cave floors. While ash deposits and fine charcoal could represent periodical burning of stable waste, they are more likely remains of expedient hearths or isolated fires.

Storage caves

Caves are useful storage facilities because they are secure locations once the entrance is walled up or otherwise barricaded. Caves are also cool and have a year-round stable temperature. This made them particularly useful resources in the days before electricity and artificial refrigeration were available. Even though electricity was introduced on rural Pelion in the 1950s, there were many areas in the region where homes and farms did not receive electrical service until much later. Before refrigeration became generally available in the countryside, meats were generally dried or salted and smoked. Vegetables and fruit could be dried, pickled or canned. However, fruit and dairy products such as milk, cheese and butter needed to be kept in the cool and dark, and many caves provided a means of refrigeration that greatly extended the life of such perishable foods.

Only a few caves on Pelion still bear evidence that they were once used for food storage. Two small caves in Kanalia have walled-up openings with doorways. The two cold storage caves are called *Garvantineika* (property of the Gavarntinas family) and *Polychroneika* (property of the Polychronis family;

KAR-31). The caves were used before World War II, but the outbuildings were constructed after the war. While these examples of privately owned caves are small storage facilities, larger cool storage caves were probably communal and thus shared by a number of dairy producers in the village or on the mountain.

Caves also provide a convenient repository for items that are too cumbersome to haul regularly between home and field. Thus, equipment is readily available in the locale where it is utilised. AGR-8 near Anemoutsa is a shallow rockshelter located in an olive grove on a high terrace and contains implements and containers used in olive cultivation and harvesting, including oil drums and bags of fertilisers. ART-2 is another field storage cave, barricaded with hammered-out oil drums and secured with a large lock. While it does not appear to be used frequently, it is crammed with plastic containers, bottles of chemicals, field tools and parts from water pumps.

Natural resource procurement – water conduits, mining and quarrying

An advantage of the Pelion villages was their ability to utilise an enormous underground water supply and the water for village use was provided by natural springs. The function of some caves, natural as well as artificial, has been concerned exclusively with the extraction or passage of water. At *Timpano* (MOU-25) on East Pelion, a conduit leads water from a waterfall into a 14 m-long tunnel carved through a limestone outcrop on the steep mountainside (Fig. 4.4)

The up to 0.7 m-wide and 0.5 m-deep conduit inside the tunnel is lined with well-laid drystone walling, which continues beyond the entrance. From the exit on the other side of the outcrop, the water is channelled through an open cemented conduit to the villages of Kissos and Agios Dimitrios and fields fur-

---

30   E.g. Acovitsioti-Hameau *et al.* 2000, 127; Kyparissi-
        Apostolika 2000.

ther down the mountain. According to informants, the tunnel was made by a Mr Aggeletos, apparently single-handedly, in the beginning of the twentieth century using hammer and chisel. The cemented components were added after the 1950s.

AFE-1 and AFE-2 are two small caves on Kalamaki Beach with low ceilings, situated just above the beach. While inconspicuous, both caves have water inside and at least during the twentieth century, people from Kalamaki village collected drinking water from the caves. AFE-2 has now collapsed, which has somewhat restricted access to the entrance.

Around Lake Karla there are several old water sources such as Tsaritsani (KAR-16) and Vania (KAR-32). They used to be productive and reliable springs, but the draining of Lake Karla meant that the water level dropped and the springs wholly or partly dried out. *Mana* (KAR-13) is near the edge of the lake in an area called *Paliochori* ("Old Village"). An artificial cave had been dug into the alluvium from which a horizontal brick-lined channel had been excavated further into the unconsolidated substrate. The opening was built in stone and mortar. Three post-Medieval ceramic sherds were collected immediately outside the cave's entrance. According

to a shepherd, young pregnant women used to go to the spring and drink the water, which they mixed with crushed limestone.

In one cave below Makrirachi on East Pelion, a small drain had been carved into the bedrock to collect seeping water from the cave wall (ZAG-5). This might have provided a sufficient supply of drinking water for a few persons sheltering in the cave.

4.5 Dwelling and short-term habitation

"Dwelling" is a commonly used term for a house, home or shelter, or for living somewhere. Archaeology and related disciplines fully recognise the importance of dwelling as a social practice,[31] but Modern use of rockshelters and caves for dwelling have rarely been subjected to scientific enquiry in Greece.[32] One

---

31   Gosden 1994; Gosden & Head 1994; Tilley 1994; Thomas 1998; Ingold 2000, 5. Anscheutz *et al.* 2001; Cloke & Jones 2001; Ashmore 2002.

32   For examples of recent cave dwellers in Europe, see Kempe 1988, 141-63.

reason may be that this practice often has been associated with people living at the margins of society (immigrants, hermits, the diseased), who have not attracted much attention. The most well-known examples of cave dwellings in Greece are the "leper caves" in Matala in South Crete, the volcanic pumice-houses of Santorini and the Turkish-speaking cave dweller community in Didimoticho in Evros. All three are cases of habitation of artificial caves, carved out of the bedrock.

Habitation in caves can take many forms depending on the purpose and the duration of the stay (Table 4.7). In most cases, caves are used temporarily and as an alternative to living in built structures. Without a doubt, some people stayed on Pelion outside the villages all year round, but caves were used for dwelling from the nineteenth century onwards only by specific subsets of society. A traditional song from Pelion tells about a brigand who would not leave the mountain during the winter out of pride and therefore "asks the sun to shed its beams on him".[33] Similar songs, mainly from the nineteenth century, recount the hardship and suffering of those brigands who chose to live on the mountain outside the villages. A rational way to keep safe and warm during winter would be to dwell inside a cave.

The activities of brigands and other criminals on the mountain waned significantly with the annexation of Thessaly to the Greek state in 1881. From the end of the nineteenth century, cave dwellers were mainly vagrants or recluses (frequently with a mental disorder) who had withdrawn from society, and seasonal workers or family groups belonging to an ethnic minority group (e.g. Gheg or Tosk Albanians). Ethnic minority groups, in particular, became more sharply defined following the territorial expansion of the Greek state after 1881, when Greece entered a "period of ideological rigidity by seeking internal enemies and marginalising minor-

ity groups".[34] This was in contrast to the situation under the Ottoman *millet* system, which had drawn part of its strength from the considerable variation in local dialects and languages, and the fact that large segments of people belonged to different ethnic groups. The attempts of the young Greek state to define citizenship in terms of religion and place of origin led to the marginalisation of various ethnicities.[35] However, these groups of Ghegs or Tosk Albanians continued to play a significant role in the countryside as seasonal labour, and would occupy caves or abandoned structures close to fields or construction sites.

The identification of dwelling caves at Pelion was only made possible by the combination of archaeological observations and ethnographic data. The ethnographic cave dwelling indicators were first-level accounts by informants who have stayed in caves themselves, and indirect information from people referring to caves used as dwelling sites (e.g. MIL-3). Archaeological cave-dwelling indicators, if one can speak of such, are tangible signs of functions that have been transferred – partly or in full – from a permanent residence to a cave. Such indicators are at best ambiguous and in most cases, it is impossible to establish archaeologically whether a cave was inhabited over an extended period, or indeed the character of the habitation. Activity areas (such as for sleeping, food preparation and consumption) within the cave may or may not be archaeologically detectable and domestic artefacts can be present or absent.

Sleeping arrangements may be conceived of as one of the clearest indications of dwelling, but Galanidou has rightly warned that "sleeping is one of the most elusive aspects of human activity in the archaeological record".[36] An example from the sur-

---

33 Liapi 2006, 275-6.

34 Michailidis 2006, 160.

35 On Arvanites and Aromanians in Greece, see the studies by Bintliff 2003 and Kahl 2004.

36 Galanidou 2000, 250.

Type	Cave characteristics and primary function	Duration of habitation	User profile	Examples
Refuge	Concealed entrance; often located on the margin of the agricultural territory of villages	Short-term (one day to weeks)	Mixed-sex, elders, children	MOU-12 AFE-7
Shelter for migrant workers	Near villages and roads; used as a seasonal home-base	Seasonal (months)	Men only or mixed-sex family units	MIL-3
Shelter for recluses	Small; used primarily for sleeping	Transient to long-term (days to months)	Male and female	MIL-23
Isolation of diseased		Long-term (months)	Male and female	ART-6
Pastoral shelter	Used for overnight stays in connection with pastoral movements	Transient (hours to one day)	Dominantly male	KER-9
Hermitage		Long-term (years)	Exclusively male	KER-5

Table 4.7. Characteristics of different types of habitation encountered on Pelion.

vey serves to illustrate this point: a shepherd from Kanalia told us that they (the shepherds) used to have resting places in many localities. When they wanted to rest they would collect branches and arrange them for a bed (*giataki*) and then sleep on the branches wrapped in their large shepherd-coats. Before they built the drystone wall in front of KAR-35, they would arrange *giatakia* inside the rockshelters. More recently, some shepherds used fleece blankets to form beds that can easily be rolled up and transported. In the single case where a bed was actually found (AGR-8), it appeared to be used as an occasional gate for a field storage facility in front of a small rockshelter. While these do not represent cases of dwelling, the ephemeral character of such sleeping arrangements does not necessarily differ significantly from those in dwelling caves. In fact, our data indicates that dwelling caves are rarely modified and that architectural elements or site furniture may not be present at all. Such constructions are redundant to some cave users, whether they are vagrants, herders or field hands, because of their migratory behaviour.

While site furniture may be absent and sleeping arrangements undetectable at dwelling caves, one would expect to come across debris resulting from what are commonly termed domestic activities. These represent a broad spectrum of activities and include processing, cooking and consumption of food, sleeping, social interaction and entertainment. Objects from the Pelion caves are more frequently associated with such domestic activities, rather than with production or religious activities. Examples are artefacts related to personal use (hairclip, toothbrush, razor, towel, clothes, shoes), entertainment (radio, newspaper, toy) and consumption and temporary storage (glasses, ceramic vessels, plates, conserves cans, bottles, snack wrappings). However, domestic artefacts are often used in transitory situations simply to make short stays in the caves more comfortable. Although such artefacts on their own do not indicate dwelling, specific objects may contribute to instilling a temporary sense of home and belonging in transient cave users.

While it is not easy to point to general characteristics of dwelling caves on Pelion, dryness, light and size play a basic role in the choice of a cave for habitation. However, the case of MIL-22 (see above) shows that each of these factors must be given a considerable range. The location could be a principal factor in the choice of habitation caves, both in terms of accessibility and in terms of proximity to a nearby settlement. With the exception of hermitages, dwelling caves are found within the agricultural territory of the villages, and close to roads or major paths below Milies ("Argyraki", MIL-3, is a particularly good example of this). On Pelion, this means that known dwelling caves are found on the foothills and lower to mid-altitude, not above 600-700 m.a.s., which is the upper altitude of most villages (no dwelling caves were identified above 445 m.a.s.).

Short-term habitation on Pelion takes many forms but usually conforms to one of two patterns: 1) brief or prolonged single stays at caves during unusual events such as war, conflict and epidemics; or 2) frequent use of caves for economic activities, but with base in a village. In either case, the duration of the stay does not extend beyond a season and only a few functions of a permanent residence have been transferred to the cave. Any structures present are usually animal-related; discernible divisions of space are rare; no permanent hearths or decorative items are found, and most food remains are "expedient".

Despite refinement of analytical techniques, the inability of archaeology to detect the signatures of ephemeral short-term use of caves has long been recognised by prehistoric archaeologists. Several of the ethnographically confirmed cases of short-term use on Pelion are related to World War II use as civilian refuges or partisan hideouts, and in most cases these uses have low archaeological visibility despite lasting for several days or weeks. Short stays in caves in connection with bad weather or during agro-pastoral field work are also likely to go unnoticed. As these places have been used by one or a few persons with little equipment for some hours, not much waste producing activity can be expected – with the exception of the consumption of food and drink.

Vagrants and social recluses

Vagrants have habitually used caves, and they often maintain a degree of mobility in their lives in which caves and rockshelters become naturally integrated. A news release from early 2011 recounts the case of a 65-year-old vagrant, who was found dead in a cave with an attached hut-construction near Agios Vlassios. The vagrant had wrapped himself tightly in a blanket, probably in an attempt to resist the cold. For several years, he would roam the Lechonia-Agria-Volos road begging and collecting garbage. At night, he would return to his shelter.[37]

In 2009, the survey team learned of an individual, originally from Kissos, who suddenly reappeared in the village after having been absent for decades. The man, presumably a shepherd, had chosen to live on his own high on the mountain with a few animals. Upon his unexplained return, he had difficulties communicating and adjusting to modern life. What is interesting about this case in connection to the survey is that the man may have sought out caves in order to survive the winter on Pelion.

An identified dwelling cave that belonged to a female social recluse nicknamed Marouko (MIL-22) is located in the gorge southeast of Agia Triada. Marouko started living in the cave because she was on poor terms with her family, and she inhabited it from the 1920s/1930s and thereafter for 40-50 years (although possibly not permanently during the whole period). Local villagers passing on a path close to the cave would sometimes leave bread and caramels for Marouko. An old woman at Agia Triada, who lived by the Moutzouris train track, shared with us her recollections of growing up with Marouko, stealing eggs from the family hen pen or wandering in the area looking for food. Marouko's cave is a surprisingly small cavity with a maximum height of 1.73 m and a mere 4-5 m² of living space (Figs 4.5 & 4.6). A small area outside the cavity offers some protec-

_____

37   Online article published at www.taxydromos.gr.

Fig. 4.5. Entrance of MIL-22.

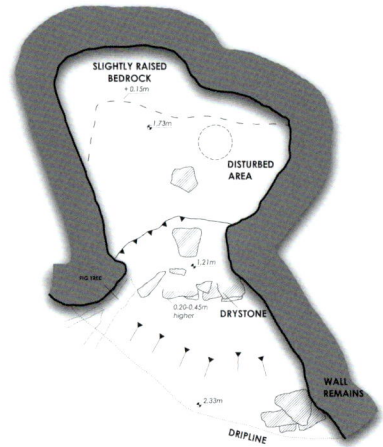

Fig. 4.6. Mapped surface of MIL-22.

tion before the drip line a bit further out. A few flat, stacked stones are the possible remains of a small retaining wall intended to keep the area in front of the cavity level. Extremely compacted sediment inside the cavity is the only indication of the cave's use as a human dwelling.

Local accounts refer to several rockshelters close to the AGET cement factory at Goritsa outside Volos being used as short-term habitations by homeless vagrants and as temporary shelters for local shepherds. Inside one of these (POR-2) is a mixture of Modern artefacts. Mapping of all the debris allowed for the identification of three tentative activity areas: 1) a sleeping- or resting-place associated with carpets, a duvet and personal effects such as shoes and a lighter – many cigarette packages and butts show that smoking also took place in this area, and a scatter of cigarette butts near the entrance was possibly thrown there from the resting-place; 2) a hearth area with charcoal, a burnt magazine and animal dung; 3) an area near the entrance with many almond shells. The practice of sitting at a cave entrance gazing and eating figs and almonds was emphasised to the survey team on many occasions. Fig trees were so common in front of cave entrances that they often could be used to find caves that were otherwise hidden from open view.

Immigrants and migrant workers

Seasonal workers often migrate long distances to find work and rarely construct permanent features. On three occasions, the survey team come across recently abandoned open-air camps in the forest, where Albanian family groups had lived while the men took on temporary employment in local agriculture or construction work.

Argyraki Cave (MIL-3) below Milies functioned up to the last decades of the nineteenth century as a dwelling for a community of allegedly 150 Albanian Gheg. The Ghegs are one of the two largest Albanian ethnic groups, residing north of Shkumbin river.[38] These men probably worked as construction workers on the Volos–Milies railway or as field hands in the fruit orchards around Milies. Approximately 90% of Ghegs were farmers,[39] and it was relatively easy for them to adapt in the arduous life on Pelion as their homeland was a mountainous region.[40] The cave and the field around it belonged to the Argyraki family of Milies, who probably also employed or rented

38 Albanian people are divided into the Ghegs and Tosks. The Tosks are the second group of Albanian people, occupying southern territories (Vickers 1997, 5).
39 Carleton 1977.
40 Coon 1977.

the cave to the Albanian seasonal workers. The cave is large and dry, and while the relatively small opening does not let in a lot of light, it keeps the temperature in the cave constant. The living space is between 140-160 m², which would hardly have allowed 150 individuals to stay in the cave at the same time. The pottery remains from the cave floor represent a mixture of shapes, including cooking pots (tsoukali), jugs and various open shapes used for serving.[41] Remains of domestic animals comprise donkey, cattle, sheep, goat and chicken. At least some of these represent food remains and a bone of red deer and a bone of sheep/goat – both chopped to extract the marrow – provide positive evidence of consumption.[42]

Two informants recounted that during the 1930s and the 1990s, families of Albanian origin stayed around an abandoned quarry with cave-like features outside the small hamlet of Vrochia, northwest of Kato Lechonia. One informant came to Pelion in 1937 when he was six years old. He lived with his family in stone huts around the Ntamari cave (ART-1), which was used mainly for storage of wood. The area was then a small hamlet with 100 people or more working at the olive groves of Agios Georgios. The quarry was actively exploited until the 1940s. After that, the quarry cave was used for keeping animals. A second cave was used as a refuge and goat pen.

Massive rockfall and casually stacked stones partly blocks the opening to the artificial cave. The material residues point to the presence of a small family group, who probably stayed there during the 1990s. Discarded objects include broken furniture, remains of clothing, a plastic toy car, a battery, a speaker cover for a radio, a broken ceramic water jug, a chicken bone and parts of mass-produced objects, some of which have been secondarily modified by cutting (e.g. a plastic knife handle and a pen). A

pile of stacked stones in the back of the cave shows an attempt to clear the floor. A concentration of charcoal against the eastern cave wall is the remnant of an expedient hearth. A handful of birdshot-shells and scattered goat faeces indicate later use of the cave as hunting stand and animal shelter.

The use of the quarry caves by families of Albanian origins within approximately 60 years of each other represents an interesting example of human movement patterns in a dynamic spatiotemporal context. If one adds to these two cases the presence of Ghegs in the region as early as the nineteenth century, we can probably speak of a repeated phenomenon within the context of seasonal work and migration.

Isolation and refuge

Stays in caves were sometimes of a particularly poignant nature, for example in cases of isolation of plague-diseased individuals.[43] Thessaloniki and other cities were hit hard by plagues during the seventeenth and eighteenth centuries, with individual outbreaks killing thousands of people.[44] Constant movement between the villages and the city extended the range of the epidemics as local peasants went to the cities to sell their wares. Some would have transferred the plague upon returning to the village. It probably did not help the situation that warm summers and a humid climate around Karla offered the plague bacillus a near-ideal environment in the lethal months from April to July. Plagues sometimes ravaged the villages; following such outbreaks, entire settlements would remain abandoned for a long time.[45] Pelion would have felt the effects of such epidemics, and would have experienced some demo-

---

41   Vroom, this volume (Chapter 2.7).

42   Panagiotidou, this volume (Chapter 2.3).

43   For a Cretan example of isolation of pest-diseased animals in caves, see Faure 1964, 216.

44   Mazower 2004, 114.

45   Makris 1982, 268.

graphic disturbance as people fled to the mountain. As there was no overall governmental response to the epidemics, each community took its own measures. A shallow rockshelter outside Agios Lavrendios (ART-6) was allegedly used to isolate members of the village who suffered from the plague. According to a local informant, the villagers would hoist down food and water to them from the cliff above, to avoid contamination.

Caves and rockshelters were used for short-term habitation in conflict situations, most notably during World War II and the Civil War. This troubled period in the history of Pelion has many unknowns and it was a violent time that led to many deaths. It is not surprising, therefore, that there are many stories about individuals and families who hid out in caves to avoid discovery by military patrols. Partisans stayed high on the mountain to raid, making headquarters in caves and moving from cave to cave to avoid detection.

4.6 Spiritual and ritual uses

Caves have played a prominent role in myths and legends throughout recorded history, but myths associated with caves are inherently intangible and elusive and bear little relation to the archaeological resource. Yet it was one of the defined objectives of the ethnohistoric research to include these kinds of narrations, mythical and spiritual, in our analysis. This broadened perspective adds to the biography of a cave and provides justification for a rather wide set of uses.[46] In addition, there is a significant relation between caves as religious sites and material culture, e.g. churches built in caves, religious graffiti and artefacts.

At times, both mythical and religious values are entangled in the local narratives. MIL-1 (*Kentavron* cave) outside Milies demonstrates in an ample

way the dynamic role that myth and religion play in the biography of a site. At the entrance to the cave, one discerns an imprint in the shape of a horseshoe on the horizontal bedrock. Pertaining to local narratives, this is a trace that Saint Taxiarchis' horse left upon the rock while rushing to the rescue of a pilgrim. Another tradition recounts that one night during the Ottoman period, perhaps the night before the celebration of the archangel Taxiarchis, a terrible noise was heard and a blinding light appeared by the precipice. It was Archangel Michael, riding his horse, and with his sword he opened a narrow path alongside the steep rock face. When the villagers went to the location the next day, they discovered his icon in an opening in the rock. There, they built a shrine and later a small stone chapel. The traces of the horseshoe and holes from the stick can still be seen on the rough stone path. A story at the same location from 1902 tells of a child who went to the chapel with an offering of wheat and bread. The villagers saw the child fall from the steep cliff, but after a few minutes the child climbed back up again unharmed, still holding in his hand the wheat and the bread. It is intriguing that this location, so steeped in Christian mythology, is today also associated with Chiron providing reasoning to the locals, intermingling the mythological and the religious with contemporary needs. It is also worth mentioning that pagans, esoterics and pseudo-scientific groups have published a number of articles and websites concerning Chiron and the Kentavron Cave.

Pelion caves are not only associated with myths and legends, but also operate occasionally as religious spaces. Several activities can be described as religious, namely the use of caves for ancient or Christian ritual and/or the possible presence of votive deposits, and the use of caves as burial vaults. Ritual activity seems to be confined to deep caves as these have obvious mysterious, deathly and womb-like associations. Human bones collected from four caves (KAR-8, VOL-3, MOU-1 and MOU-16) are most likely parts of disturbed burials from the His-

---

46   Kefalliniadi 1961; Platakis 1977.

torical period (Prevedorou, this volume).[47] They were in all cases associated with Late Helladic and Hellenistic-Roman ceramics that can be interpreted as remains of grave offerings. However, these are mainly coarse and plain wares for cooking, serving food, transportation and storage activities, and there are few decorated wares with the exception of a few painted sherds and Roman red-gloss ware.[48] It is possible that simple burials and occasional offerings to various nature deities took place in caves during the Hellenistic-Roman periods and at least some of the sherds found in the deeper zones of the caves (e.g. MOU-1, MOU-7, and MOU-16)[49] could represent such activities, with coarse wares containing a small offering of meat, grain, wine or oil.

Mountains were regarded as the natural dwelling places of the Nymphs (Hesiod, Theogony 129-30) and several mountains had sacred places, altars and temples for other deities.[50] Arvanitopoulos made a brief excavation in a cave below the Plaka summit of Ossa, north of Spelia. During his investigation, a number of dedications to the mountain nymphs, pottery from the fourth to the third century and fragments of terracotta figurines were recovered.[51] No formal, sacred caves have been identified on Pelion and there are no signs of the worship of the original Thessalian goddess Enodia (it is still questionable what her exact characteristics are, but later she was associated with Artemis and other classical Greek goddesses). There are traces of her worship at Pagasai and mostly Pherai, where a temple dedicated to her was discovered.[52] Noticeably, ceramic material from the Archaic and Classical periods appears to be missing from the Pelion caves.

Historical sources, however, support the notion of ritual cave use in historical Thessaly. The sons of nobles used to visit the cave of Chiron on Pelion, evidently at an initiatory moment in their lives.[53] Buxton provides the following citation from the Hellenistic geographer Herakleides Kretikos 2.8:[54]

> On the peak of the oros there is the so-called 'cave of Cheiron', and a shrine of Zeus Aktaios. At the rising of the Dog Star, the time of the greatest heat, those among the citizens who are most notable and in the prime of their lives, having been chosen in the presence of the priest, climb up to the cave, clad in thick new fleeces – so cold is it on the mountain.[55]

Arvanitopoulos excavated the sanctuary of Zeus Akraios and Chiron mentioned in the passage at the beginning of the last century, within the fenced-in military area on the summit of Pliassidi.[56] Zeus Akraios was already known as one of the main deities of the Magnesians, worshipped on Pelion.[57] The remains consisted of a peribolos, two temples and a stoa. Votive pottery and weapons were recovered at the sanctuary and the finds suggest that it was used in the fifth and fourth centuries BC. A cave was located at the periphery of the sanctuary and it is possible that it served some cultic function in connection with Chiron or with the deity worshipped in temple A. However, while such "co-locational" worship of Zeus and a lesser deity or hero are known from several other mountain tops in Greece, no inscriptions or other finds at the cave can confirm a specific cult of Chiron at this site.

---

47   Human bones have also been reported in earlier reports from Tsouka at Mouresi (MOU-11) and from Alikopetra (not the Alikopetra caves AGR-4-6 investigated by PCP).

48   Voskos, this volume (Chapter 2.6).

49   In all three cases, the innermost parts of the caves appear to have been unused in the Modern period. Instead, modern pastoral activities took place at the large cave entrances.

50   Langdon 2000, 463-4.

51   Arvanitopoulos 1908-09, 243ff; 1910, 183-4.

52   Morgan 2003, 92.

53   Dowden 1999, 87.

54   Buxton 1994, 93.

55   Herakleides II, 8; IG IX-2, 1110 and 1108.

56   Arvanitopoulos 1911, 305-15.

57   Gorrini 2006, 283.

Plutarch does speak of a cult of Chiron, as a healer, in Magnesia; other sources mention a Chironidai festival in Demetrias.[58]

A cult of Zeus Meilichios has been proposed for cave POR-5 on the Goritsa Hill outside Volos.[59] The cave bears the inscription on a rock near the cave entrance, *Διός Μιλιχίο* (Zeus Meilichios), dated to the third century BC or later.[60] Later uses of the site include storage of ammunition by German troops and shelter for animals, the latter evidenced by a collapsed drystone wall.

Christian uses of caves

Various explanations exist as to what triggered the use of chapels in caves and rockshelters. A popular belief is that Ottomans did not allow Orthodox Christians to build churches, despite the Küçük Kaynarca Treaty in 1774;[61] as a way of evading the ban, local Christian communities established religious spaces in caves. However, this contradicts the generally tolerant position of the Ottoman authorities towards Orthodox Christians and other faiths – especially on the almost exclusively Orthodox Pelion, which already enjoyed a number of economic and political privileges granted by the Ottoman establishment. A more believable explanation is that religious recluses would resort to caves and that some of these would later be turned into chapels. Cave chapels are usually small, one-room

religious localities, often found in the Greek countryside.[62] As with other small countryside chapels, these are visited primarily on the dedicatory saint or saints' feast day, and do not have a priest permanently attached. Some country chapels would have been acquired by monasteries and thereby gained monastic status.

One cave chapel (MOU-10) was documented during the survey in the northern part of Fakistra, south of Damouchari Bay. Historically, it is part of a tripartite cave complex consisting of a sea cave (*Panagia Megalomata*) 20 m below the chapel and a rockshelter (*Kryfo Scholeio*) next to the chapel.[63] The chapel and the rockshelter are on a steep cliff immediately above the Aegean Sea and access is only by a narrow path from the carpark high above.

Local traditions maintain that the cave of Panagia Megalomata below harboured the 'secret school' of Tsagarada, where during the Ottoman period Orthodox children could learn to read and write under the tutelage of a monk named David – since local Ottoman officials did not permit Christian priests to teach. *Kryfo Scholeio* is a reflection of the popular myth of the secrecy of Orthodox belief during the Ottoman period.[64] According to the locals, the sea cave provided shelter to both a monk and the Panayia Icon (Megalomata) during the last half of the seventeenth century. The entrance of the cave is wide enough to allow a boat through. The dry part of the cave is divided into two sizeable chambers with considerable roof height. Freshwater drips into natural rock basins, which are always filled with drinking water.

The cave chapel on the steep cliff above the sea cave is dedicated to Panagia Megalomata. The opening of the cave has been walled-up by drystone walling and has a small timber-framed window and a

58   Plu., Quaest. Conv. 647a. Gorrini 2006, 284.
59   Backhuizen 1992; Helly 2006; Adrymi-Sismani 2010
60   Te Riele 1977.
61   Treaty of Küçük Kaynarca 2011. This was a treaty signed between the Russian Empire and the Ottoman Empire after the latter was defeated in the Russo-Turkish War of 1768-74. Included in the treaty was a religious stipulation allowing Russia to represent Eastern Orthodox Christians and exacting a promise from the Sublime Porte to permanently protect the Christian religion and its churches.

62   Dubisch 1995; Hart 1992; Kenna 1976; Stewart 1991; Skoura 1993.
63   Grafiou-Nida 1971, 83.
64   See parallel from Crete in Brumfield 2002, 52.

Fig. 4.7. Drystone structures at MOU-10 (*Kryfo Scholeio*): the "monk cell" with the engraved wooden beam above the door. Below and to the left is the small chapel.

doorway (Fig. 4.7). A decayed and defective door with remains of rusting hinges has been placed temporarily inside the chapel. The floor is fitted with natural stone slabs. The cave room is divided by a simple, wooden iconostasis or chancel screen upon which are fastened several icons and candle holders. In the tiny sanctuary east of the iconostasis is the Holy Table, consisting of two niches (*prosthesis, diakonikon*) carved into the eastern wall of the cave. A drystone wall separates the two small niches, each of which is fitted with horizontal stone slabs. On these are exhibited more icons, incense burners and a few small votive offerings left there by contemporary visitors. Mortar has been used to consolidate some of the stones in the exterior wall and the drystone walling around the niches.

Many artefacts inside the chapel can be classified as religious, such as the framed icons and religious paintings (including the Virgin Mary with the Christ child, and St. George), candles and brass candle stands (*kantili*) and incense burners (*thimiato*). Others are "domestic" artefacts, such as plastic and glass bottles (for oil), a bracelet, coins, dried flowers, matches, lighters, a dustpan, etc. Although frequently visited by tourists and a few locals (especially on the celebration of Panagia Megalomata on September 8), the chapel appears to be a bit neglected.

Next to the chapel is a partly collapsed cave containing two drystone structures. Above the entrance to the best-preserved structure is an engraved beam commemorating David the monk: "[ill.] ΙΟΠΑΝΩϹΙΩΤΑΤΩΚΩ. ΔΑΒΗΔ.

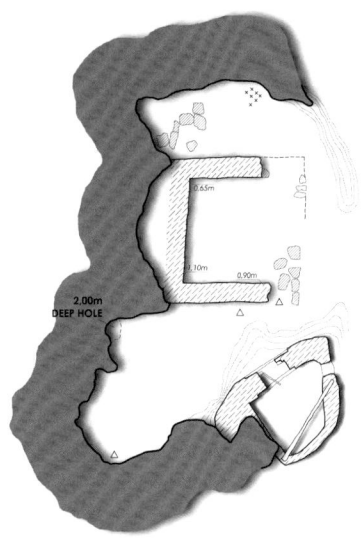

Fig. 4.8. Drystone structures inside the rockshelter of MOU-10. The engraved beam is positioned above the door of the southern structure. The adjacent cave chapel is a few metres further to the south along the cliff edge. Triangles indicate ceramic finds.

IEPOMONAXOY. ETOC. AXNS", which translates into "[cannot be read] THE VERY REVEREND ΚΩ DAVID. MONK. YEAR. 1656". Another approximately 4 x 4 m-large square structure inside the rockshelter is preserved up to a height of 1.1 m (Fig. 4.8), and with a wall thickness of 0.6 m. The east wall (with the entrance) has been demolished, but a few stones still indicate the course of the wall. Around the structure were found two post-Medieval roof tile fragments and an uncharacteristic ceramic sherd.

Apart from the case of caves as chapels and religious spaces, some monks on Pelion selected caves for hermitages. Hermitages were solitary dwellings for the exclusive use of an anchorite who wanted to live in spiritual and social seclusion, perhaps copying a saint. Such persons would occupy a single or two small caves, typically containing a cell and a small chapel.[65] Greek Orthodox monasticism has its roots in the lives of St. John the Baptist and Jesus, both of whom went out into the desert to fast, pray and battle with the devil. In the third century, several Christians followed in their footsteps, most importantly St. Anthony the Abbot, who is considered the founder of anchorite monasticism, where hermits live separately but meet on occasion to share meals and holy services. From this grew cenobitic monasticism, where the monks lived together and shared labour and worship on a daily basis. After the Ottoman conquest of Greece, St. Anthony's ascetic practices and cenobitic monasticism again gained in popularity.

The hermit caves on Pelion have been described to us as either "founder caves" (see below) or hermit cells for single occupancy. The hermitages that we know of from Pelion are therefore different from, or less organised compared to the traditional *skete*, which in monastic terminology means a branch of a recognised major monastery. A skete consists of calubae (*kalivi* = hut), the latter a collective noun meaning 1) a company of two to three hermits living in the wilderness but under the jurisdiction of a *skete*, and 2) the hut itself in which they lived. Hermits worshipped in small churches called *prosefchadia*. There they read their prayers and held morning services and vespers. Monastic communities sometimes formed around such *skete* churches, as was the case at Meteora and the late Byzantine/post-Byzantine hermitages on the limestone lakeshores of the Great Prespa Lake in northwest Greece.[66] Interestingly, the monasteries themselves attracted settlement and many of the Pelion villages developed around monasteries.[67] This is also reflected in a number of place- and village-names with religious connotations.

A small cave, *Askitario* (KER-5), has a historical connection to the nearby Sourvies monastery, which was founded in 1550. Around 1700, Ger-

65   Psilakis 1994.

66   Skouvaras 1962.
67   Synodiou 1995, 15-8.

asimos the Younger came to the monastery from Leontari (Peloponnesus) and became a leading figure in the area and an important monk at the Sourvies monastery. After working for some decades at the monastery, he began frequenting the Askitario cave and gradually became a recluse. Allegedly, he spent most of his time there from 1727 until he died in 1740, after which the Orthodox Church posthumously recognised him as a saint. No modifications have been made to the cave, but red crosses have been painted on nearby rocks to show the way to the entrance. On the cave floor were scattered candle fragments but no other visible artefacts. A small plaque that hangs inside the cave commemorates Gerasimos the Younger and there is an icon of Agios Georgios and pictures of scenes from Christ's crucifixion. At least 12 dates ranging from 1889 to 2003 are engraved on the cave walls with initials and simple crosses. A group of three visitors wrote a message in Greek on the cave wall that they visited the locality on May 3, 1889. One Dimitrios, who wrote his name and the slogan "Red Revolution" (*Kokini Epanastasi*) on May 1, 1944, was undoubtedly among the group of *andartes* who stayed at the then abandoned Sourvies monastery. One of the monks currently working on restoring the monastery showed us a nearby shallow space in the rock that Gerasimos the Younger allegedly used for sleeping. There is a good view of the area, but the little rockshelter itself would not have been practical for any activity other than resting. No cultural remains or signs of modification were found in or around the shelter.

Further north, towards Veneto, *Osios Symeon* (KER-7) is a so-called "founder cave" where the monk Osios Symeon lived as a hermit in the late sixteenth century before he established the Flamouri monastery. The cave is situated in a remote area facing the Aegean Sea on Northeast Pelion. The entrance is small and difficult to spot – despite several painted red crosses on the rock around the entrance. Inside the cave, ropes (climbing rope and ordinary

rope) have been attached to the upward sloping bedrock to facilitate access. The cave receives little natural light beyond the entrance. Inside a rock niche hangs a *kantili* and on a flat slab of rock nearby is placed a wine bottle, a plastic water bottle and a glass jar. A ruined picture frame and a brass candle holder lie on the floor. A laminated commercial postcard featuring Osios Symeon was found near the entrance alongside a plastic bag, matches and a lighter. Inside the cave, the visitor is met with a message written in dark red capital letters – "*ΙΕΡΟΝ ΣΠΗΛΑΙΟΝ ΟΣΙΟΥ ΣΥΜΕΩΝ 1593*" (The Holy Cave of Osios Symeon 1593) – and several crosses painted in the same colour. It is evident that the cave is visited rarely and possibly only by a representative from the Flamouri Monastery on the founder's nameday.

It is not clear when religious ascetics ceased to use caves on Pelion, but we know that recluses – both religiously inclined individuals and vagrants and mentally ill – continued to frequent caves on West Pelion at least until the 1970s. There are parallels between the two groups, as illustrated by the case of a cave overlooking the village of Drakeia, where the locals claim that a "crazy hermit" lived for a period (AGR-1).

Religious landmarks, such as churches, chapels and monasteries, were significant to the locals for structuring landscape perception and the limits of village life (see Chapter 3). While this "sacred geography" certainly included caves in the form of chapels or historic hermitages, caves were not generally looked upon as spiritual sites. It follows that few or no aspects of economic behaviour in caves on Pelion or the rest of Greece have a ritual dimension. The most obvious element today is the small Christian crosses that shepherds would engrave next to their initials in caves, most likely to highlight their affiliation with the Orthodox Christian faith. Another connection to religion was the coordination of the pastoral calendar with religious festivals. For instance, August 15 is the feast of the Virgin Mary and is celebrated in the Orthodox calendar by the Vlachs of Greece (Aromani).[68] A shepherd from Trikeri and his family who have followed a seasonal migratory pattern for years referred to religious festivals when asked about the timing of his moves. In these cases, seasonal abandonment of caves at high-altitude pastures would be coordinated with such schedules. Whole village communities on West Pelion would fall into a seasonal pattern of movement and migrate to settlements on the coastal lowland during the winter. In spring, the household would be loaded onto pack animals and people would go to spend their summer in the cooler villages on the mountain.

4.7 Non-economic uses of caves

Recreational

Recreational caving has been described as "the activity of entering a void, such as a cave, for the pure joy of the activity".[69] The emotive power of caving is so great that cave tourism may be seen as a twenty-first-century parallel to traditional pilgrimage and is a pattern of cave use in itself. Recreational caving may overlap with more professional caving purposes when recreational cavers add speleological activities to their cave visits or vice versa. These activities may include photography, mapping and various cave-related scientific specialties. Speleologists from Volos and Athens are active on Pelion and periodically visit some of the deeper or vertical caves, particularly Malaki (ART-8), Roumbos (AFE-3), Agios Athanasios (KAR-8), Drakou tou Skamni (MOU-19), Tsouka (MOU-11) and Drakotrypa (KAR-21). Several of these caves have been explored on earlier occasions and the results have been published as brief reports in speleological periodicals (e.g. *Bulletin of the Speleological Society of Greece*).

Less organised visits to caves are common, but there is no way to estimate how frequent they are. The beach caves on East Pelion attract a large number of visitors in the summer, as they are located directly on the shore. Other caves that are signposted and locally promoted (e.g. *Tsouka*, MOU-11) or are highly visible in the landscape also receive their share of visitors, mostly foreigners or Greek urban weekend trippers. Some of the graffiti dates from the 1990s until today were evidently made in connection with such visits and new dates appear in caves where there were no graffiti before.

The low accessibility and visibility of caves on Pelion due to the increasing vegetation cover largely explains why there is relatively little evidence that can be interpreted as a result of recreational visits. Recreational visits to caves are often only accomplished after a long search that includes strenuous navigation through the terrain and vegetation. *Lykotrypa* (AGR-1) is an insignificant but highly visible rockshelter west of Drakeia village. It looks rather alluring from the distant vantage point of the village and most visits to the shelter set off from here. The slope in front of the shelter is heavily vegetated and the shallow rockshelter is difficult to access. Two handwritten paper notes left in a rock niche by recent visitors describe in brief the difficulties of approaching it. In such a case, the process of discovery rather than the locality itself carries the attraction.

Looting

Part of the allure of finding a cave is what one might find inside it. Undocumented and illicit digging at cave sites is a well-known phenomenon on Pelion and is still taking place on the mountain today. Illicit digging is easily accomplished in caves as they are usually hidden from the view of local authorities or other parties.

World War II set the stage for a later wave of illicit digging as caves with a historical connection to

---

68   Nandris 1999, 119; Megas 1958.
69   Wilson 2005, 469.

guerrilla activities were targeted. Treasure hunters looking for wartime valuables were particularly interested in gold sovereigns issued by British agents to the Greek *andartes* or abandoned and hidden military weapons and equipment.[70] The story of the civil war gold left by the British is a well-established myth in almost all of rural Greece. Of course, cases of anyone finding anything of true value are extremely rare. Stories are still circulating about caves high on the mountain that contained weapons and other military equipment, such as swords, helmets and trumpets. A remote cave on North Pelion, *Spilia Me Kranoi* (KER-8), was a documented military storage place used by EAM/ELAS forces in the Ano Kerasia area. Several large disturbances in the cave floor clearly show illicit digging for gold sovereigns or other valuables supposedly hidden by the partisans. Local newspapers reported a state-approved, organised treasure hunt for gold sovereigns in the Ano Kerasia area in 2010.[71]

While the prospect of finding gold sovereigns must have been a strong motivating factor in many cases, it is not always clear what prompted digging in specific caves. Some cave sites were disturbed because looters expected to find ancient movable heritage. We learned from our informants that one group of looters has 'specialised' in the search for antiquities and employs sophisticated techniques to reach their target. The most elaborate evidence of well-planned looting is from *Agios Athanasios* (KAR-8) near Lake Karla. Here, a fuel-driven generator, makeshift electrical lightning and fresh-air pipes showed a considerable investment in time and equipment. Within and around the looting trench in the cave were found Mycenaean and early Historic period ceramics and human bones from a possible, disturbed burial.

Although not a principal aim of the project, such disturbances were recorded when encountered.

While our records do not provide accurate information on the rate or intensity of treasure hunting, it was clear that several caves had been subjected to illicit digging with visible effects on the cave floors. In some cases, the excavation tools were discovered next to the disturbances.

Despite being situated on an almost vertical rock wall 10 m above ground, a small cave 1 km north of Kala Nera (MIL-13) was subjected to illicit digging during August/September 2007. The recent date was indicated by shopping receipts discovered below the cave and information relayed by local informants. Below the cave, the looters had left a plastic bag containing handmade, coarse ware ceramics joined with travertine rock, evidently from the cave deposits (Voskos, this volume, Cat. Nos 134-8). A stick-and-rope ladder had been made to access the cave and near the entrance was found a mattock and evidence of digging.

Illicit excavation and destruction of cave sites bring to the fore not just the motives of those implicated in such things, but also their social, educational and economical position, which generates the motives.[72] However, the looting of cave sites on Pelion is far from what Hollowell has termed "subsistence digging" carried out by impoverished local people who need to gain a livelihood.[73] Most illicit digging on Pelion is treasure hunting by locals for whom the economic temptation of artefact looting is less significant than the thrill of the search and possibility of discovery. It is not always clear why certain caves would be targeted for looting, but the individuals involved would most certainly be familiar with local stories and legends concerning treasures and other relevant aspects of past cave use.

The hunt for buried treasure in caves has obvious downsides. It inevitably involves digging that causes extensive damage to archaeological resources that might be in the affected area; it damages the layers

---

70   Freris 1986, 116.
71   www.taxydromos.gr

72   Durrans 1994, 66.
73   Hollowell 2006.

of sediments that enable geologists to unravel the geological history of the cave; finally, it leaves the interior of the cave unsightly and disfigured. The responsible cultural authorities do not have the resources to control, preserve and protect the countless cave and rockshelter sites within their jurisdiction; in many cases have the scale and nature of the destruction caused by looting activities remained undocumented.

4.8 Caves in a changing economic landscape

PCP rested from the beginning on the premise that caves represent an extended network of cultural sites and landscapes that exist outside the immediate habitation and cultivation zones of villages and fields. In other words, the Pelion caves and the records associated with them provide an alternative view (although fragmented) of the extent and variety of human activity that took place in the countryside away from the main settlements. We know that rural communities including both villagers and semi-nomadic pastoralists have traditionally depended on their non-cultivated landscapes as part of their means of making a living from the countryside,[74] and that this included the use of caves as both multi- and special-purpose sites. Our basic proposition is therefore that caves have not been, and were not perceived as, marginal sites, but were at the core of a comprehensive understanding of local economic history.

A comprehensive study of settlement structure, social organisation and subsistence patterns on the mountain and the adjacent lowland would require a thorough investigation into the social and economic landscape of Pelion in the post-Ottoman era. Unfortunately, our resources did not permit us to treat these aspects with the degree of detail that they deserve, which is why we have chosen to focus instead

on major social and economic transformations. The region's socio-economic history is to a great extent reflected in the census figures for the Pelion villages provided by the National Statistical Service of Greece (to be referred to as NSSG 1961, etc.). These figures allowed us to follow the demographic developments in individual villages as well as in different areas on the mountain from 1881 until today.

While it is apparent that life on Pelion has undergone significant changes since Thessaly's annexation to Greece, it is difficult to record these changes in the landscape outside the villages. A further problem is that censuses are based on interview and questionnaire data obtained from permanent households, and, therefore, vagrants, migrant workers and transhumant shepherds (often the primary cave users) may be omitted from census returns. We wanted to show how our fieldwork may supplement the historical data and be used to explore effects of social and economic change in the non-cultivated landscape. For instance, the pastoral community had its own complexities and was involved in husbandry activities that were largely independent of life in the villages. Pastoral artefacts and structures at the caves were shaped by the necessity of this cultural and economic system to exploit its changing circumstances in the most optimal way. Knowledge about the response of cave users to economic and social change is, therefore, an important supplement to village population censuses when drawing a picture of economic and social transformation on Pelion.

The following is an effort to summarise the economic and social developments on Pelion during the last 130 years and to place caves within the narrative as part of these changing conditions.

From 1881 until the end of the 1940s

It would not be an overstatement to say that Pelion's economic profile had gone through remarkable transformations throughout the Ottoman period,

---

74   Forbes 1996, 191.

before 1881.[75] In contrast to the lowland, the economy on the mountain was characterised by minimal agricultural activity contrasted with specialised handicrafts, industrialised manufacturing and a range of maritime ventures. Piracy, predatory raids and the occasional conflicts that occurred during the period had not prevented the Pelion villages from becoming vital commercial centres with strong ties to the outside world.

There was no exodus of any Muslim population from the mountain after the annexation of Thessaly to the Greek state, as only Christians populated the Pelion villages. Lowland Lechonia did have a Turkish minority and the settlements on the Thessalian Plain lost a significant number of Muslim inhabitants, but the latter had already begun their departure during the conflict in 1878. Little has remained to attest to Pelion's Ottoman heritage, mainly echoes of Turkish placenames (such as Agios Vlassios, former name Karambas) and the remains of a Turkish tower in Palaikastro. In the Greek census of 1907, 3,519 people stated Islam as their religion and they were all residents of Thessaly.[76] In comparison, the total population of the region was 381,279.[77]

In 1881, the Greek Financial Diaspora, with the backing of the Greek government, had acquired large areas of Thessalian land from the departing Ottoman landholders. It was hoped that the new owners would invest in their large estates (*tsiflikia*) and introduce a series of initiatives that would modernise agricultural production. Little happened, however, in terms of making cereal production more effective; landholders preferred and continued the long-held tradition of renting out their land for winter grazing to semi-nomadic pastoralists.[78] In the summer, many of the latter would return with their flocks to permanent settlements in the mountains in Epirus and Macedonia. The pastoralists would make large profits by selling wool to Central European buyers, with the support of the Ottoman authorities, who taxed this trade. Such seasonal movements were stifled with the annexation of Thessaly to the Greek state, as brigands harassed flocks taken across the border.

In 1896, the Greek government started the expropriation of land, which was completed in 1917 with the "Agricultural Reformation". In 1917, 11,900 square kilometres were expropriated, 43.7% being pasturelands. At the beginning of the twentieth century, only 22.4% of the total area of Thessaly was cultivated.[79] At the same time of land expropriation, scrublands were given to shepherds and nomads started to settle permanently in certain areas.[80]

After 1881, the population inflow to the lowland settlements along the Pagasetic Gulf found work in the labour-intensive cash crop industry and in other emerging economic areas, e.g. industry, iron forging, tobacco factories and ceramic workshops. Sherds of ceramic vessels, presumably from workshops in Volos, show that local production included inexpensive water vessels for local consumption (see Vroom, this volume). The new population consisted of both local landless Greeks and seasonal or permanent migrants from other parts of Greece.[81] Graffiti in the Pelion caves suggests that the practice of writing dates or initials emerged in the last decades of the nineteenth century. While this could be the product of leisure walkers and explorers, it may also relate to the influx of newcomers, some undoubtedly literate,

---

75 Skouvaras 1959; Makris 1982.
76 This seems to be a small number for the whole of Greece, but perhaps can be explained by the fact that Macedonia is not yet included.
77 NSSG 1909.

78 In Thessaly during Ottoman times, the development of mountain animal husbandry was closely linked with cereal cultivation on the plains, where herds grazed fallow lands of the large *tsiflik* estates. See also Islamoglou-Inan 1994; Vergopoulos 1975.
79 Vergopoulos 1975.
80 Sivignon 1975; Vergopoulos 1975.
81 Bennett 1988, 220.

who found employment in the pastoral sector. Like the Albanian shepherds many years later, they were perhaps keen to set their mark in the new cultural landscape, even if it was just initials, a date or a figure carved in a cave wall. Two of the earliest dates that can be attributed to possible pastoral caves are from the edge of Lake Karla, where shepherds grazed large flocks of animals.[82]

The most comprehensive large-scale construction project in Pelion's history was the construction of the Volos–Milies train line only a decade after the annexation of Thessaly to the Greek state. The project was carried out in several stages from 1892 to 1903 and required a large number of building specialists and workers. Some of the earliest post-Ottoman evidence of cave use on Pelion could be related to these construction works. Locals from Milies told us that 'Ghegs' inhabiting the Argyraki cave (MIL-2) worked on the train line but that they also – and perhaps to a larger extent – became involved in cultivation and olive gathering.[83] Finds of remains of cooking pots, dishes, jugs, iron nails and animal bones are probably the remains of these migrant workers and they underline that the cave was fully integrated into the production zone of Milies and was part of the daily economic activities there (see Pantzou & Papadopoulos, this volume).

Agri- and arboriculture remained an important source of income and prosperity for the Pelion villages as they continued to supply the growing urban market in Volos. Olives can grow up to a height of c. 750 m.a.s. and vines up to a height of c. 1200 m.a.s., which meant that a broad belt of lowland and slopes on both sides of the mountain were highly suitable for growing grapes and olives. However, as travellers in the mid-nineteenth century observed, the villages

of Pelion had different levels of local self-sufficiency in terms of agricultural products. There was little land under cereal cultivation on Pelion and only the former municipalities of Makrinitsa and Viviis (Karla's area) grew some wheat.[84] Production was insufficient to meet local demands in the large mountain villages and it was necessary to obtain wheat from the lowland. The Pelion economy must, therefore, be seen in connection with developments on the agrarian markets in Volos and elsewhere in Thessaly. Pastoral production and the small-scale, stationary pastoralism that characterised most villages on the central part of the mountain would probably have been geared towards direct local consumption to a higher degree than cereal production.

The trading network between Greece and the old Ottoman world underwent serious transformations in the nineteenth century, as former commercial centres that had relied on caravans and sailing ships declined and suffered population losses. Volos, in contrast, was located in a rich agricultural region and had been able to provide commercial and financial services to Pelion and other parts of the surrounding countryside. It experienced rapid growth that was supplied mainly by the declining centres and less by a rural demographic surplus. The smallest Pelion villages (e.g. Pouri, Kerasia, Veneto, Anilio, Lambinou, Xourichti and Makrirachi) seem to have been the least affected by de-population, while larger villages with semi-industrial production lost more inhabitants.

In the four decades after the annexation of Thessaly to Greece, different regions on Pelion experienced shifts in population numbers in diverse ways. A distinction can be made between general developments on the mountain and developments that are specific to some villages. There is a tendency for villages with-

---

82   Such dates may also simply have been left by Pelioretes visiting the caves.

83   Ghegs were members of a north Albanian ethnic community that often worked on Pelion as immigrant seasonal workers. They probably abandoned the area on the annexation of Thessaly to Greece.

84   Sivignon 2009, 463. Today, the few preserved threshing floors on Pelion are also found mainly on the northern part of the mountain, as indicated by local hiking maps (see Haratsis 2003).

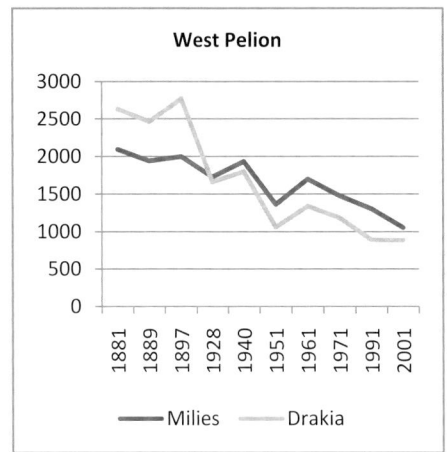

West Pelion

— Milies   — Drakia

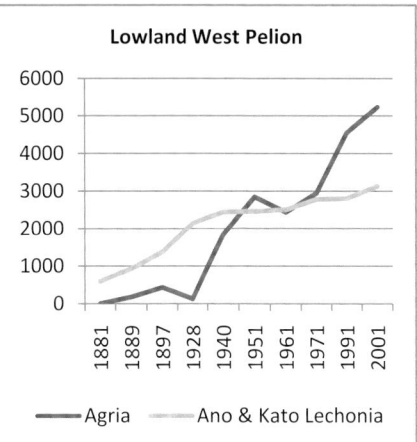

Lowland West Pelion

— Agria   — Ano & Kato Lechonia

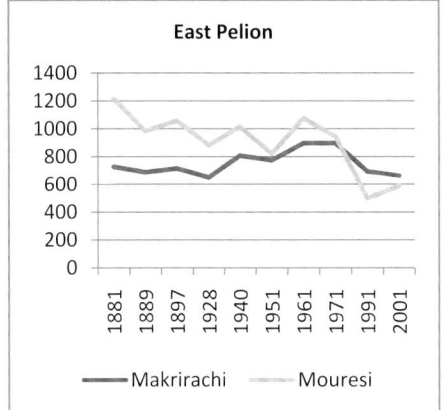

East Pelion

— Makrirachi   — Mouresi

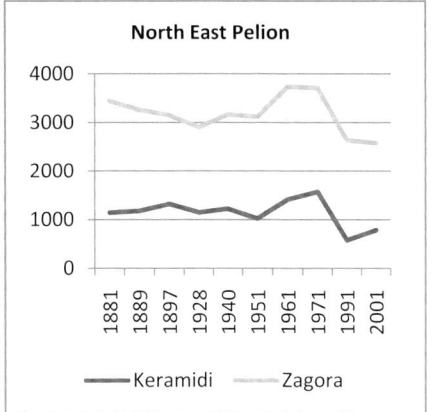

North East Pelion

— Keramidi   — Zagora

Fig. 4.9. Population trends on Pelion from 1881-2001, exemplified by eight villages in various regions on the mountain.[87]

in the same area to display similar population trends. This is hardly surprising since there would have been close economic ties and kinship inter-dependencies between neighbouring settlements.

On East Pelion, Zagora had been a prominent hub in the old commercial sailing network and its downturn had begun many years earlier. Its 15.6% population decline from 1881 (3,442 inhabitants) to 1928 (2,902 inhabitants) is merely the last development in a long-term trend. Despite an overall decline, most villages on the Aegean side never saw large shifts in the number of inhabitants during this interval (Fig. 4.9). Kissos experienced a more significant loss of population (39.8%) between 1881 and 1928 than any other East Pelion village.[85] Although

unsupported, this was perhaps related to a gradual decline in mining activity in the central part of East Pelion, where mines were actively exploited from the early twentieth century until the early 1940s (MOU-3, MOU-18, MOU-26, MOU-27, MOU-28). Even in the early nineteenth century, Pelioretes had begun to migrate and excelled as merchants and bankers. These expats came from both sides of the mountain. Evidence of their wealth and contributions are the mansion-like buildings that can still be seen on Pelion today. Several Pelioretes chose to travel to Egypt, mostly from the second half of the nineteenth century.[86]

---

85   The negative population trend at Kissos continued into the 1950s.

86   Konstantinidi 1936; Vlachakis 2003-05.

87   Data from NSSG 1881-2001.

On West Pelion, some villages climb to historic highs in the census of 1897, which they never reached again. At the end of the nineteenth century, there was a general downward trend in population in most of the large villages on the western side of the mountain. Makrinitsa (inclusive of Koukourava and Staghiates) saw a particularly steep drop in population after the end of the nineteenth century. The population almost halved from 1897 (5,083 inhabitants) to 1928 (2,608 inhabitants). Population decline was also noticeable for Portaria, Katochori, Drakia, Milies, Agios Georgios and Agios Lavrentios. Kanalia is an interesting exception. In this lowland village, there was a gradual increase in population from the time of the annexation of Thessaly to the Greek state until the 1970s, when the number started to drop from 2,035 in 1971 to 1,213 in 2001. This may be related to the draining of the lake. An informant told us that after the lake was drained, some people moved to nearby Kato Kerasia and took up agriculture and pastoralism.

In West Pelion, population numbers climbed significantly in some settlements along the Pagasetic Gulf, such as Lechonia. The village had been growing steadily since 1881, but picked up the pace in the beginning of the twentieth century when many of the other villages started slumping. Even Agria followed this slump but saw major growth after 1928. Based on interviews, particularly in Lechonia, this growth was attributed to the fertile land and cultivation of vegetables and fruit. The rapid population growth in the lowland settlements was analogous with similar developments in Volos and reflected the success of an emerging urban economy.

The influx of new inhabitants after the massive exchange of population following the long war period (1912-22) was little felt on Orthodox Christian Pelion, but in Volos the sudden labour surplus gave impetus to the emerging industry such as the Aget Iraklis cement plant between Volos and Agria, which had operated since 1924, and the EPSA soft drinks factory, which opened at Agria the same year. The 14,000 or so refugees from Asia Minor mainly found refuge in the area that today is known as Nea Ionia.[88] The slow fall in rural population numbers merely continued throughout and after this decade. Then things started to change. Extensive land reforms (laws of 1917-26) and large investments in land amendment and drainage ended up transforming both the landscape and the land tenure system of Northern Greece. Many villages experienced a rise in population numbers between the 1928 and 1940 censuses. The developments seemingly included the uncultivated landscape: an increase in cave use at the advent of the 1930s can perhaps be substantiated by the fact that a growing number of graffiti dates are carved in an increasing number of caves (Fig. 2.8.20). The dates left in caves during the 1930s more than triple in comparison to previous decades. Moreover, a new pattern not previously seen on Pelion is that closely spaced dates (from the same or succeeding years) are left in the same pastoral caves. In some cases, this could point to continuous use by the same or perhaps cooperating shepherds.[89]

The post-war period to the 1980s

After World War II and the Civil War, international emigration and migration to urban and lowland areas caused a population decline on Pelion that ranged at individual villages from slight to considerable. Probably as an effect of the displacements caused by the

88   Konstantaras-Statharas 1994.
89   Shepherds were generally illiterate and there is a possibility that the dates and initials were left by hikers and independent visitors. However, the engravings are likely from shepherds as most of these caves are not obvious targets for visitors, such as KAR 10, KAR-20 and MAK-23. MIL-1 is perhaps an exception as it may have had some influx of visitors, but it is most certainly also a pastoral cave. Moreover, many dates are made with a knife-carving technique that requires good practice and available time, both indicative that the carvers were shepherds.

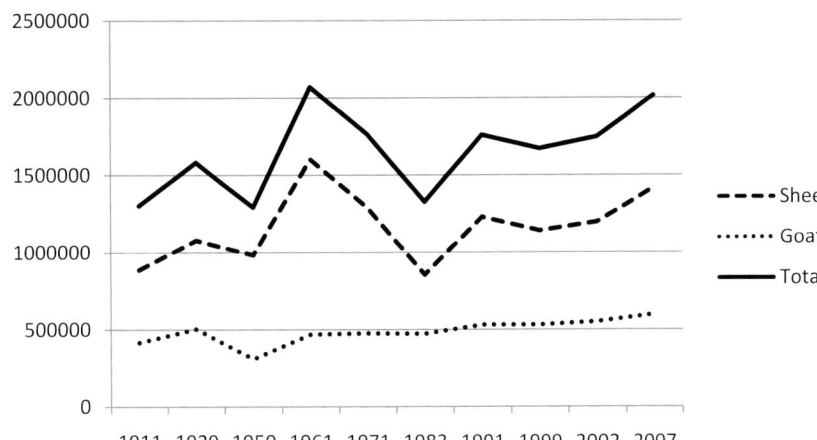

Fig. 4.10. Sheep and goat populations in Thessaly from 1911-2007.[93] The number of Thessalian goats has been on a modest but constant rise since 1950. Sheep husbandry topped during the 1960s and after a decline, the number of Thessalian sheep is once more on the rise, due in part to EU subsidies.

long war period, Pelion villages lost more than 14% of their population during the 1940s, but in 1961, the census numbers had returned to their pre-war level.

The post-war period saw an expansion and modernisation of the pastoral economy caused by increased demand on Greek and foreign markets for pastoral products, primarily cheeses and "Volos olives". The region became more closely integrated into national and global commercial markets. This resulted in a stronger influence on fluctuating prices for agro-pastoral products (especially sheep cheeses), which significantly altered the social organisation of village production.

These transformations went hand in hand with technical improvements and industrialisation of production that included the introduction of cars, tractors, mechanical milking machines and locally cultivated or industrially produced animal fodder.[90] Also, new construction materials were introduced, such as concrete building blocks, cement render, corrugated iron and elenit.[91] These materials were gradually introduced in the countryside and were found in connection with pastoral shelters in and

around caves. These advances had their effects on the organisation of agro-pastoral production.

The nomadic way of life largely disappeared in Greece after the war. Migratory sheep flocks were reduced by 27.5% from 1965 to 1984, while sedentary flocks in the same period increased by 15.6%.[92] Although the total number of sheep (not goats) also declined in Thessaly in this period (Fig. 4.10), sheep and goat farming remained an important rural activity. The animals were kept on tracts of land that were not attractive for agriculture, such as mountainous areas on North Pelion or around Lake Karla.

The loss of population in the rural communities on Pelion as a whole accelerated after 1970 and reached an all-time low in the 1991 census. For instance, in 1991, the steady population of Mouresi (including Agios Ioannis and Damouchari) was 499. This is down from the post-war period high in 1961 when the community had 1,076 inhabitants.[94] Other villages that experienced a steep decline in population numbers included Lambinou, Glafyra and Kanalia. The latter village on Lake Karla had been subject to major social and economic changes when the draining in 1962 caused the disruption of traditional

90 Informants from West Pelion vividly told us about the appearance of the first taxi in Milies.

91 For instance, elenit were extensively used as roofing materials in the 1960s and 1970s. It was abandoned in the 1980s when the presence of asbestos minerals was confirmed in elenit sheet roofs.

92 Gidarakou & Apostolopoulos 1995, 58.

93 Source: National Statistical Service of Greece (NSSG).

94 NSSG 1962, 45; NSSG 1994, 175.

lifestyles connected to activities on the lake. The disappearance of Lake Karla as an important watering place and the emergence of large tracts of land for cash crops would have changed local pasturage patterns and the scheduling of stays in the rockshelters around the lake.

Other settlements on the coastal lowland of West Pelion, such as Aghria, Kato Lechonia and Kala Nera, were expanding and their populations continued to grow during the 1970s and 1980s.

In these conditions of socio-economic transformation and declining populations in the mountain villages, some areas that were used for agro-pastoral production in earlier times were abandoned; the caves around them fell out of use and their associated structures allowed to decay. Caves located further away were forgotten. Fig. 2.8.20 shows a significant fall in the number of dates made in caves – and the number of caves with engraved dates – during the 1970s and 1980s. If a lower number of dates can be taken as evidence of less activity in and around the caves, it is hardly surprising that partial or complete abandonment of these marginal production sites followed in the wake of economic and social changes. Departing shepherds meant that structures were poorly maintained or abandoned altogether and that fewer artefacts would be deposited in the caves during these decades. With cheap construction materials newly available, shepherds increasingly chose to construct new shelters and abandon their long-term use of caves and rockshelters.

The contemporary period (c. 1990-the present)

Pelion has not been isolated from or unaffected by globalised economic, political and social actualities. Although the mountain was always affected by influences of the outside world, it is clear that the changes of recent years are of a different sort and on a larger scale. The process of globalisation is ongoing and affects the communities on the mountain in different ways. In a real way, it marks the end of certain traditional ways of life and the inclusion of the region's rural inhabitants in a larger global system. Herders have long ceased to be isolated while on the mountain; they coordinate their activities with their family and friends via mobile telephone and other electronic devices. When we asked a shepherd how we could find him on a specific part of the mountain the following week, he replied as a matter of course that we could arrange it over Skype as he was usually online in the evening.

While this little incident illustrates an important observation, namely that the local pastoral economy are ready to embrace changes,[95] some of the recent challenges faced by traditional animal husbandry have fundamentally transformed a way of life. Herders and villagers on Pelion continue to change many of their routines in an attempt to adjust to rapidly shifting social and economic conditions during the last three decades. This transpires both from the informant interviews and from archaeological observations around the mountain. The entry of Greece in 1981 into what is now the European Union meant that subsidies and the opportunity to participate in regional development programmes became available, and Thessaly quickly became one of the top recipient regions in Greece of farm subsidies through the EU's Common Agricultural Policy (CAP).

Opening up the Greek agricultural markets also had negative effects. Agropastoral production experienced an increased dependency on subsidies that went hand in hand with a reduction in traditional, family-based labour. The pastoralist system was subjected to increased competition and demands for modernisation and intensification. The continued use of export subsidies and the promotion of high production levels would depress prices on the international market and sometimes lead to the dumping of agricultural produce (Fig. 4.11). Local herders

---

95   Brochier *et al.* 1992, 54.

Fig. 4.11. Victims of subsidies: apples dumped next to an orchard near Agios Lavrentios, West Pelion (2006).

also complained to us that the profit margin for dairy products is low.

The decline in the rural population around Pelion, as well as a change in agricultural methods favouring more intensive but less widespread land use, lowered the pastoral pressure on the landscape.[96] Over the last few decades, this has led to changes in vegetation dynamics and to an increase in forest cover.[97] Local perception today views the mountain as "closed", referring to the many overgrown fields and stone-paved tracks.[98] Similar developments have been documented throughout Europe.[99]

While the rich vegetation has worked as an asset for the growing tourist industry on an increasingly greener Pelion, the shrubbery has also added to the changes in the economic landscape outside the villages. Abandoned fields and agricultural terraces around the villages, which were until recently cleared

for upland fruit tree cultivation, are now less and less likely to be used as they are covered by thick blankets of bracken fern and other types of vegetation. Informants told us that areas around the villages used to be much less vegetated because people had many animals that were grazing the wild vegetation. These elements of the landscape fell out of the economic system once they grew over and can only play a role again once the vegetation has been cleared. The increase in grasses and foliage lead to a larger risk of uncontrolled forest fires, such as those in 2007 on South Pelion, which burned out a band of agricultural land and forest from Lambinou/Syki across the middle of the peninsula to near Afissos.

This wild vegetation naturally had an effect on the accessibility of the caves. Dense foliage covering most of the mountain (except the southern part and the foothills around Lake Karla) means that many localities – and the tracks that lead to them – are today unknown, except to some locals. For instance, the entrance of ZAG-5 below Makrirachi village was visible in the past, but thick blankets of thorny climbers today cover the entrance. Several caves around Agios Lavrentios have disappeared under vegetation to the degree that they can no longer be spotted even when

---

96 Bradshaw & Hannon 2004; Zervas 1998; de Rancourt *et al.* 2006; Hadjigeorgiou & Zervas 2009.
97 Ispikoudis & Chouvardas 2005, 155.
98 See Pantzou & Papadopoulos, this volume.
99 Williams 2000, 29-30; Bradshaw 2004.

Figs 4.12. & 4.13. Two generations of shepherds in front of their shepherd's huts (*mitato*), outside Kanalia (left) and above Makrinitsa (right). There is more than half a century between their dates of construction.

one stands in front of them. The coastal stretch east of Tsagarada between Makrilitharo and Kryfo Scholeio has several caves that "were used", according to a local informant. Today, the area has grown largely inaccessible apart from a narrow footpath.

Rapid contraction of the pastoral system, less reliance on transhumance and the centralisation of cheese production had consequences for the organisation of pastoral activities and through that for the use of caves. The general picture from the mountain is that most animal caves have fallen out of regular use. This can be traced archaeologically through recent portable artefacts at the caves, which imply general abandonment during the 1980s and the beginning of the 1990s. Many others did not contain any material items indicating recent (economic) use, and it is rare to find pastoral structures around caves that have been constructed or significantly modified

more recently than the 1980s. Only occasional and spontaneous occupation by free-ranging animals can now be observed at these sites.

Production is now done in a central location, often open-air pastoral stations, and less often "at home", whether in a house or a cave. Large goat/sheep caves with easy access, like MAK-3 (Alafoklisi) and MOU-1 (Tsounaga), continue to be used occasionally, but cheese-making operations have been moved to locations closer to the village.

Rather than utilising caves and rockshelters for overnight stays, most shepherds now prefer open-air stations, which are more spacious, have more facilities, and can be constructed at locations that are more convenient (Table 4.8). It is our impression that especially smaller caves and rockshelters have been abandoned, perhaps because these were

Cave shelters	Open-air *mandri*
Low-investment in terms of structures	Locationally flexible
Cool in summer and warm in winter	Functionally flexible (easy to modify)
Convenient for mobile herders and vagrants	More comfortable

Table 4.8. Advantages related to the economic uses of caves vs. open-air sites. The switch to pastoral open-air stations is not necessarily a difficult one and perhaps more than anything else depends on the scheduling of activities and on available building materials in any one area.

often associated with rapidly declining small-scale household pastoralism. A shepherdess told us that whereas in the past every family had their flock of sheep in Vyzitsa, today only three are actively engaged in pastoralism. This was quite a common pattern occurring in other villages as well.

Artefacts from the caves that can be dated to the last two decades and tied directly to pastoralism were few, with new requirements and standards occasionally reflected in scattered finds of vaccination syringes, medicine vials and bottles and bones of domestic animals (one a goat carcass with a plastic ear

	Socio-economic development		
	Traditional pastoralist/ farming economy (until the end of the 1940s)	Post-war period, increasing industrialisation, expanding market for pastoral products, tourism, abandonment of countryside (1950s–1980s)	Contemporary (c. 1990-the present)
Characteristics of local cave use	Caves are used extensively for pastoral purposes, as a shelter for sheep and goats. Shepherds sometimes stay overnight. Caves near villages are used for cool food storage. Alternative use as refuge during times of unrest.	General decline in the regular use of caves and rockshelters from the 1970s onwards (increased use of open-air sites). Use of caves by nomadic shepherds in steep decline. Electrical refrigeration means that caves are no longer used for cold storage.	New uses of caves, such as treasure hunting, exploration, excavation, heritage interest. Pastoralists have mostly turned to the use of huts or open-air pens supported by constructed shelters. Immigrant shepherds use caves.
Archaeological effects	Continuous construction, use and maintenance of cave structures. Domestic debris is deposited in and around caves (e.g. broken water jars) Refuse deposition, primarily ceramics and organic materials, happens mainly in connection with economic activities.	The introduction of corrugated iron, cement and plastic. Few graffiti dates. Useable building materials around abandoned caves are removed or allowed to decay. Structures around caves fall into disrepair or are plundered for potential building material.	Continued collapse and decay of structures around caves. Vegetation growth around caves. Discard around caves from occasional pastoral use, but otherwise increasingly unrelated to economic activities. Due to a sharp decline in predator population, modern folds are delimited by lightly built wood and wire fences to keep the animals in.
Caves as culturally perceived sites	Caves are used as landmarks and are captured in social discourses.	Caves may be remembered in terms of events in the recent past that have meaning to the speakers. Many caves acquired names that are handed down but whose associated stories are not. Few people know the literal meaning of cave names.	Locational and historical information about caves falls out of individual and collective memory as primary users have moved away or died. "Rediscovery" of some caves and their associated stories by cave tourists.

Table 4.9. Three-stage model for the evolution of cave use since the end of the eighteenth century, and its associated behavioural and archaeological transformations.

tag). Side by side with these objects are artefacts of a more traditional character, such as a shepherds' staff or a hand-carved goat collar of wood. Clothes, shoes and low-level deposits of scattered rubbish occurred regularly, but it is not possible to say with certainty whether pastoral users left such remains in the caves.

However, a decline in cave use and household pastoralism does not necessarily indicate a parallel decline in overall pastoral capacity. This can perhaps most clearly be seen at Lake Karla, where, despite significant changes in the physical, social and economic landscape around the lake, the pastoral capacity has remained almost unchanged. An informant from Kanalia told us that the village in the "old days" had roughly 15,000 goats and sheep. Today, according to the informant, the figure is 14,000 goats and sheep, plus 1,500 cows.

Today, the traditional pastoral collaborative networks from the last few centuries have ceased to exist and pastoral enterprises are organised and run by individuals or by families who negotiate and maintain relationships with dairy companies, local municipalities and governmental organisations. Similar developments have been recognised elsewhere in the Mediterranean.[100] In the last few decades, some shepherds have preferred to hire Albanian workers during much of the year to free up time to pursue other economic activities. While it has not been apparent in the material culture deposited in the caves, the influx of foreign immigrant workers is visible in the non-Greek names engraved on the bedrock in and around the caves. An increase in engraved dates in caves especially on the northern part of the mountain (*Gidospilia*, *Spilia*, *Bourdovanou*, Apostolis' Shelter and *Alafoklisi*) can be connected to developments during the 1990s when Albanian immigrants again found employment as field hands and shepherds in the Greek countryside. Many portable artefacts of recent date, such as iron- and plastic

containers and packaging materials from food and drink products, can perhaps be associated with this 'renaissance' in the pastoral function of caves (Table 4.9).

It is not only practices in the pastoral sector that are changing. The most recently available census from 2001 has showed, with a few exceptions, a slight upturn in population numbers from 1991. There has been a general recovery all over the mountain, not least due to the increasing role of Pelion as a tourist destination. This has to some degree led to a decline in employment within the agricultural sector – or a shift to part-time employment during the tourist off-season.[101] Another phenomenon that merits attention is the gradual repopulation and regeneration of Pelian villages by young people leaving urban centres.

Tourism meant more building activity. Generous definitions of rural settlement limits during a public assessment in 1985 led to a horizontal expansion of the settlements as people started building on land outside the traditional village limits. A building boom started on Pelion in the late 1990s and contractors started buying and dividing larger plots of land in order to increase the allowable area for building.[102] Villages that were once tightly nucleated now tend to expand laterally. New roads are built and access to the mountain from Volos is made easier. Some fear that these developments will lead to an "urbanised" Pelion that could potentially endanger the ecosystems on the mountain. Such apprehension is understandable as part of the image that is sustained in the tourist business is Pelion as a green and nature-dominated mountain. Its mythical past involves healing plants and herbs; agribusinesses today produce an abundance of fruit, and herb mar-

---

100   Mientjes 2004, 184-5.

101   Labour demands remain extremely high at the time of the apple harvest in the autumn. During this time, villages such as Zagora now have significant dormitory functions and provide housing for many mainly Albanian immigrant workers.

102   Tratsa 2010.

kets and roadside stands sell herbs to passers-by. A mental connection is made to myths involving herb-collecting centaurs with healing powers.

One of the first to foresee the tourist potential of caves on Pelion was the Austrian "father of Greek ecotourism" Alfons Hochhauser (1906-81). Hochhauser foresaw the prospects of caves in tourism and was the first to develop ecotourism on Pelion. He was fascinated by the myth of Thetis and Pileas' wedding taking place in a Pelian cave.[103] However, while there are plenty of mythological references that involve caves, caves on Pelion have not yet been commercially developed and promoted as tourist attractions, the deep and "mysterious" Tsouka Cave at Mouresi being the only exception.[104] Neither do we know of any structured outdoor educational and recreational adventure programmes that involve exploring caves. A specialised segment of the tourist masses goes for caving, which is becoming increasingly popular. These developments are part of a relatively new trend on Pelion during the last decade, where efforts are made to establish mythical links, identify tourist/heritage attractions, preserve hiking tracks and so on.

---

103   Akrivos 2010, 141-91.
104   A commercial show cave would be any cave that is developed and operated for profit in order to exhibit its natural features to the public.

Chapter 5

Synthesis and perspectives

N. Andreasen, N. Pantzou & D. Papadopoulos

Greek summary

Αυτό το κεφάλαιο επιχειρεί να χτίσει τις κατά βάση σησύγχρονες 'βιογραφίες' τεσσάρων σπηλαίων. Ο σκοπός των 'βιογραφιών' είναι να δώσει παραδείγματα του βαθμού της λεπτομέρειας με την οποία η ερευνητική μας μέθοδος θα μπορούσε να προσφέρει πληροφορίες για τις μεμονωμένες θέσεις. Παρόλη την ανεπαρκή χρονολόγηση, είναι δυνατόν να διακριθούν ορισμένα κοινά χαρακτηριστικά αναφορικά με την πυκνότητα, την ποικιλία και τη σχετική ηλικία των αντικειμένων που βρέθηκαν. Αυτό μπορεί να βοηθήσει στην αναγνώριση μεταβλητών που ευθύνονται για την επιλογή συγκεκριμένων σπηλαίων.

Σε αυτό το κεφάλαιο καθορίζονται επίσης τα αποτελέσματα των ποσοτικών αναλύσεων των δεδομένων που συνελέγησαν κατά την διάρκεια της έρευνας πεδίου. Διερευνώνται διάφορες περιβαλλοντικές μεταβλητές που θα μπορούσαν να είναι σχετικές με την επιλογή των σπηλαίων, συμπεριλαμβανομένου του υψομέτρου, της τοποθεσίας, της όψης της εισόδου του σπηλαίου, της απόστασης από τον οικισμό και του τύπου σπηλιάς.

Τα υλικά κατάλοιπα από τα σπήλαια αποτελούν ένα δυσανάλογα μεγάλο τμήμα των σωζόμενων αρχαιολογικών δεδομένων για πολλές προϊστορικές περιόδους. Ως εκ τούτου, η άποψη για τη σύγχρονη χρήση των σπηλαίων βασίζεται, σε ένα αρχαιολογικό αρχείο που προέρχεται από στρωματογραφίες σπηλαίων. Οι σύγχρονες πρακτικές θεωρούνται κατάλοιπα παλαιότερων πρακτικών, αλλά τα αποτελέσματα του ερευνητικού προγράμματος μας έδειξαν

ότι αυτή η άποψη είναι μη αντιπροσωπευτική όσον αφορά το εύρος και τον χαρακτήρα των δραστηριοτήτων που έλαβαν χώρα σε σπήλαια στην εποχή μας.

Οι επιπτώσεις στην ερμηνεία των αρχαιολογικών δεδομένων που έχουν οι παρατηρήσεις μας σχετικά με τη χρήση των σπηλαίων στην εποχή μας, μπορούν να συνοψιστούν ως εξής:

- Δεν βρήκαμε επαρκή στοιχεία για να υποστηριχθεί η χρήση των σπηλαίων ως χώρων μόνιμης κατοίκησης. Μόνο σε εξαιρετικές περιπτώσεις ταυτοποιήσαμε σπήλαια που χρησιμοποιήθηκαν ως τόποι κατοίκησης για μεγάλο χρονικό διάστημα.
- Λίγα αρχαιολογικά στοιχεία σχετίζονται με κάποια αλλαγή στην χρήση των σπηλαίων, διότι οι οποιεσδήποτε αλλαγές που συνέβησαν ήταν συχνά εφήμερες και δεν άφησαν πολλά ίχνη.
- Μικτές πρακτικές θα συνεπάγονται ενίοτε διαφορετικές παράλληλες χρήσεις ενός σπηλαίου (π.χ. κατοικία/αποθήκευση) ή αντικατάσταση μίας χρήσης από κάποια άλλη μέσα σε σύντομο χρονικό διάστημα. Το γεγονός αυτό μπορεί να δημιουργήσει ένα είδος οριζόντιας στρωματογραφίας, στην οποία αρκετές διαδικασίες της σύγχρονης εποχής έχουν παραγάγει συγκεκριμένα μοτίβα υλικών καταλοίπων.

Οι συμβατικές διαιρέσεις ανάμεσα σε αγροτικό και αστικό, παραδοσιακό και μοντέρνο δεν ισχύουν για την σύγχρονη Ελλάδα και πιθανώς δεν ίσχυσαν ποτέ. Το Πήλιο εμπλεκόταν ανέκαθεν, σε μικρότερο ή μεγαλύτερο βαθμό, σε κοινωνικά, οικονομικά και πολιτικά συστήματα πέρα (συχνά αρκετά παραπέρα) από τα όρια των μεμονωμένων χωριών και της ίδιας της περιοχής.

Η αξιολόγηση σπηλαίων και βραχοσκεπών απλώς και μόνο από πλευράς των πρακτικών χρήσεών τους ή της οικονομικής τους σημασίας, μειώνει την αξία που έχουν αυτές οι θέσεις μέσα στο τοπίο και το ιδεολογικό σύστημα, μέσα στο οποίο λειτουργούν. Το ερευνητικό πρόγραμμα υποδεικνύει ότι οι σπηλιές συνδέουν τους ανθρώπους με το ευρύτερο τοπίο. Αρκετές από τις σπηλιές που ερευνήσαμε μπορούν να κατηγοριοποιηθούν σαν ιδεο-τεχνικά τέχνεργα, επειδή περιέχουν κατασκευές σαν ερημητήρια και γκραφίτι που συνδέουν αυτές και τους χρήστες τους με εθνικές και κοινωνικές ομάδες. Τα σπήλαια συμμετέχουν επίσης στο σύγχρονο/πρόσφατο ιδεολογικό τοπίο. Ως σύμβολα της Ελληνικής αντίστασης κατά την διάρκεια του 2ου Παγκοσμίου Πολέμου και του Ελληνικού Εμφυλίου, εκπροσωπούν μνημειώδη τοπόσημα της ιστορίας του έθνους.

5.1 Cave biographies: Case studies

Throughout the project, we have approached the topic of cave use by using both archaeological and ethnographic methods, and it is through this mixture of approaches and data that we have attempted to disentangle the connections between cave and land use, animals, agriculture and people. While the retrieval of multiple classes of evidence from any one cave may be somewhat optimistic, the exercise strongly underlines the importance of integrating diverse but complementary lines of investigation. Pre-excavation survey of cave sites ought to routinely examine evidence of the most recent phases of use and consider them part of the locality's history. Excavations significantly disturb the microenvironment of caves and may lead to the loss of surface deposits that can be informative of later periods. Once excavated, the status of a cave site changes permanently and it rarely reverts to earlier uses.

The following represent our attempts at constructing the primarily Modern "life histories" of four caves. In contrast to the regional outlook of the preceding chapters, the purpose of the "biographies" is to give examples of the degree of detail with which our research method could provide information about individual sites.

One may argue that because we cannot be sure about the precise date of the surface assemblages or about the chronological contemporaneity of their various components, then the findings of the material studies presented in Chapter 2 have only limited relevance. We admit the scarcity of datable material and diagnostic elements limits the utility of these analyses, but our intention was never to perform a detailed and exhaustive analysis of individual caves and rockshelters. This would have required a project on an entirely different scale. While a stronger focus on individual sites likely would have brought to light more facets of their specific uses and their users, our goal has been to provide a conceptual background for attempting to think through neglected categories of archaeological evidence in caves. Despite poor chronological control, we can discern certain patterns in the density, variety and relative age of the cultural material. This may help identify variables that account for the selection of particular caves. Some were obviously selected because the topogra-

Fig. 5.1. Horseshoe-shaped impression.

Fig. 5.2. Artificial hole and cross (visible 10 cm to the right of the hole) in the bedrock.

phy offered expansive views of and easy access to a wide variety of resources within the immediate area.

Case 1 – Linking the cave to the myth: Chiron's Cave (MIL-1)

Kentavron or "Chiron's Cave" is found west of Milies by following a small uphill path, which starts from the rail tracks close to the iron bridge. The cave lies on the south side of the Kakoskali Gorge, which drains a large area on the southwest slope of the central Pelion massif and discharges at the coast south of Kala Nera. The gorge is heavily wooded and access can be difficult.

Because the cave has not been excavated, nothing is known about its early use. A probable Late Helladic rim sherd of an open vessel collected from the back of the cave is the only evidence of activity before the nineteenth century. Broken tiles inside and below the cave indicate that there was a roofed structure in the past, but the tiles cannot be dated with any precision.

It is only in the late eighteenth century when the historian Grigorios Konstantas mentions the "Cave of Taxiarches" that it is possible to follow the site's history in more detail. In his description, Konstantas

mentions a "hoof-print" on the bedrock in front of the cave, which he ties to the story of Saint Taxiarchis. According to local legend, the saint's horse left the mark upon the rock while he was rushing to the rescue of a pilgrim. There is a mark on a horizontal rock face in front of the cave but this resembles a medieval heater shield rather than a horse's hoof (Fig. 5.1). On the small plateau in front of the cave several holes are also drilled into the bedrock (Fig. 5.2).

The date "1902" has been scratched into the bedrock at Kentavron no fewer than three times, more frequently than any other year. While it is difficult to establish the significance of graffiti dates, it coincides temporally with an occurrence that allegedly took place on the rock ledge above the cave. A story from 1902 tells about at child who went to the chapel above the cave with an offering of wheat and bread.[1] The villagers witnessed the child fall from the steep cliff, but after a few minutes, he miraculously climbed up again unharmed, still holding in his hand the wheat and the bread. It is not hard to imagine different scenarios where one might choose to memorialise such an event. Other dates and initials from the first dec-

---

1   Firiki, http://firiki.pblogs.gr/tags/cheiron/gr.html

ade of the twentieth century (1908, 1909) show at least occasional visits to the cave. It is not certain how the cave was used in the years up until World War II, when visitors/users again engraved dates and initials (1936, 1938 and 1940); we know that Mr Vergozisis, father of Apostolis, used it as a goat pen until 1948. Five years later, a 16-year-old visitor engraved his age and initials (N.K.E. 1953).

Petrocheilou, a well-known speleologist, made a survey of Kentavron in 1974 and published a report in which she speculated about the historic and touristic value of the cave. She argued that the walls and the ceiling had been "artificially altered in ancient times" and that this supported her theory that the cave was associated with the worship of Chiron or, curiously, with his personified existence. An examination of the rock wall during the survey showed no apparent signs of marks or cuts, but the local bedrock erodes in natural, rectangular chunks that could be mistaken for human manipulation.

A sketch plan published with her article shows that the frontal enclosure at this time consisted of two separate walls. One was a thick, semi-circular wall with a central entrance in front of the cave. The other wall ran inside the adjacent rockshelter and turned at a 90-degree angle to meet the rock wall (Fig. 5.3). This wall was thinner and equipped with a door.

It is unclear whether the cave was in use at the time of Petrocheilou's visit, but four years earlier, in the spring of 1970, a resident of Vyzitsa carved his initials (M.E.K.) on the cave wall. The fact that the same initials reappear on the wooden door in the dry wall could indicate that M.E.K. owned the locality or used it regularly to shelter a flock of animals. Some locals in Milies know the cave at this time as *Stani* (Sheep Pen). Petrocheilou mentions that animals are kept in Milies during the winter, so the cave was most likely used as a pen outside the cold season. It is significant that people in neighbouring Vyzitsa would tell us that the cave belonged to the territory of their village and were aware of the myths associated with it. The word "Vyzitsa" is engraved in the bedrock in front of the drystone wall. The possibility exists that the cave was on the limit of the two villages and that user rights would change back and forth between Milies and Vyzitsa over time (see also Chapter 3).

The two Germans Walter Hausmann and Wolfgang Jöchle visited the cave in 1981 and seven years later published an article about their discoveries. It is a bizarre yet fascinating example of how even "scientific" and research-driven visits and narratives have integrated the dominant Centaur myth. The authors included in their article two photos that show the course of the drystone walls as Petrocheilou recorded them some years earlier. Sometime after the visit of Hausmann and Jöchle, the walls were re-routed and built into a single feature (See Appendix 1). A part of the straight wall was demolished along with a part of the semi-circular wall in front of the cave (Fig. 5.4). The demolished parts were used for building a new part that combines the two walls, thus producing one long wall. The drystone wall is extremely well constructed and it appears to have been made by a person skilled in construction. The wall follows the topography of the bedrock and it is built in some parts to a height of more than two metres. Vertical extensions in the form of corrugated iron plates were added to the southern end of the wall, which covers the opening of the cave. Three old telephone poles from the Milies rail line were placed on top of the wall to act as supports for an angled roof structure built from rafts and probably overlaid by sheets of plastic (Figs 5.5 & 5.6). Within the walled-in area, a thin dividing wall was made from scrap wood and parts of wooden transport boxes. An analysis of the animal dung shows that goats/sheep were kept in the cave at least during spring or early summer; a discarded plastic injection syringe and veterinary medicine bottles show that vaccination was one of the activities carried out at the cave. The find of a spark plug outside the cave could indicate that a portable generator was used to provide light at some point.

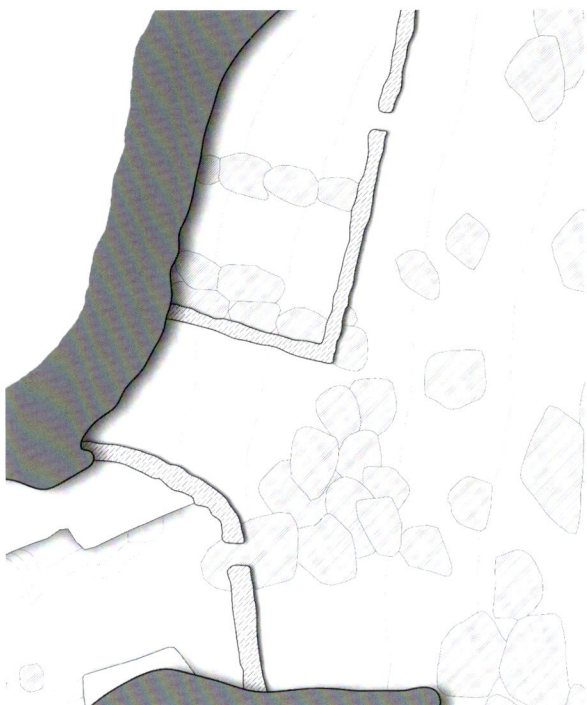

Fig. 5.3. Course of the drystone walls as they appeared until c. 1980s. Part of redrawn sketch after Petrocheilou (1974).

Fig. 5.4. The still visible cornerstones of the straight section of the first phase of the wall are integrated into the wall as it appears today.

The cave was used as an animal fold probably until sometime in the 1980s, after which it was abandoned and decay set in. The roof rafts, which originally spanned the distance from the drystone wall to the bedrock, had been placed to extend the protection provided by the overhang outside the cave. Once

Fig. 5.5. The stalling area with the collapsed roof beams.

Fig. 5.6. Telephone poles as part of the wall/roof construction.

Fig. 5.7. The combined wall as it appears today.

the covering material was removed, they started to gradually rot and collapse (Fig. 5.5). Gradually, the narrow path that leads to the cave became overgrown and access to the cave became difficult. A single graffiti date from 1995 was the only evidence of visitation in this period.

In the beginning of the 2000s, the speleology association HERON 'rediscovered' the cave and cleared the vegetation from the old path. Graffiti from 2000, 2002 and 2005 showed a new influx of visitors, mostly day-trippers and tourists. Discarded pages from a school notebook told us that these visitors included students from Volos. Finally, the Pelion Cave Project documented the cave in 2006 and 2007 and collected artefacts from its surface. Although total abandonment of the site is obvious, the drystone wall is still perfectly preserved (Fig. 5.7). Most recently, a revisit to collect dung samples for analysis under the Pelion Cave Project revealed that a graffiti date had been added in 2008.

Case 2 – Bourdovanou (MOU-7)

Bourdovanou is situated immediately north of the Agios Stefanos district in Tsagarada.[2] It is a north-west-facing limestone cave that possesses a commanding view over the Chalorema gorge with Mouresi village opposite. Today, shrubs and young trees partly obscure the view from the entrance. The cave is 5.75 m wide at the opening and c. 33 m deep. It is irregularly tunnel-shaped and becomes increasingly narrow towards the back. The ceiling steadily lowers moving into the cave, with a height of less than one metre at the back. A fine, dusty sediment covers the cave floor, which is littered with large and small rocks. Burrowing animals have caused some disturbances of the surface sediment in the inner part of the cave.

---

2   The Pelion Cave Project surveyed the cave on 18.9.2008.

The presence of ceramic sherds inside the cave points to its use as early as the Late Hellenistic–Early Roman period. With a concentration of sherds edible sea snails and limpets were also found, which must be interpreted as human food waste. The concentration points either to the presence of a more intensely occupied area of the cave or to a more intense disturbance in this area, which brought artefacts to the surface from underlying levels. The character of the use of the cave during the Early Historical period is unknown, but the marine invertebrates and varying pottery shapes show that food preparation and consumption took place at least occasionally.

Bourdovanou may have gone through several cycles of use and abandonment. The strongest indications of this are the retaining wall and the low remains of two derelict drystone walls at the opening of the cave (Fig. 2.2.11). The likelihood that the two latter walls are contemporaneous is small as they seem to fulfil the same purpose, namely to form a barrier at the entrance in connection with pastoral activities.

The cave was used in the Modern period as a goat-pen. Two glazed-ware sherds and a graffiti date of 1929 are evidence of activity in the nineteenth–twentieth centuries. One sherd is from a large basin used in food preparation, serving and eating, and the other is from a jug (see Vroom, this volume). As there is no water source immediately nearby, the cave user was required to make trips to and from springs and/or wells to collect drinking water. This included the use of water jugs. While purely speculative, perhaps one of the derelict walls can be tied to this period of use, along with the date of 1929. Four animal bones of cattle, goat and wild boar cannot be dated. A tooth was probably shed by a wild boar during a visit to the cave.

Beyond a semi-circular retaining wall in front of the cave is a steep slope that continues towards the bottom of the valley (Fig. 5.8). While it prevents erosion of the area in front of the cave, the retaining wall also supports a lath fence forming a 14 m² animal enclosure in front of the cave's drip line (Fig. 5.9). A

Fig. 5.8. Retaining wall outside the drip line.

Fig. 5.9. Remains of lath fence enclosure.

vertical post and thin undressed stems of local trees to the right of the entrance show that the small plateau in front of the cave could be closed off from the small path that leads to the locality. This would have formed a fenced-in area outside the drip line of c. 40 m², with a further c. 80 m² inside the cave. The fence is not suitable for pigs, so goats would have been the most probable livestock sheltered in the cave.

Two 1999 dates were engraved in the southern cave wall and one panel, bordered by engraved dots, read as follows:

ΚΧΖ
ΕΤΟΣ
1999 χρονια 17
οκτωβριου 14

While highly tentative, the engraving suggests that the cave user "ΚΧΖ" first arrived at (or started using) the locality in 1982. In 1999, seventeen years later, he commemorates his first visit to the cave by writing his initials and date. The absence of other recent dates and the presence of a built wooden structure points to ownership of a single user or a family group.

Recent use as a goat-pen can be substantiated by the presence of wooden transport pallets in the back of

the cave, likely used as a portable gate or barrier. A decomposing ram carcass and the decaying wooden enclosure and young bush vegetation in front of the cave are signs that both humans and livestock have recently abandoned the site, most likely within the last 5-10 years. It is striking that no refuse was found that could be dated to within the last half century. A shepherd who we met on top of the slope told us that he used the structure until some years ago, but that he now kept his goats in a built shelter. Based on his testimony, we became convinced that he probably made the 1999 graffiti.

Case 3 – Alafoklisi *(MAK-3)*

Alafoklisi is a large cave accessible by dirt road from the Fytoko settlement above Nea Ionia. The cave lies at 526 m.a.s. inside a creek to the left of the road. When entering the creek, it is initially shallow and insignificant but it quickly widens and it becomes framed by imposing vertical rock walls. The cave is deep and informants told us that it could shelter more than 200-300 goats. This estimate appears to be reliable as the inner part of the cave has about 100 m² of level floor space and the outer part has an additional 80-90 m², which is within the drip line (Fig. 5.10). The name of the cave hints at it being a sanctuary for deer, and according to informants

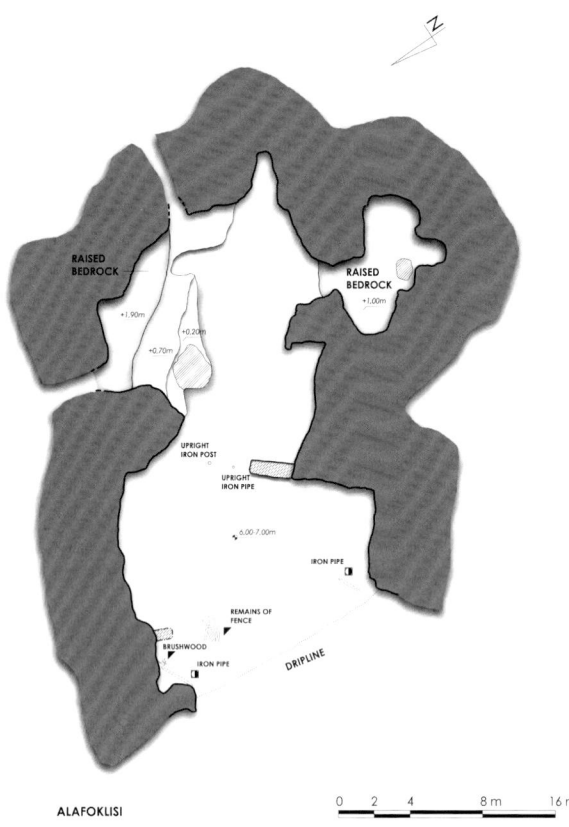

Fig. 5.10. *Alafoklisi* (MAK-3).

from Agios Georgios and Makrinitsa, wild deer would occasionally find shelter in the cave when the weather was bad, and hunters would rest there, too.

The sparse remains of a rubble wall are found to the right of the west-facing entrance. Iron pipes (some still vertically embedded in the ground), scattered brushwood, laths and tangled bundles of metal wire show that the shelter functioned as a pen, with animal access controlled by fences. A further indication that shepherds would use the site regularly for resting their animals is a battered fodder/water zinc trough. On the sloping ground outside the shelter are discarded pieces of clothing and a boot, but no other artefacts were found. An iron pole and a wooden staff are embedded in the bedrock, perhaps to hang things out of reach from nibbling livestock.

A transhumant goatherd that we met by chance close to the cave told us that he had rented grazing land from the municipality throughout the summer but that he was now preparing to take his flock south to Trikeri. He would pass by the cave during grazing rounds and had used it occasionally for shelter. That shepherds have used the cave intensively in the past is evident from the numerous graffiti, which include many engraved names and dates near the opening (1914, 1927, 1940, 1952, 1955 (x3), 1956, 1957, 1961 and 1991). The last name, Gagas ('ΓΚΑΓΚΑΣ'), appears several times. This name belonged to one of the major clans of Arvanitovlachs in Thessaly, of which a settlement was based in Almyros.[3] It also appears that there is a geographic connection between the name Gagas and the Vlach village of Vlasti between Kozani and Kostaria.

A little further to the south, and following the left side of the gorge, is an overhang or a shallow, collapsed rockshelter. There are many names, anthropomorphic drawings and dates on the bedrock around it. Both the preponderance of dates from the 1950s–1960s and the engraved name 'ΓΚΑΓΚΑΣ' are reminiscent of *Alafoklisi* and is evidence that several sites in the area were used by the same shepherds, particularly in the post-war period.

Based on this evidence, it seems likely that Vlach shepherds frequented the cave for an extended period during the mid-1900s.

Case 4 – Argyraki (MIL-3)

Argyraki is situated at the end of a small olive grove just off the gorge that runs from Milies to Kala Nera. The easy access to the cave and its proximity to a major transport route as well as the village of Milies make the location extremely attractive for a range of purposes (Fig. 5.11).

3   Sylogos Vlahon Eparhias Almyrou,
     http://www.almyros.vlahoi.net/almiros.htm

The entrance to the cave is 2 m wide and is easily barricaded by a wall or a gate. That it was indeed fenced off in the past is suggested by the remains of an old drystone wall just to the left of the entrance and a disintegrating gate found towards the rear of the cave. The latter is constructed from three upright posts with horizontal laths tied onto the posts with metal wire.

The cave is dry and provides good shelter, although artificial light is required aside from a small part of the day when the sun shines directly into the cave. The floor today consists of desiccated dung sediments mixed with clasts eroded out from the cave wall. There are numerous small and large disturbances in the cave floor characteristic of intrusive, burrowing animals. From the sections in these burrows we could observe a significant sediment layer interspersed with consolidated ash layers. The ash deposits could indicate either the presence of a hearth or burning of dung deposits on the cave floor.

The cave floor measures around 160 m² but only approximately 100 m² is proper living space, with a suitable roof height and at a distance from the entrance. Local narratives tell us of 150 Ghegs living simultaneously in the cave, but a more realistic estimate would be 2 or 3 extended family units, perhaps totalling 30 persons.

According to Milies inhabitants, the Ghegs worked during the Ottoman period in the fields around the cave and built many of the drystone terraces below Milies (see Chapter 3). An informant estimated that they left around the 1880s, but it is highly possible that they stayed until the beginning of the twentieth century (or returned for a second period), since there are accounts of the Ghegs' contribution to the construction of the train line that arrived at Milies in 1903. Their (seasonal) presence may thus have covered several decades around the turbulent time when Thessaly was annexed to the Greek state.

The Gheg occupation of the cave fit well with the date given to the pottery assemblage recovered

Fig. 5.11. Looking towards the entrance from inside *Argyraki* (MIL-3).

from the cave floor (see Vroom, this volume). An assemblage of animal bones could be contemporaneous with the ceramics. It contains a range of domesticates, such as donkey, cattle, sheep, goat and chicken. A single red deer bone and a sheep/goat bone have been chopped, probably for marrow extraction. Chopping marks are only inflicted on bones when the animal has already been cooked and it shows that cave users brought selected parts of meat (with bones) to the cave for consumption (see Panagiotidou, this volume). A disintegrating 9 cm-long hand-forged iron nail (Cat. No. 620) would also fit with a date at the end of the nineteenth century.

Graffiti and initials just inside the entrance tell us of visits to the cave in 1952 ("B.Γ.") and 1952 ("E.A."). As these dates are unlikely to represent tourist visits, they could be graffiti of herders using the cave. The disused lath gate and the extensive faeces deposits in the cave indicate its use as an animal pen.

Evidence from the most recent decades is sparse and consists of two graffiti dates from (1997 and 2000) and four fragments of bright green bottle glass (Cat. No. 592). During our survey of the cave in 2008, we noticed that we were the first to disturb a fine sediment layer on top of the cave floor. Evidently, some time had passed since the last visit to the cave. On this occasion, we surveyed the cave and collected artefacts from the floor. A re-visit in 2009 revealed that the owner or user currently keeps a horse in the cave.

5.2 Quantitative analysis of the survey results

Introduction

This part sets out the results of quantitative analyses of the data collected during the course of the field survey. In these analyses, we defined archaeological caves as in Chapter 1, i.e. as caves known to have contained artefacts, cultural deposits or deliberately deposited human remains. This category includes caves that provide recent historical evidence of human activity, such as modifications of cave structures and Modern carvings and graffiti.

The analyses include comparisons of the proportion of archaeological caves on West and East Pelion and by Lake Karla, associations of cave archaeology with landscape and cave geomorphological variables, associations of cave archaeology with internal sediment condition – and the development of separate predictive models for the two study regions based on the identified correlations and associations of cave archaeology with the variables measured in the cave surveys. A separate study of the associations between finds categories is also presented.

Several environmental variables possibly pertaining to cave selection were measured, including altitude, location, entrance aspect, distance to the settlement and cave type. The results showed that caves were typically located at an elevation below 300 m.a.s. Most caves were natural rock formations with only one entrance. The caves were often close to human residences and less than 100 m from a water source.

Cave sites with cultural remains

Within the three areas, there is significant variation between the proportions of archaeological caves. The percentage of caves that have been recorded as containing archaeological material is higher around Lake Karla than in the rest of West Pelion (Table 5.1). This may reveal a genuine spatial pattern of greater archaeological potential in the caves around the lake, and could reflect the intensity of agropastoral activity in this area.

Identification of pastoral use in comparison to other uses was more frequent around Lake Karla than on both West and East Pelion. The ethnographic fieldwork confirms that pastoral cave use was not so

	Archaeological caves	Total no. of caves	% of archaeological caves
West Pelion	64	83	77.1%
East Pelion	46	55	83.6%
Karla foothills	20	21	95.2%

Table 5.1. Total number of caves surveyed and proportions of archaeological caves in three areas of Pelion.

frequent on East Pelion. This is hardly surprising as there is limited available grazing land on most of the wooded and steep Aegean side of the mountain.

Use of caves as a dwelling was attested in six cases – and only on West Pelion. A tentative explanation could be that the inland side of the mountain is better connected to the infrastructure and the urban centre of Volos, as well as possible sources of income in the greenhouses, fields and small-scale industry of West Pelion.

Relation between archaeological caves and altitude

There is no clear relationship between caves with archaeological evidence and altitude (Fig. 5.12). This lack of correlation is repeated both across the mountain as a whole and within individual regions, and may suggest that altitude is not a significant factor in relation to cave use on Pelion.

Nearly 42% of the caves are on the coast and in the lowlands, below 100 m.a.s., and half of these are concentrated around the eastern edge of Lake Karla. The remaining are mostly beach caves and other coastal caves along the eastern (Aegean) side of Pelion, especially along the limestone coast of the Mouresi and Afetes municipalities.

Caves on the coast and lowland frequently show signs of use. It is possible that caves located at lower altitudes are more visible within the surrounding landscape, and are hence more likely to have been investigated by the PCP. However, the same argument would also apply to past human activities – lower-altitude caves may have seen a more intense use in this and the latter part of the last century due to the greater passage of people in the lowlands. Apart from beach caves, localities below 300 m.a.s. are the richest in cultural remains.

We also note that some archaeological caves (or parts of caves) were used in prehistory for burial or votive deposition, rather than for occupation or

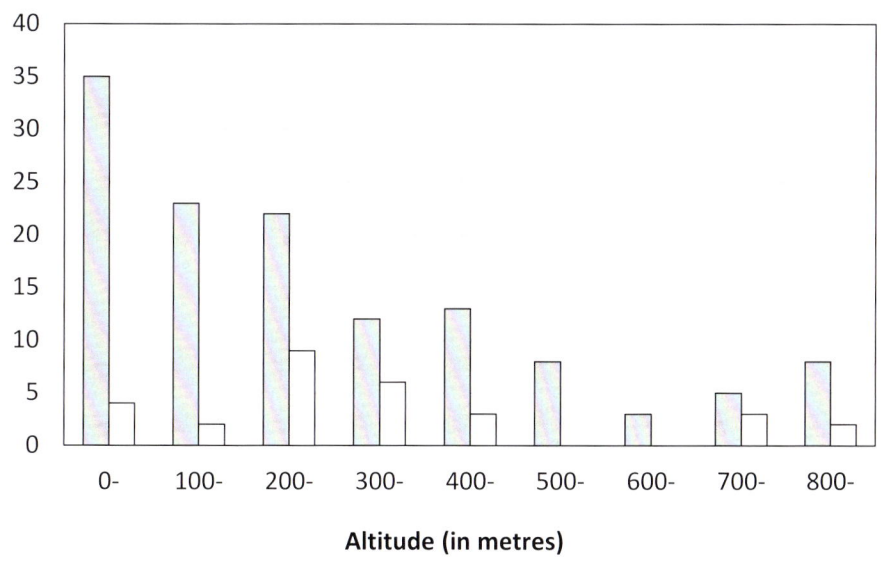

□ Caves with archaeological evidence

□ Caves with no archaeological evidence

Fig. 5.12. Relationship of archaeological caves to altitude.

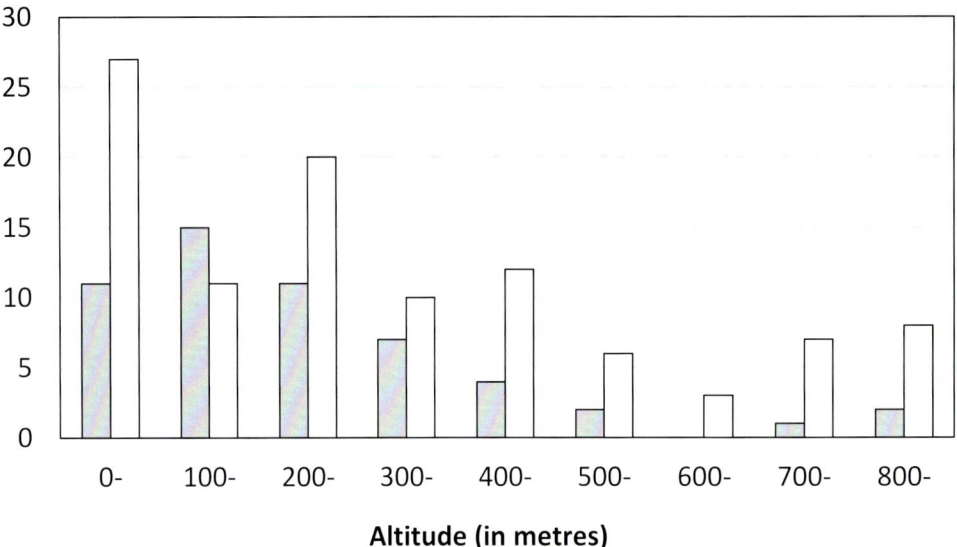

Fig. 5.13. Relationship of drywall structures to altitude.

☐ Caves with drystone structures   ☐ Caves with no drystone structures

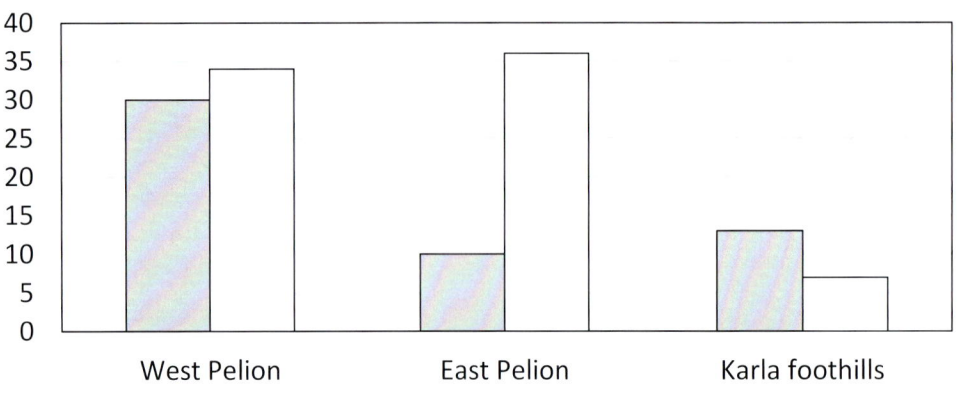

Fig. 5.14. Proportion of archaeological caves with drystone structures on West and East Pelion and by Lake Karla.

☐ Caves with drystone structures
☐ Caves with no drystone structures

subsistence-related activities. There is no evidence from the survey that high altitude locations were preferred for these types of activities.

Built features are more frequently associated with lower altitude caves and there is a slight non-significant negative trend for decreasing frequency of built structures with altitude (Fig. 5.13). Drystone structures are particularly common at Lake Karla,

where almost two-thirds of all archaeological caves are associated with a drystone wall or the remains of such a wall (Fig. 5.14). Drystone walling is much less common on Aegean Pelion, where it occurred at only ten caves out of forty-six. Since there is a close connection between drystone walling and animal husbandry, we believe this pattern supports the observation that animal husbandry was historically

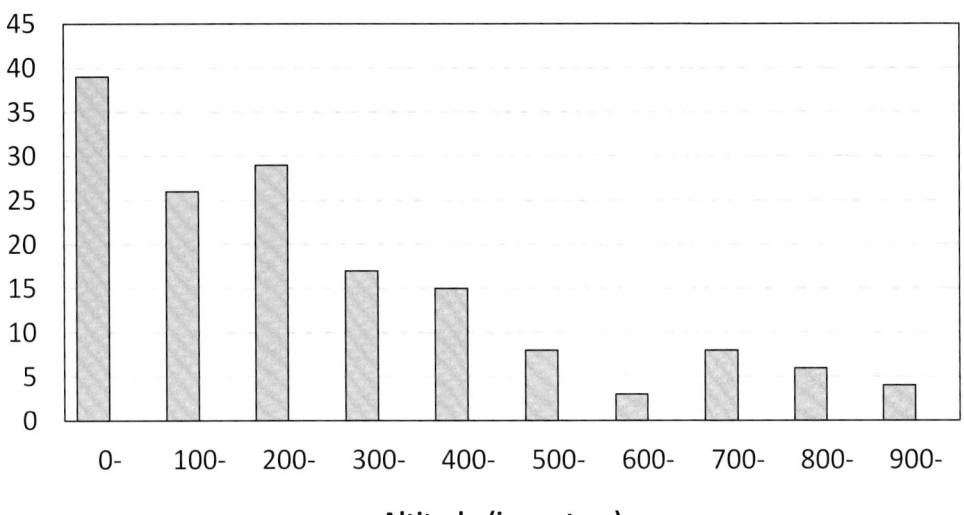

more common along Lake Karla and some areas of West Pelion.

The average altitude for archaeological caves is lower on East Pelion (290 m.a.s.) than on West Pelion (362 m.a.s.). The reason for this is that caves occur close to or on the rocky Aegean coast (particularly at Mouresi and Afetes municipalities), while coastal caves below 100 m.a.s. are almost absent along the Pagasetic Gulf where the slope is gentler, building activity more intense and the conditions more suitable for farming. The most noteworthy low-altitude cave is Malaki (ART-8), which is a solution cave in a limestone formation at the foot of a hill some distance from the coast. Otherwise, Quaternary marine and continental deposits dominate most of the lowlands along the Gulf.

Over half of the caves are situated within the low altitude belt (100-500 m.a.s.; Table 5.2). They are either below or at the same altitude as most of the mountain villages on Pelion. Most caves registered by the Ephorate prior to the project were below 500 m.a.s. 18.8% of the archaeological caves were found at mid-altitude (500-1000 m.a.s.). The caves in the upper part of this altitude interval have fewer artefacts (less than 10). Four caves above 900 m.a.s. were all from West Pelion. While only two villages on East Pelion extend beyond 500 m.a.s., seven larger villages on West Pelion are at a higher altitude with some parts of Makrinitsa reaching 800 m.a.s.

No caves (with or without archaeology) were discovered on the mountain at high altitude (above 1000 m.a.s.). With the exception of Chania, there

Zone	M.a.s.	No. of caves					
		W Pelion		E Pelion		Karla	
1. Coast & Lowland	0-100	1	*1.6%*	17	*38.6%*	17	*85%*
2. Low altitude (foothills)	100-500	49	*76.6%*	17	*38.6%*	3	*15%*
3. Mid-altitude	500-1000	14	*21.8%*	10	*22.8%*	0	*0%*
4. High altitude	1000-1600	0	–	0	–	0	–

Table 5.2. Distribution of archaeological caves grouped by altitude.

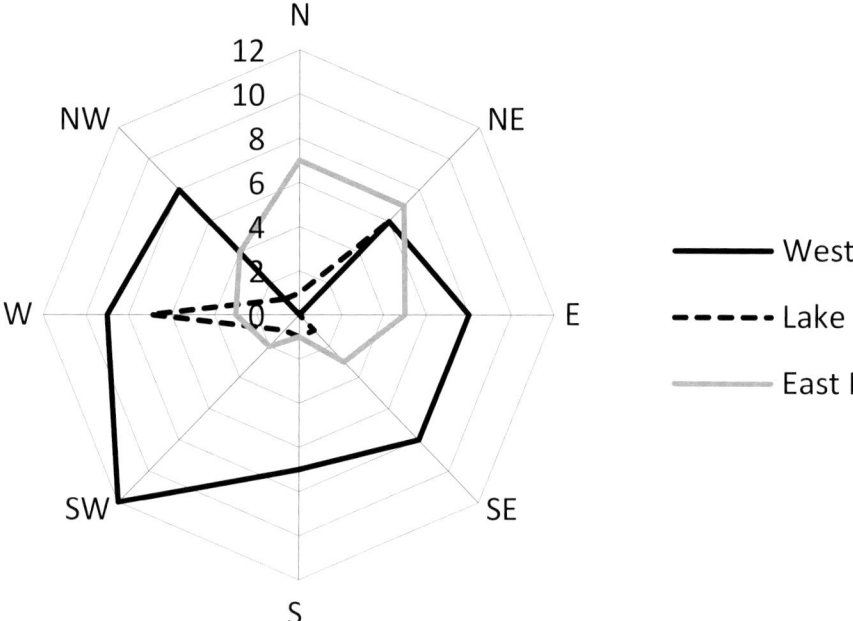

Fig. 5.16. Distribution of entrance directions on West and East Pelion and by Lake Karla.

are no villages on the central and highest area of the mountain. We expect that caves at mid- and high altitude are underrepresented in our survey since the upper parts of the mountain are becoming increasingly inaccessible, as vegetation is encroaching mainly on the old fields and paths that are situated above the villages.

Relation between archaeological caves and aspect of cave entrance

Aspect (direction faced when looking out of the cave entrance) was potentially an important factor in the selection of caves for past human activities. Aspect affects the nature and quantity of light entering the cave, it controls the exposure of the entrance to prevailing winds and precipitation and it determines the visual prospects and viewshed from the cave entrance. Aspect also relates to the surrounding area from which the cave may be visible. For instance, if a Pelion village and its orchards were located on the south-facing slopes on the north side of a gorge, then north-facing caves on the opposite (i.e. south) side of the gorge might be more noticeable to such a

community. Finally, as regularities in entrance orientation are characteristic of built structures, the same factor might have influenced the selection of caves for particular activities in the past.

Combining the three regional cave datasets, there is a noticeable difference in the distribution of entrance aspects. On West Pelion, there is a tendency for archaeological caves to have entrances facing towards the southwest quadrant, with a deficit of north-facing archaeological caves. East Pelion is characterised by north- and northeast-facing caves.

At Lake Karla, there is a tendency for archaeological caves to have entrances facing either towards the west or towards the northeast, while few caves have northern or southern aspects. The presence of archaeological caves at Karla with entrances facing northeast is partly attributable to the inclusion of two caves situated at the natural mound of Delichani Magoula. This landscape feature forms prominent scarps that face towards the northeast.

The dataset may contain an obvious bias towards southwest- and northeast-facing caves because Pelion is situated in an northwest–southeast direction,

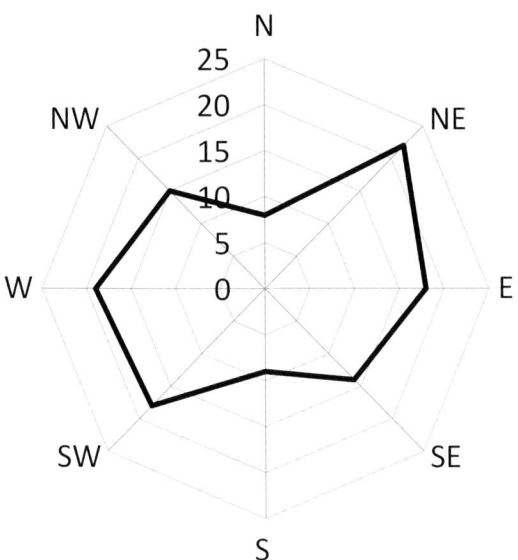

Fig. 5.17. Distribution of entrance directions of all archaeological caves on Pelion.

which should yield a higher number of southwest-facing caves on the inland side and a higher number of northeast-facing caves on the Aegean side. Other factors also play a role: most of the Pelion caves are within karst areas in which stream sinks have developed. Principal drainage directions, in particular, affected the orientation of limestone exposures, and hence the directions of cave entrances. Also, in all three areas the orientations of many cave passages reflect tectonically controlled jointing in the carbonate and metamorphic bedrocks (Vaxevanopoulos, this volume).

The patterns recorded are not suggestive of an avoidance of activity in caves that face towards a certain direction (even though there is a predominance of northeast-, northwest- and southwest-facing caves in the dataset as a whole), or of a preference for using caves that face in a specific direction. A possible preference for caves with west-facing entrances as burial sites in the prehistoric or early historic periods may reflect the general favouring of these orientations in the entrances of prehistoric monuments such as tombs.

Relation between archaeological caves and size of cave

A strong and statistically significant relationship exists between the recorded presence of archaeological evidence and the size of the cave, as expressed variously by cave entrance size (i.e., width multiplied by height) and by cave depth. On Pelion, the relationship is primarily with cave entrance height and with cave depth, as cave entrance width is not associated with the presence of archaeological remains in this region.

Cave entrance size is strongly associated with the presence of archaeological evidence, as is cave depth. There are several competing explanations for the pattern of association between measures of entrance size and the recorded presence of archaeological evidence: 1) large caves can hold more sediment, therefore increasing the probability of their containing archaeological material; 2) large caves may be more noticeable or more attractive to antiquarians and archaeologists, increasing the likelihood of excavation and recovery of archaeological material; and 3) large caves may have been more noticeable or attractive to people in the past, enhancing the likelihood of deposition of archaeological material.

However, it is inevitable that some of the cave sites that are currently classified in our database as 'non-archaeological' either do, in fact, or once used to contain archaeological material, but this is a supposition that can only be tested by future intrusive ground-based investigations.

Relation between archaeological caves and ground slope

In all of the surveyed areas, there were only weak and non-significant relationships between the presence of cultural material in caves and the general topographical slope, as assessed by ground slopes below the caves. This indicates that the presence of archaeological evidence is not related to the general topographic gradient surrounding the cave.

However, the ground slope immediately outside the cave did show a relationship with the presence of archaeological material – in both of the districts surveyed, archaeological evidence was more likely in caves where the area immediately outside the entrance was either horizontal or had a negative slope, i.e. sloped down approaching the cave entrance. As with the size of the cave entrance, the presence of a level or even raised area immediately outside the cave entrance may be indicative of past archaeological excavation; nonetheless this feature could also be an indicator of human activity at the cave site in the more remote past.

As with the ground slope outside the cave, the ground slope inside the cave entrance also showed a slight trend in relation to archaeological evidence, with archaeological caves being associated with a more positive slope, i.e. ground sloping down into the cave. The trend is understandable in the same manner as the ground slope outside the cave, perhaps because of removal of deposits from inside the cave passages to spoil tips outside the cave entrance, either in antiquity or more recently during excavation by archaeologists or cavers.

Slope did seem to be a factor affecting some sites. In some cases, there was a need for flatter areas outside the cave entrance, and where caves occur on steeper slopes, such areas were created by constructing a retaining wall. Drystone retaining walls are self-draining and easy to construct, and were therefore ideal for this purpose. In all cases, these gravity walls hold the earth through their own weight. If not carefully constructed, they may topple relatively easily, as the internal leverage of the earth is high.

Relation between archaeological caves and condition of internal cave deposits

The presence and nature of internal sedimentary deposits were recorded in a descriptive fashion in the cave survey database. This information was assigned to three categories of deposit condition: 0 = sediments absent or present in minimal or residual quantities; 1 = in situ sediments in a damaged or eroding condition and 3 = in situ sediments present in a stable or accruing condition. The distribution of each deposit condition was determined for archaeological and non-archaeological caves in the two study areas. A similar pattern emerged in both study areas: the majority of archaeological caves had sediments in damaged or eroded condition, with a smaller number having stable or accruing deposits and few archaeological caves having no or minimal sediments. The difference in proportions of archaeological versus non-archaeological caves for the three sediment condition categories was statistically significant for Pelion. The pattern can be understood as reflecting the fact that caves with little or no sediment are unlikely to produce archaeological finds.

Relation between archaeological caves and proximity to water

Water is a vital commodity for subsistence and pastoral activities, and proximity to water is often incorporated into models of archaeological site location. On Pelion, water may be more accessible due to a prevalence of springs and surface watercourses. Distance to surface water is therefore not a clear locational factor and would in any case be difficult to put a figure on, as springs and streams are widespread on the mountain.

5.3 Experiences and perspectives

Archaeologists have been interested in the contents of caves for a variety of reasons, and cave archaeology has a long and distinguished tradition in Greece. Nevertheless, upon reviewing the sparse literature on Modern cave use in Greece, we must conclude that archaeology has avoided analyses of such sites and that the research potential that these sites offer has generally been ignored. At the same time, there have been few anthropological or ethnoarchaeological studies undertaken on contemporary cave use.

To improve this situation, a core aim of the Pelion Cave Project was to demonstrate the decisive role played by the perspective offered by regional ethnoarchaeological landscape study in documenting the variability of cave use with an emphasis on the processes of migration, industrialisation and urbanisation.

In the preceding chapters, we have examined the secular and spiritual significance of caves within the cultural and physical topography of a single mountain. Our research provides for the first time a detailed insight into the status of caves in the region as local, living heritage, as well as an insight into the diverse use of caves by earlier generations of people on Pelion, particularly during the course of the past two centuries.

The survey provided virtually snapshot views of the situation at the caves and, consequently, studies of change over a longer period at each site were not possible. On the other hand, the investigation gave completely new insights into other important aspects, for instance in the area of rural graffiti. In addition, with regard to the methodological part of the project, it became clear that the mountain, spanning a well-defined and manageable geographical region, was well suited to the majority of practical objectives planned for the project.

Archaeological implications

Our project was designed to address Modern cave sites as a potentially valuable resource for archaeological knowledge. As is the case with Prehistoric caves, archaeological evidence from Modern caves provides a glimpse of past societies' cultural understanding of natural places in the landscape.

Material culture from caves constitutes a disproportionately large part of the surviving archaeological record for many Prehistoric periods. This, ironically, has led to a situation where the view on Modern cave use is based on the archaeological record derived from cave sequences. Modern practices are viewed as remnants of older practices – but the results from our project have shown that such a view is deeply unrepresentative of the range and character of activities that took place in caves in the Modern period.

The implications that our observations on Modern cave use hold for the interpretation of the archaeological record can be briefly summarised as follows:

- Looking to the ethnographic record, we found little evidence to support the expectation that caves served as permanent home bases. There are only a few indicators that people would replace a built residence with a cave, and this usually only happened in special circumstances.
- While caves are convenient places in the landscape for a wide variety of purposes, they represent only a relatively small fraction (albeit situationally variable) of the locations used outside the villages.
- Our study shows that the Modern landscape use of both caves and rockshelters on Pelion was highly flexible. Caves transcend traditional rural spatial divisions in that they occur as features of the built village, the fields around it and the area outside the agricultural territory.
- There appeared to be little *archaeological* evidence for changes in cave use; the changes that occurred were often ephemeral and would leave few traces behind.
- Mixed practices would mean in some cases that different uses could occur simultaneously in a cave (e.g. dwelling/storage), or be replaced by another within a short time interval. This can produce a sort of horizontal stratigraphy in which several processes in modern times have produced their own particular patterns of material remains.

Ethnographic implications

One of the important ideas to come out of our project is that it is the individual that breathes social life into our data on cave use. The individual offers us the chance to expand our theoretical and analytical horizons with respect to the evidence. Mainly we have come to discover that the roles of the individual in relation to cave use are flexible. Through the ethnographic component of the project, we have seen that it is in a story's power to disrupt some of the traditional taxonomies of pastoralists, vagrants, partisans and villagers, and view in a more nuanced way those practices that unfolded around caves during the last 130 years.

Similarly, we can say that the conventional dichotomies of rural and urban, traditional and modern, etc. are not valid for contemporary Greece and probably never have been. Pelion has always been involved, to a greater or lesser extent, in social, economic and political systems beyond (and often far beyond) the boundaries of the individual villages and the region itself.

We have been able to compare the assemblages from scores of caves closely associated in space and in time, and to extend our conclusions from the level of the isolated, individual cave to an entire region. As is the case for the Prehistoric period, Modern cave use can only be understood in the context of the entire social and economic patterns of the investigated region.

As showed in the previous chapter, the evolution of cave use since the end of the eighteenth century and its associated behavioural and archaeological transformations could be broadly described in a three-stage model (Table 4.9).

To assess caves and rockshelters merely in terms of their practical function or economic significance diminishes these sites in terms of the significance they hold within the landscape and the ideological system in which they operate. Our survey indicates that caves connect people to the wider physical and ideological landscape. Several of the caves we investigated can be classified as "ideo-technic artefacts" in that they house built structures such as hermitages and contain graffiti that connect them and their users to ethnic and social groups. Caves also feature in the recent ideological landscape; as symbols of the Greek resistance during World War II and the Greek Civil War, they represent monumental landmarks in the nation's history.

Challenges for the future

This volume has been an attempt to make links back into Pelion's recent past, using archaeological and ethnographic evidence. The intention was to go beyond the view of recent cave occupation as an unrepresentative example of Greek rural life, centred upon marginal and seasonal economic activities. We argue that many of the concepts prevalent in the interpretation of data derived from archaeological cave and rockshelter sites assume a degree of fixedness in the social use of landscape and consumption and discard of material culture items that finds few contemporary parallels.

We believe there is a great potential for systematic mapping of Modern cave use and for a fully integrated, community-based archaeological ethnography of Modern cave use practices. We also hope that our approach demonstrated the potential of historical archaeology and ethnoarchaeology for social issues arising from the emergence of modernity in Greece, and will be of value to researchers who are seeking inspiration for how historical archaeology can operate in a Greek context. To some extent, we offer a proposal for later historical archaeology in Greece that expands and facilitates innovative and unfamiliar ways of looking at the last centuries of Greek rural history. We would like to think that this project is representative of modern scholarship, which is increasingly crossing disciplinary lines and recognising the validity of examining social and cultural phenomena in more than a narrow archaeological or other disciplinary manner. [4]

---

4 Buchli & Lucas 2001.

Bibliography

ACOVITSIOTI-HAMEAU, A., J.-É. BROCHIER & P. HAMEAU 2000
'Témoignages et marqueurs du pastoralisme actuel en Grèce: une ethnographie des gestes et des restes et les applications archéologiques corrélées', *Ethnologia* 6-7 (1988-89), 93-135.

ADRYMI-SISMANI, V. 2010
'Αρχαίες θέσεις και μνημεία της Μαγνησίας', *Εν Βόλω* 37-38, 17-8.

AKRIVOS, K. 2010
Ποιος θυμάται τον Αλφόνς. Athens.

ALEXANDERSEN, V. 1978
'Approximale Furchen bei dänischen mesolithischen und neolithischen Molaren', *Deutsche Zahnhärztliche Zeitschrift* 33, 213-5.

ALEXANDROU, C. 2014
'Nazi operations in northern Pelion during the final stages of World War II', in *Lake Karla Walking Guide*, S. Dodouras, I. Lyratzaki & T. Papayannis (eds.), Mediterranean Institute for Nature and Anthropos, Athens, 73-7.

ALT, K.W. & C. KOÇKAPAN 1993
'Artificial tooth-neck grooving in living and prehistoric population', *Homo* 44, 5-29.

ALT, K.W. & S.L. PICHLER 1998
'Artificial modifications of human teeth', in *Dental anthropology: Fundamentals, limits and prospects*, K.W. Alt, F. W. Rösing & M. Teschler-Nicola (eds.), Vienna/New York, 387-416.

ANDREASEN, J.O. & J. KUROL 1997
'The impacted first and second molar', in *Textbook and color atlas of tooth impactions*, J.O. Andreasen, J.K. Petersen & D.M. Laskin (eds.), St. Louis, 197-218.

ANDREASEN, N.H., N. PANTZOU & D.C. PAPADOPOULOS 2009
'The Pelion Cave Project – an ethno-archaeological investigation of the human use of caves in the Early Modern and Modern period in East Thessaly', *Proceedings of the Danish Institute at Athens* 6, 177-89.

ANGEL, J.L. 1944
'Greek teeth: Ancient and modern', *Human Biology* 16, 283-97.

ANSCHEUTZ, K.F., R.H. WILSHUSEN & C.L. SCHEICK 2001
'An Archaeology of Landscape: Perspectives and Directions', *Journal of Archaeological Research* 9, 157-211.

ARNOLD, C. 1997
Steel pots: The History of America's Steel Helmets. San Jose, California.

ARROYO, M.A.B. & G.M.R. MORALES 2007
'La Fragua Cave, a seasonal hunting camp in the lower Asón valley (Cantabria, Spain) at the Pleistocene-Holocene transition', *Anthropozoologica* 42(1), 61-84.

ARSENIOU, L. 1977H
Θεσσαλία στην Αντιστάση. Athens.

ARVANITOPOULOS, A.S. 1910
Ανασκαφαί καί ερευναι έν Θεσσαλία κατά τό ετος 1910. *Πρακτικά της Αθήναις Αρχαιολογικης Εταιρείας (ΠΑΕ)* 1910, 168-264.

ARVANITOPOULOS, A.S. 1911
Ανασκαφαί καί ερευναι έν Θεσσαλία κατά τό ετος 1911. *Πρακτικά της Αθήναις Αρχαιολογικης Εταιρείας (ΠΑΕ)* 1911, 280-356.

ASDRACHAS, S.I. 2003
Ελληνική Οικονομική Ιστορία ΙΕ' – ΙΘ αιώνας, Athens, 403-22.

ASDRACHAS, S.I. 2005
Greek Economic History, 15th–19th centuries. The Historical Review (Volume II), Athens.

ASHMORE, W. 2002
'Decisions and Dispositions: Socializing Spatial Archaeology: Archaeology Division Distinguished Lecture 99th AAA Annual Meeting, San Francisco, CA, November 2000', *American Anthropologist* 104(4), 1172-83.

BACKHUIZEN, S.C. 1992
A Greek City of the Fourth century BC. Biblioteca Archaeologica 10, Rome.

BAERENTZEN, L., J. IATRIDES & O. SMITH (EDS.) 1987
Studies in the History of the Greek Civil War, 1945-1949. Copenhagen.

BAKALAKI, A., 1997
'Students, Natives, Colleagues: Encounters in Academia and in the Field', *Cultural anthropology: journal of the Society for Cultural Anthropology* 12(4), 502.

BASS, W.M. 1995
Human osteology: A laboratory and field manual, Columbia.

BASU, P. 1997
Narratives in a landscape: Monuments and memories of the Sutherland Clearances. MSc dissertation, University College London.

BATZIOU-EFSTATHIOU, A. 1999
'Το νεκροταφείο της Νέας Ιωνίας (Βόλου) κατά τη μετάβαση από την ΥΕ ΙΙΙΓ στην ΠΓ εποχή', in *Η Περιφέρεια του Μυκηναϊκού Κόσμου, Α΄ Διεθνές Διεπιστημονικό Συμπόσιο, Λαμία 1994*, E. Froussou (ed.), Lamia, 117-30.

BENNETT, D.O. 1988
'The poor have much more money: changing socio-economic relations in a Greek village', *Journal of Modern Greek Studies* 6(2), 217-43.

BERMÚDEZ DE CASTRO, J.M. & J.L. ARSUAGA 1983
'L' usure anormale du collet de la dent chez les populations préhispaniques des Canaries', *L' Anthropologie* 87, 521-33.

BERMÚDEZ DE CASTRO, J.M. & P.-J. PÉREZ 1986
'Anomalous tooth-neck wear in North African Mesolithic populations', *Paleopathology Newsletter* 54, 5-10.

BERMÚDEZ DE CASTRO, J.M., J.L. ARSUAGA & P.-J. PÉREZ 1997
'Interproximal grooving in the Atapuerca-SH hominid dentitions', *American Journal of Physical Anthropology* 102, 369-76.

BERNARD, H.R. 2000
Social research methods: qualitative and quantitative approaches. Thousand Oaks, California.

BERRIZBEITIA, E.L. 1989
'Sex determination with the head of the radius', *Journal of Forensic Sciences* 34, 1206-13.

BERRYMAN, H.E., D.W. OWSLEY & A.M. HENDERSON 1979
'Non-carious interproximal grooves in Arikara Indian dentitions', *American Journal of Physical Anthropology* 50, 209-12.

BINFORD, L.R. 1964
'A consideration of archaeological research design', *American Antiquity* 29(4), 425-41.

BINFORD, L.R. 1981
Bones: Ancient Men and Modern Myths. New York.

BINTLIFF, J. 2003
'The ethnoarchaeology of a 'passive' ethnicity: The Arvanites of Central Greece', in *The Usable Past. Greek Metahistories*, K.S. Brown & Y. Hamilakis (eds.), Lanham-Boulder, 129-44.

BOAZ, N.T. & F.C. HOWELL 1977
'A gracile hominid cranium from upper member G of the Shungura Formation, Ethiopia', *American Journal of Physical Anthropology* 46, 93-108.

BOESSNECK, J. 1969
'Osteological differences between sheep (*Ovis aries* Linné) and goat (*Capra hircus* Linné)', in *Science in Archaeology: a survey of Progress and Research*, D. Brothwell & E. Higgs (eds.), New York, 331-58.

BONFIGLIOLI, B., V. MARIOTTI, F. FACCHINI, M.G. BELCASTRO & S. CONDEMI 2004
'Masticatory and nonmasticatory dental modifications in the Epipalaeolithic necropolis of Taforalt (Morocco)', *International Journal of Osteoarchaeology* 14, 448-56.

BOOCOCK, P.A., C.A. ROBERTS & K. MANCHESTER 1995A
'Maxillary sinusitis in Medieval Chichester, England', *American Journal of Physical Anthropology* 98, 483-95.

BOOCOCK, P.A., C.A. ROBERTS & K. MANCHESTER 1995B
'Prevalence of maxillary sinusitis in leprous individuals from a Medieval leprosy hospital', *International Journal of Leprosy and other Mycobacterial Diseases* 63, 265-8.

BOSI, D. 2006
M33 Analisi di un Elmo: trattato tecnico sull'elmetto italiano della seconda guerra mondiale.

BOSSI-NOGUEIRA, E. 1975
L'Elmetto Italiano 1915-1971. Milan.

BOSSI-NOGUEIRA, E. 1991
History dell'elmetto Italian. Milan.

BRADSHAW, R.H.W. 2004
'Past anthropogenic influence on European forests and some possible genetic consequences', *Forest Ecology and Management* 197, 203-12.

BRADSHAW, R.H.W. & G.E. HANNON 2004
'The Holocene structure of north-west European forest induced from palaeo-ecological data', in *Forest Biodiversity: Lessons from History for Conservation*, O. Honnay, K. Verheyen, B. Bossuyt & M. Hermy (eds.), Oxford, 11-25.

BRAIN, C. K. 1981
The Hunters or the Hunted? Chicago.

BRICKLEY, M. 2006
'Rib fractures in the archaeological record: A useful source of sociocultural information?', *International Journal of Osteoarchaeology* 16, 61-75.

BROCHIER, J.E., P. VILA, M. GIACOMARRA & A.T. TAGLIACOZO 1992
'Shepherds and sediments: geo-ethnoarchaeology of pastoral sites', *Journal of Anthropological Archaeology* 11, 47-102.

BROOK, I. 2009
'Sinusitis', *Periodontology 2000* 49, 126-39.

BROOKS, A. & S. ADCOCK 1999
Dry Stone Walling – a practical handbook.

BROTHWELL, D.R. 1963
'The macroscopic dental pathology of some earlier human populations', in *Dental anthropology*, D.R. Brothwell (ed.), New York, 271-88.

BROWN, T. 1991
'Interproximal grooving: Different appearances, different etiologies. Reply to Dr Formicola', *American Journal of Physical Anthropology* 86, 86-7.

BROWN, T. & S. MOLNAR 1990
'Interproximal grooving and task activity in Australia', *American Journal of Physical Anthropology* 81, 545-53.

BRUMFIELD, A. 2000
'Agriculture and rural settlement in Crete, 1669-1898', in *A Historical Archaeology of the Ottoman Empire: Breaking New Ground*, Uzi Baram & L. Carroll (eds.), New York, 37-78.

BUIKSTRA, J.E. & D.H. UBELAKER (EDS.) 1994
Standards for data collection from human skeletal remains: Proceedings of a seminar at the Field Museum of Natural History. Fayetteville.

BULL, G. & S. PAYNE 1982
'Tooth eruption and epiphysial fusion in pigs and wild boar', in *Ageing and Sexing Animal* bones *from Archaeological Sites* (BAR British Series 109), B. Wilson, C. Grigson & S. Payne (eds.), Oxford, 55-71.

BULTRINI, N. 2006
Adrian: La storia e il mito dell'elmetto della Grande Guerra. Chiari.

BURSIAN, C. 1862
Geographie von Griechenland. Leipzig.

BUXTON, R. 1994
Imaginary Greece – the contexts of mythology. Cambridge.

ÇAKIRLAR, C. 2009
'To the shore, back and again. Achaeomalacology of Troia', *Studia Troica* 18, 59-86.

CAMERON, C.M. 1991
'Structure abandonment in villages', *Archaeological Method and Theory* 3, 155-94.

CAMPBELL, T.D. 1925
Dentition and palate of the Australian Aboriginal, Adelaide.

CANTI, M.G. 1997
'An investigation of microscopic calcareous spherulites from herbivore dungs', *Journal of Archaeological Science* 24, 219-31.

CAPUTO, R. 1996
'The active Nea Achialos Fault System (Central Greece): Comparison of geological, morphotectonic, archaeological and seismological data', *Annali di Geofisica* 39(3), 557-74.

CAPUTO, R. & S. PAVLIDES 1993
'Late Cainozoic geodynamic evolution of Thessaly and surroundings (central-northern Greece)', *Tectonophysics* 223, 339-62.

CASTAÑEDA, Q.E. & C.N. MATTHEWS (EDS.) 2008
Ethnographic archaeologies reflections on stakeholders and archaeological practices. Lanham, MD.

CATLING, H.W. 2009
Sparta: Menelaion I: The Bronze Age. Vols I–II. British School at Athens, Supplementary Volume 45. London.

CHANG, C. 1981
The archaeology of contemporary herding sites in Didyma, Greece. Unpubl. PhD dissertation, Anthropology Department, State University of New York at Binghamton.

CHANG, C. 1992
'Archaeological Landscapes. The ethnoarchaeology of pastoral land use in the Grevena Province of Greece', in *Space, Time, and archaeological Landscapes*, J. Rossignol & L. Wandsnider (eds.), New York & London, 65-90.

CHANG, C., 1993
'Pastoral transhumance in the southern Balkans as a social ideology. Ethnoarchaeological research in northern Greece', *American Anthropologist* 95, 687-703.

CHANG, C. 1999
'The ethnoarchaeology of pastoral sites in the Grevena Region of Northern Greece', in *Transhumant pastoralism in Southern Europe – Recent perspectives from Archaeology, History and Ethnology*, L. Bartosiewicz & H.J. Greenfield (eds.), Budapest, 133-44.

CHANG, C., 2000
'The material culture and settlement history of agropastoralism in the Koinonos of Dhidhima. An ethnoarchaeological perspective', in *Contingent countryside.*

Settlement, economy, and land use in the southern Argolid since 1700, S.B. Sutton (ed.), Stanford, California, 125-40.

CHANG, C. & H.A. KOSTER, 1986
Beyond bones: toward an archaeology of pastoralism. New York.

CHANG, C. & H.A. KOSTER, 1994
Pastoralists at the periphery: herders in a capitalist world. Tucson.

CHANG, C. & P.A. TOURTELLOTTE 1993
'Ethnoarchaeological survey of pastoral transhumance sites in the Grevena region, Greece', *Journal of Field Archaeology* 20(3), 249-64.

CHARISI, D., C. ELIOPOULOS, V. VANNA, C. KOILIAS & S.K. MANOLIS 2011
'Sexual dimorphism of the arm bones in a modern Greek population', *Journal of Forensic Sciences* 56, 10-8.

CHATZILAZARIDIS, L.I. 2000
Προϊστορικά βραχογραφήματα στη Β. Ελλαδα [Prehistoric rock engravings in North Greece]. Unpubl. PhD dissertation, University of Thessaloniki.

CHERRY, J. 1988
'Pastoralism and the role of animals in the Pre- and Protohistoric periods of the Aegean', in *Pastoral economies in Classical antiquity*, C.R. Whittaker (ed.), Cambridge, 6-34.

CLAASSEN, C. 1998
Shells. Cambridge Manuals in Archaeology, Cambridge.

CLOKE, P. & O. JONES 2001
'Dwelling, place, and landscape: an orchard in Somerset', *Environment and Planning A* 33(4), 649-66.

CLOSE, D.H. (ED.) 1993
The Greek Civil War: studies of polarization, 1943-1950, New York.

COON, C.S. 1977
'The Highland Ghegs', in *A Reader in Cultural Anthropology,* C.S. Coon (ed.), Huntington, New York.

CREIGHTON, O.H. & J.R. SEGUI 1998
'The ethnoarchaeology of abandonment and post-abandonment behaviour in pastoral sites: evidence from Famorca, Alacant province, Spain', *Journal of Mediterranean Archaeology* 11(1), 31-52.

DAVIDSON, A. 1981
Mediterranean Seafood, London.

DE RANCOURT, M., N. FOIS, M.P. LAVIN, E. TCHAKERIAN, & F. VALLERAND 2006
'Mediterranean sheep and goats production: An uncertain future', *Small Ruminant Research* 62(3), 167-79.

DELAMOTTE, M. & E. VARDALA-THEODOROU 1994
Κοχύλια από τις ελληνικές θάλασσες. Kifisia.

DESBOROUGH, V.R.D'A. 1972
The Greek Dark Ages. London.

DIETRICH, B.C. 1974
The Origins of Greek Religion. Berlin.

DIMITROPOULOS, D. 2010
'On the Settlement Complex of Central Greece: An Early Nineteenth-Century Testimony', *The Historical Review / La Revue Historique* 7, 323-46.

DOULGERI-INTZESILOGLOU, A. 1992
'Εργαστήρια κεραμεικής Ελληνιστικής εποχής στην αρχαία πόλη των Φερών', in *Διεθνές Συνέδριο για την Αρχαία Θεσσαλία (στη μνήμη του Δημήτρη Ρ. Θεοχάρη), Βόλος 1987,* Ε. Kypraiou (ed.), Athens, 437-46.

DOULGERI-INTZESILOGLOU, A. 1994
'Τα Ελληνιστικά λυχνάρια των Φερών', in *Γ΄ Επιστημονική Συνάντηση για την Ελληνιστική Κεραμεική,* Πρακτικά, Athens, 363-88.

DOWDEN, K. 1989
Death and the maiden: girls' initiation rites in Greek mythology. London.

DROUGOU, S. & G. TOURATSOGLOU 1994
'Τα χρονολογημένα σύνολα Ελληνιστικής κεραμεικής από τη Μακεδονία', in *Γ΄ Επιστημονική Συνάντηση για την Ελληνιστική Κεραμεική,* Πρακτικά, Athens, 128-37.

DROUGOU, S. 1988
'Τα πήλινα κτερίσματα του μακεδονικού τάφου στην πλατεία Συντριβανίου Θεσσαλονίκης', *Αρχαιολογική Εφημερίς* 127, 71-94.

DUBISCH, J. 1995
'The Church and the Annunciation of Tinos and the domestication of institutional space', in *Constructed Meaning: Form and Process in Greek Architecture,* E. Pavlides & S.B. Sutton (eds.), Special Issue of *The Modern Greek Studies Yearbook* 10/11:271, 389-418.

DURRANS, B. 1994
'Theory, profession, and the political role of archaeology', in *Archaeological approaches to cultural identity,* S.J. Shennan (ed.), London & New York, 66-75.

ECKHARDT, R.B. & A.L. PIERMARINI 1988
'Reply to Formicola', *Current Anthropology* 29, 668-70.

ECKHARDT, R.B. 1990
'The solution for teething troubles', *Nature* 345, 578.

EDGEWORTH, M. 2006
Ethnographies of archaeological practice: cultural encounters, material transformations. Lanham, MD.

EFSTRATIOU, N. 1999
'Pastoralism in highland Rhodope: archaeological implications from recent observations', in *Transhumant pastoralism in Southern Europe – recent perspectives from Archaeology, History and Ethnology*, L. Bartosiewicz & H.J. Greenfield (eds.), Budapest, 145-58.

ELIOPOULOS, C. 2006
The creation of a documented human skeletal reference collection and the application of current aging and sexing standards on a Greek skeletal population. Unpubl. PhD dissertation, University of Sheffield.

EUROPEAN COMMISSION 2009
Natura 2000 in the Mediterranean Region, Belgium. http://ec.europa.eu/environment/nature/info/pubs/ docs/biogeos/Mediterranean.pdf [accessed 8.5.2010]

FARDI A., A. KONDYLIDOU-SIDIRA, Z. BACHOUR, N. PARISIS & A. TSIRLIS 2011
'Incidence of impacted and supernumerary teeth -a radiographic study in a North Greek population', *Medicina Oral, Patología Oral y Cirugía Bucal* 16, 56-–61.

FAURE 1964
Les fonctions des cavernes crétoises. Paris.

FEUER, B. 1992
'Mycenaean Thessaly: our present state of knowledge', in *Διεθνές Συνέδριο για την Αρχαία Θεσσαλία (στη μνήμη του Δημήτρη Ρ. Θεοχάρη), Βόλος 1987*, Ε. Κυpraiou (ed.), Athens, 286-7.

FLOOD, J. 1997
Australian Aboriginal use of caves, in *The Human Uses of Caves*, C. Bonsall & C. Tolan-Smith (eds.), British Archaeological Reports International Series 667, Oxford, 193-200.

FORBES, C.M.H. 1976
'Farming and foraging in prehistoric Greece: a cultural ecological perspective', in *Regional Variation in Modern Greece and Cyprus: Toward a Perspective on the Ethnography of Greece*, M. Dimen & E. Friedl (eds.), Annals of the New York Academy of Science 268, New York, 127-142.

FORBES, H.A. 1996
'The uses of the uncultivated landscape in modern Greece: a pointer to the value of the wilderness in antiquity?' in *Human Landscapes in Classical Antiquity. Environment and Culture*, G. Shipley & J. Salmon (eds.), London, 68-97.

FORBES, H.A. 2007
Meaning and identity in a Greek landscape: an archaeological ethnography. Cambridge & New York.

FORD, D., & P. WILLIAMS 2007
Karst Hydrogeology and Geomorphology. West Sussex, England.

FORMICOLA, V. 1988
'Interproximal grooving of teeth: Additional evidence and interpretation', *Current Anthropology* 29, 663-4.

FORMICOLA, V. 1991
'Interproximal grooving: Different appearances, different etiologies', *American Journal of Physical Anthropology* 86, 85-7.

FOTIADIS, G.
Χαρακτηριστικά στοιχεία της βλάστησης και της χλωρίδας του Πηλίου, http://www.iama.gr/ethno/ faskomilo/Fwtiadis.pdf [accessed 05.09.2008]

FRAYER, D.W. & M.D. RUSSELL 1987
'Artificial grooves on the Krapina Neanderthal teeth', *American Journal of Physical Anthropology* 74, 393-405.

FRAYER, D.W. 1991
'On the etiology of interproximal grooves', *American Journal of Physical Anthropology* 85, 299-304.

FRERIS, A.F. 1986
The Greek Economy in the Twentieth Century. London.

FREUND, G. 1968
'Review of D. Theocharis, Die Anfänge der thessalischen Vorgeschichte. Ursprung und erste Entwicklung des Neolithikums' (1967, translation from Greek), *Quartär* 19, 415-8.

FURUMARK, A. 1941
Mycenaean Pottery: Analysis and Classification. Stockholm.

GALANIDOU, N. 2000
'Patterns in caves: Foragers, horticulturists, and the use of space', *Journal of Archaeological Anthropology* 19, 243-75.

GALLIS, K. 1996
'The Neolithic world', in *Neolithic Culture in Greece*, G.A. Papathanassopoulos (ed.), Athens, 23-37.

GALLOWAY, A. 1999
Broken bones: Anthropological analysis of blunt force trauma, Springfield.

GIALIS, S. & C.S. LASPIDOU 2014
Lake Karla and the contradictory character of Greek Environmental Policies: A brief historical overview. IWA Regional Symposium on Water, Wastewater and Environment: Traditions and Culture. Patras, Greece, 22-24 March 2014.

GIANNOPOULOU, M. & S. DEMESTICHA 1998
Tskalaria. Τα εργαστήρια αγγειοπλαστικής της περιοχής Μανταμάδου Λέσβου. Athens.

GIDARAKOU, I. & C. APOSTOLOPOULOS 1995
'The productive system of itinerant stockfarming in Greece', *Medit* 3, 56-63.

GISAKIS, I.G., F.D. PALAMIDAKIS, E.-T.R. FARMAKIS, G. KAMBEROS & S. KAMBEROS 2011
'Prevalence of impacted teeth in a Greek population', *Journal of Investigative and Clinical Dentistry* 2, 1-8.

GLEGLE, L. 2009
'The Ottoman domination, the chifliks and the association of rural movements with the national matter after the 17th century', *ΘΕΣΣΑΛΙΑ, Thessaly. History and Culture of Thessaly*, 495-9.

GORECKI, P.P. 1991
'Horticulturalists as hunter-gatherers: Rock shelter usage in Papua New Guinea', in *Ethnoarchaeological Approaches to Mobile Campsites: Hunter-Gatherer and Pastoralist Case Studies* (International Monographs in Prehistory), C.S. Gamble & N.A. Boismier (eds.), Ann Arbor, MI, 237-62.

GORRINI, M.E. 2006
Healing heroes in Thessaly: Chiron the Centaur. *Αρχαιολογικό Έργο Θεσσαλίας και Στερεάς Ελλάδας*, Πρακτικά επιστημονικής συνάντησης Βόλος 27.2-2.3.2003, Vol. I, Volos, 283-95.

GOSDEN, C. & L. HEAD 1994
'Landscape – A Usefully Ambiguous Concept', *Archaeology of Oceania* 29, 113-6.

GOSDEN, C. 1994
Social Being and Time. London.

GOULD, R.A. (ED.) 1978
Explorations in ethnoarchaeology. Albuquerque.

GOULD, R.A. 1980
Living archaeology. Cambridge.

GRAFIOU-NIDA, G. 1971
'Σπήλαιον Μεγαλομμάτας Πηλίου', *Bulletin of the Greek Speleological Society* 11(3-4), 83.

GRIGSON, C. 1982
'Sex and age determination of some bones and teeth of domestic cattle: a review of the literature', in *Ageing and Sexing Animal* bones *from Archaeological Sites* (BAR British Series 109)*, B. Wilson, C. Grigson & S. Payne (eds.), Oxford, 7-23.

GROVER, P.S. & L. LORTON 1985
'The incidence of unerupted permanent teeth and related clinical cases', *Oral Surgery, Oral Medicine, Oral Pathology* 59, 420-5.

GUNN, J. 2004
Encyclopedia of Caves and Karst Science, New York.

HAAS, J.N., S. KARG & P. RASMUSSEN 1998
'Beech leaves and twigs used as winter fodder: examples from Historic and Prehistoric times', *Environmental Archaeology* 1, 81-6.

HADJIGEORGIOU, I. & G. ZERVAS 2009
'Evaluation of production systems in protected areas: Case studies on the Greek Natura 2000 network', *Options Mediterraneennes Series* A 91, 101-1.

HALSTEAD, P. 1985
'A study of mandibular teeth from Romano-British contexts at Maxey', in *Archaeology and Environment in the Lower Welland Valley, 1. East Anglian Archaeology 27,* F. Pryor & C. French (eds.), 219-24.

HALSTEAD, P. 1996A
'The development of agriculture and pastoralism in Greece: when, how, who, what?', in *The origins and spread of agriculture and pastoralism in Eurasia*, D.R. Harris (ed.), London, 296-309.

HALSTEAD, P. 1996B
'Skoteini, Tharrounia: The Cave, the settlement and the cemetery (by Adamantios Sampson)', *American Journal of Archaeology* 100, 179-80.

HALSTEAD, P., 1998
'Ask the Fellows who Lop the Hay : Leaf-Fodder in the Mountains of Northwest Greece', *Rural history* 9(2), 211-34.

HALSTEAD, P. & J. TIERNEY 1998
'Leafy hay: an ethnoarchaeological study in NW Greece', *Environmental Archaeoology* 1, 71-80.

HALSTEAD, P., P. COLLINS & V. ISAAKIDOU 2002
'Sorting sheep from the goats: morphological distinctions between the mandibles and mandibular teeth of adult Ovis and Capra', *Journal of Archaeological Science* 29, 545-3.

HAMILAKIS, G. & A. ANAGNOSTOPOULOS 2009
'Introduction. Public Archaeology, Special Volume', *Archaeological Ethnographies* 8(2-3), 65-87.

HAMILAKIS, Y. & A. ANAGNOSTOPOULOS 2009
'What is Archaeological Ethnography?' *Public Archaeology* 8(2), 65-87.

HAMILAKIS, Y., A. ANAGNOSTOPOULOS & F. IFANTIDIS 2009
'Postcards from the Edge of Time: Archaeology, Photography, Archaeological Ethnography (A Photo-Essay)', *Public Archaeology* 2, 283-309.

HARATSIS, N. 2003
A Hiker's Guide to Mount Pelion. Volos.

HART, L.K. 1992
Time, Religion, and Social Experience in Rural Greece. Lanham, MD.

HARTWEG, R. 1945
'Remarques sur la denture et statistiques sur la caries en France aux époques préhistorique et proto-historique', *Bulletins et Mémoires de la Société d' Anthropologie de Paris* 6, 71-113.

HAYES, J.W. 1972
Late Roman Pottery. London.

HAYES, J.W. 1985
'Sigillate orientali', in *Enciclopedia dell'Arte Antica Classica e Orientale II. Ceramica Fine Romana nel Bacino Medieteraneo (Tardo Ellenismo e Primo Impero)*, Roma, 1-96.

HAYES, J.W. 1991
'Fine wares in the Hellenistic world', in *Looking at Greek Vases*, T. Rasmussen & N. Spivey (eds.), Cambridge, 183-202.

HAYWARD, P.J., G.D. WIGHAM & N. YONOW 1995
'Molluscs (Phylum Mollusca)', in *Handbook of the marine fauna of North-West Europe*, P.J. Hayward & J.S. Ryland (eds.), Oxford, 484-628.

HELLENIC NATIONAL METEOROLOGICAL SERVICE
http://www.hnms.gr [accessed 05.09.2008]

HELLY, BR. 2006
'Un nom antique pour Goritsa?', *Πρακτικά 1ου Αρχαιολογικού Έργου Θεσσαλίας και Στερεάς Ελλάδας, Πανεπιστήμιο Θεσσαλίας, Βόλος, 27.2-2.3.2003, Βόλος 2006*, 145-69.

HEURTLEY, W.A. & T.C. SCEAT 1930-31
'The tholos tombs of Marmariane', *Annual of the British School at Athens* 31, 1-55.

HEYDEN, D. & P. GENDROP 1975
Pre-Columbian architecture of Mesoamerica (History of world architecture). New York.

HILLSON, S. 1996
Dental Anthropology. Cambridge.

HOLLOWELL, J. 2006
'Moral arguments on subsistence digging', in C. G. *The Ethics of Archaeology – Philosophical Perspectives on Archaeological Practice*, C. Scarre & G. Scarre (eds.), Cambridge, 69-93.

HOLTORF C. 2006
'Studying archaeological fieldwork in the field: Views from Monte Polizzo', in *Ethnographies of Archaeological Practice: Cultural Encounters, Material Transformations*, M. Edgeworth (ed.), Lanham, 81-94.

HOURMOUIADIS, G., P. ASIMAKOPOULOU-AT-ZAKA & K.A. MAKRIS 1982
The story of a civilization – Magnesia. Athens.

HOWLAND, R.H. 1958
'Greek lamps and their survivals'. *The Athenian Agora*, Vol. IV. Princeton.

ΨΑΧΝΟΥΝ ΘΗΣΑΥΡΟ ΣΤΗΝ ΚΕΡΑΣΙΑ! ΧΕΚΙΝΗΣΕ Η ΕΚΣΚΑΦΗ ΜΕ ΜΠΟΥΛΝΤΟΖΑ. NOVEMBER 2, 2010
http://www.taxydromos.gr. [accessed 20.11.2010]

I.G.M.E. 1978
Geological Map of Greece 1: 50.000, Ayia-Panayia Ayias Sheet.

I.G.M.E. 1978
Geological Map of Greece 1: 50.000, Volos Sheet.

I.G.M.E. 1978
Geological Map of Greece 1: 50.000, Zagora-Syki Sheet.

IATRIDES, J., (ED.) 1981
Greece in the 1940s: A Nation in Crisis. London.

INGOLD, T. 1993
'The Temporality of the Landscape', *World Archaeology* 25(2), 152-74.

INGOLD, T. 2000
The Perception of the Environment: Essays in Livelihood, Dwelling, and Skill. London.

INTZESILOGLOU, B. 2000
'Η Θεσσαλία κατά την Ελληνιστική εποχή. Το ιστορικό πλαίσιο', in *Ελληνιστική Κεραμεική από τη Θεσσαλία*, Ε. Kypraiou (ed.), Athens, 11-5.

IOANNOU, I. 1964
'Το σπήλαιο "Κωστας"', *Bulletin of the Greek Speleological Society* 7(8), 217-20.

ISLAMOGLOU-INAN, H. 1994
State and Peasant in the Ottoman Empire: Agrarian Rower relations and Regional Economic Development in Ottoman Anatolia during the 16th century. New York.

ISPIKOUDIS, I. & D. CHOUVARDAS 2005
'Livestock, land use and landscape', in *Animal production and natural resources utilisation in the Mediterranean mountain areas*, Georgoudis A, Rosati A, Mosconi C. (eds.), 151-7, The Netherlands: EAAP Publication No. 115.

JACOBSHAGEN, V., S. DURR, F. KOCKEL, K.O. KOPP & G. KOWALCZYK 1978
'Structure and Geodynamic Evolution of the Aegean Region', in *Alps, Apennines, Hellenides: Geodynamic investigation along geotraverses by an International group of geoscientists* (Scientific report – Inter-Union Commission on Geodynamics; no. 38), H. Closs, D.H.Roeder & K. Schmidt (eds.), Stuttgart, 537-564.

JOHNSEN, D.C. 1977
'Prevalence of delayed emergence of permanent teeth as a result of local factors', *The Journal of the American Dental Association* 94(1), 100-6.

JONCHERAY, J.P. 1976
Essai de Classification des Amphores: Découvertes lors de Fouilles Sous-Marines. Fréjus: Cahier d'archéologie sub-aquatique. *Journal of European Studies* 28(3), 217-29.

KAHL, T. 2004
'Minorities in Greece – historical issues and new perspectives', *Jahrbücher für Geschichte und Kultur Südosteuropas* 5 (2003), 205-19.

KALLINTZI, K., & M. CHRYSSAPHI 2010
'Κεραμική της Ύστερης Αρχαιότητας από τα Άβδηρα', in *Κεραμική της Ύστερης Αρχαιότητας από τον Ελλαδικό Χώρο (3ος - 7ος αιώνας μ.Χ.)*, Πρακτικά επιστημονικής συνάντησης, Θεσσαλονίκη 12-16 Νοεμβρίου 2006, Τόμος Β, D. Papanikola-Mpakirtzi & N. Kousoulakou (eds.), Thessaloniki, 386-401.

KAMIZIS, D., A. STROULIA & K.D. VITELLI 2010
'From Franchthi Cave to Kilada: reflections on a long and winding road', in *Archaeology in situ : sites, archaeology, and communities in Greece*, A. Stroulia & S. Sutton (eds.), Lanham, MD, 397-435.

KATSANEVAKIS, S. 2007
'Growth and mortality rates of the fan mussel Pinna nobilis in Lake Vouliagmeni (Korinthiakos Gulf, Greece): a generalized additive modeling approach', *Marine Biology* 152, 1319-31.

KEFALLINIADI, N. 1961
Τα σπήλαια της Νάξου και οι θρύλοι των. Athens.

KEMPE, D.R.C. 1988
Living Underground: A History of Cave and Cliff Dwelling. London.

KENNA, M. 1976
'Houses, Fields and Graves: Property and Ritual Obligation on a Greek Island', *Ethnology* 15, 21-34.

KERNEY, M.P. & R.D. CAMERON 1979
A field guide to the land snails of Britain and North-west Europe. London.

KILIAS, A., G. NASTOS, G. FALALAKIS & D. MOUNTRAKIS 1995
'Tertiary extension and exhumation of the HP/LT Makrynitsa metamorphic core complex in the Pelion Mountain (Eastern Thessaly)', *Annales Géologiques des pays Helléniques 1e série* 38, fasc. A (1998), 163-85.

KITROEFF, A. 1983
'The Greeks in Egypt, 1919-1937: Ethnicity and Class', *Journal of the Greek Diaspora* 10(3), 5-15.

KNOBLAUCH, H. 2005
Focused Ethnography, Forum Qualitative Sozialforschung / Forum: Qualitative Social Research [Online Journal], 6, Art 44.

KOFOS, E. 1977.
Ο Ελληνισμός στην περίοδο 1869-1881. *Ιστορία του Ελληνικού Έθνους.* Τόμος ΙΓ', Athens, 339-40.

KOLIOPOULOS, J.S. 1981
'Shepherds, Brigands, and Irregulars in Nineteenth Century Greece', *Journal of the Hellenic Diaspora* 8, 41-53.

KONSTANTARAS-STATHARAS, D. 1994
Το Χρονικό της Νέας Ιωνίας, 1924-1994. Nea Ionia.

KONSTANTINIDI, A. 1936
Οι Πηλιορείται εν Αιγύπτω, τόμος 1ος, εκδ. συλλόγου Ζαγοράς, Αλεξάνδρεια, Αύγουστος 1936, 18.

KORRE-ZOGRAPHOU, K. 2000
Τα κεραμεικά του Τσανάκ Καλέ 1670-1922, Athens.

KORRE-ZOGRAPHOU, K. 2001
Λεύκωμα. Κεραμεικόν εμπόριο... Athens.

KOUKOULI -CHRISSANTHAKI, C. 1996
'Macedonia – Thrace', in *Neolithic Culture in Greece*, G.A. Papathanassopoulos (ed.), Athens, 112-6.

KYPARISSI-APOSTOLIKA, N. (ED.) 2000.
Theopetra Cave: 12 years of excavation and research, 1987-1998. Proceedings of the international conference, Trikala, 6-7/11/1998. Institute for Aegean Prehistory, Athens.

LAGIA, A., E. PETROUTSA & S. MANOLIS 2007
'Health and diet during the Middle Bronze Age in the Peloponnese: the site of Kouphovouno', in *Cooking up the past: Food and culinary practices in the Neolithic and Bronze Age Aegean*, C. Mee & J. Renard (eds.), Oxford, 313-28.

LAMB, H.H. 1977
Climate: Present, Past and Future. Volume 2. Climatic History and the Future. London.

LANGDON, M.K. 2000
'Mountains in Greek religion', *The Classical World* 93(5), 461-70.

LAURENT, P.E. 1830
An introduction to the study of ancient geography. Oxford & London.

LAZAROS, A. 1977
Η Θεσσαλία στην Αντίσταση. Athens.

LEAKE, W.M. 1835
Travels in Northern Greece IV. London.

LESSA, A. & N. GUIDON 2002
'Osteobiographic analysis of Skeleton I, Sítio Toca dos Coqueiros, Serra de Capivara National Park, Brazil 11,060 BP: First results', *American Journal of Physical Anthropology* 118, 99-110.

LEWIS, M.E., C.A. ROBERTS & K. MANCHESTER 1995
'Comparative study of the prevalence of maxillary sinusitis in Later Medieval urban and rural populations in Northern England', *American Journal of Physical Anthropology* 98, 497-505.

LIAPI, K. 2006
Το δημοτικό τραγούδι στη Μαγνησία I. Volos.

LOUPASAKIS, C. & N. NIKOLAOU 2007
Safety assessment and protection of monuments located in caves – a case study from the city of Volos, Greece. *Bulletin of the Geological Society of Greece* Vol. XXXX. Proceedings of the 11th International Congress, Athens, May 2007.

LOVELL, N.C. 1997
'Trauma analysis in paleopathology', *Yearbook of Physical Anthropology* 40, 139-70.

LUBELL, D. 2004
'Are land snails a signature for the Mesolithic – Neolithic transition', *Documenta praehistorica* 31, 1-24.

LUKACS, J.R. & R.F. PASTOR 1988
'Activity-induced patterns of dental abrasion in prehistoric Pakistan: Evidence from Mehrgarh and Harappa', *American Journal of Physical Anthropology* 76, 377-98.

MACKRIDGE, P. 2009.
Language and National Identity in Greece, 1766-1976. Oxford.

MAGNITOS, N.I. 1860
Περιήγησις. Η τοπογραφία της Θεσσαλίας και Θετταλικής Μαγνησίας. Athens.

MAIS, K., R. SEEMAN & N. SIMEONIDIS 1978
'Βραχογραφίες σε σπήλαια της περιοχής Αλιστράτης Σερρών', *Δελτ. Ελλ. Σπηλαιολ. Εταιρίας* [Bulletin of the Greek Speleological Society] 15(1), 71-7.

MAKRIS, K.A. 1982
'Post-Byzantine and Modern Magnesia', in *The Story of a Civilization – Magnesia*, G.C. Hourmouziadis, P. Asimakopoulou-Atzaka & K.A. Makris (eds.), Athens, 177-275.

MALAKASIOTI, Z. 2004
'Ελληνιστικά λυχνάρια της Άλου', in *ΣΤ΄ Επιστημονική Συνάντηση για την Ελληνιστική Κεραμική*, D. Zaphiropoulou (ed.), Athens, 89-109.

MALL, G., M. HUBIG, A. BÜTTNER, J. KUZNIK, R. PENNING & M. GRAW 2001
'Sex determination and estimation of stature from the longbones of the arm', *Forensic Science International* 117, 23-30.

MARAN, J. 2007
'Emulation of Aeginetan pottery in the Middle Bronze Age of coastal Thessaly: regional context and social meaning', in *Middle Helladic Pottery and Synchronisms, Proceedings of the International Workshop held at Salzburg, October 31th –November 2nd, 2004,* F. Felten, W. Gauss & R. Smetana (eds.), Vienna, 167-82.

MARCUS, G. 1998
Ethnography through Thick and Thin. Princeton, NJ.

MARGARITIS, G. 2000
Ιστορία του Ελληνικού εμφυλίου πολέμου1946-1949, Vol. 1. Athens.

MARZETTI, P. 1996
Combat helmets from around the world. Parma.

MARZETTI, P. 2003
Elmetti: Helmets, Parma.

MATOS, V. 2009
'Broken ribs: Paleopathological analysis of costal fractures in the human identified skeletal collection from the Museu Bocage, Lisbon, Portugal (Late nineteenth to

Middle twentieth centuries)', *American Journal of Physical Anthropology* 140, 25-38.

MAZOWER, M. 1993
Inside Hitler's Greece. The experience of occupation, 1941-44. New Haven.

MAZOWER, M. 2004
Salonica – City of Ghosts. London.

MEE, C., H.A. FORBES & M.P. ATHERTON 1997
A rough and rocky place : the landscape and settlement history of the Methana Peninsula, Greece : results of the Methana Survey Project, sponsored by the British School at Athens and the University of Liverpool. Liverpool.

MEGAS, G. 1958
Greek Calendar Customs. Athens.

MEHRA, P. & H. MURAD 2004
'Maxillary sinus disease of odontogenic origin', *Otolaryngologic Clinics of North America* 37, 347-64.

MÉZIÈRES, A.-J.-F. 1854
Mémoires sur le Pélion et l'Ossa (Archives des Missions Scientifiques et Litteraires 3), Paris.

MERRETT, D.C. & S. PFEIFFER 2000
'Maxillary sinusitis as an indicator of respiratory health in past populations', *American Journal of Physical Anthropology* 111, 301-18.

MESKELL, L. 2005
'Archaeological ethnography: Conversations around Kruger Park', *Archaeologies* 1(1), 81-100.

MESKELL, L. 2007
'Falling Walls and Mending Fences: Archaeological Ethnography in the Limpopo', *Journal of Southern African Studies* 33(2), 383-400.

MICHAILIDIS, I.D. 2006
'The formation of Greek citizenship (19th century)', in *Citizenship in historical perspective*, S.G. Ellis, G. Halfdanarson & A.K. Isaacs (eds.), Pisa, 155-62.

MIENTJES, A.C. 2004
'Modern pastoral landscapes on the island of Sardinia (Italy). Recent pastoral practices in local versus macro-economic and macro-political contexts', *Archaeological Dialogues* 10.2, 161-90.

MORGAN, C. 2003
Early Greek states beyond the polis. London & New York.

MOUNTJOY, P.A. 1986
Mycenaean Decorated Pottery: A Guide to Identification (Studies in Mediterranean Archaeology 73). Göteborg.

MOUNTRAKIS D.M. 1985
Geology of Greece. Thessaloniki.

MOUNTRAKIS, D.M. 1986
'The Pelagonian zone in Greece. A polyphase-deformed fragment of the Cimmerian Continent and its role in the geotectonic evolution of the Eastern Mediterranean', *Journal of Geology* 94, 335-47.

MURRAY, P. & C. CHANG 1981
'An ethnoarchaeological study of a contemporary herder's site', *Journal of Field Archaeology* 8, 372-81.

MURRAY, P. & P.N. KARDULIAS 1986
'A Modern-Site Survey in the Southern Argolid, Greece', *Journal of Field Archaeology* 13(1), 21-41.

NANDRIS, J.G. 1999
'Etnoarchaeology and Latinity in the mountains of the southern Velebit', in *Transhumant pastoralism in southern Europe – recent perspectives from archaeology, history and ethnology*, L. Bartosiewicz & H.J. Greenfield (eds.), Budapest, 111-31.

NANOS, G.D. & G. DIANELOS 2011
'Energy analysis for apple growing in Pelion, Central Greece', *The Annals of "Valahia" University of Targoviste* 11, 4-8.

NANOU-SKOTINIOTI, A. 1988
'Τα τοπωνύμια της Μακρινίτσας', *Thessaliko Imerologio* 13, 121-44.

NATIONAL STATISTICAL SERVICE OF GREECE (NSSG) 1884
Population Census 1881. Athens.

NATIONAL STATISTICAL SERVICE OF GREECE (NSSG) 1890
Population Census 15-16th April 1889. Athens.

NATIONAL STATISTICAL SERVICE OF GREECE (NSSG) 1897
Population Census 5-6th October 1896. Athens.

NATIONAL STATISTICAL SERVICE OF GREECE (NSSG) 1909
Population Census 27th October 1907. Athens.

NATIONAL STATISTICAL SERVICE OF GREECE (NSSG) 1935
Population Census 15-16th May 1928. Athens.

NATIONAL STATISTICAL SERVICE OF GREECE (NSSG) 1950
Population Census 16th October 1940. Athens.

NATIONAL STATISTICAL SERVICE OF GREECE (NSSG) 1955
Population Census 7th April 1951. Athens.

NATIONAL STATISTICAL SERVICE OF GREECE (NSSG) 1962
Population Census 19th March 1961. Athens.

NATIONAL STATISTICAL SERVICE OF GREECE (NSSG) 1975
Population Census 14th March 1971. Athens.

NATIONAL STATISTICAL SERVICE OF GREECE (NSSG) 1985
Population Census 5th April 1981. Athens.

NATIONAL STATISTICAL SERVICE OF GREECE (NSSG) 1994
Population Census 17th March 1991. Athens.

NICOD, J., M. JULIAN & E. ANTHONY 1996
'A historical review of man-karst relationships: Miscellaneous uses of karst and their impacts', *Rivista Geografica Italiana* 103, 289-338.

NIKOLAOU, E. 2004
'Οι πήλινοι λύχνοι από το βόρειο νεκροταφείο της Αρχαίας Δημητριάδος', in *ΣΤ΄ Επιστημονική Συνάντηση για την Ελληνιστική Κεραμική*, D. Zaphiropoulou (ed.), Athens, 47-60.

NIXON, L., 2006
Making a landscape sacred : outlying churches and icon stands in Sphakia, southwestern Crete. Oxford.

NIXON, L. & S. PRICE 2001
'The Diachronic Analysis of Pastoralism through Comparative Variables', *Annual of the British School at Athens* 96, 395-424.

NTINA, A. 2010
'Πρωτοχριστιανική κεραμική από τις Φθιώτιδες Θήβες', in *Κεραμική της Ύστερης Αρχαιότητας από τον Ελλαδικό Χώρο (3ος - 7ος αιώνας μ.Χ.), Πρακτικά επιστημονικής συνάντησης, Θεσσαλονίκη 12-16 Νοεμβρίου 2006,* Τόμος Β, D. Papanikola-Mpakirtzi & N. Kousoulakou (eds.), Thessaloniki, 563-79.

NÚÑEZ JIMÍNEZ, A. 1987
Geografía y Espeleología en Revolucion. Havana.

OLIVER, J. & T. NEAL (EDS.) 2010
Wild signs: Graffiti in Archaeology and History (Studies in Contemporary and Historical Archaeology 6), BAR International Series 2074.

ORTNER, D.J. 2003
Identification of pathological conditions in human skeletal remains. San Diego.

PALMER, A.N. 2007
Cave geology. Dayton, Ohio.

PANHUYSEN, R.G.A.M., V. COENED & T.D. BRUINTJES 1997
'Chronic maxillary sinusitis in Medieval Maastricht, The Netherlands', *International Journal of Osteoarchaeology* 7, 610-4.

PAPACONSTANTINOU, M.F. 2000
'Μακεδονικού τύπου αμφορείς από το νοτιοανατολικό νεκροταφείο της Λαμίας', in *Ελληνιστική Κεραμεική από τη Θεσσαλία*, E. Kypraiou (ed.), Athens, 193-203.

PAPADEMETRIOU, E. 2005
Modern Glazed Pottery of Cyprus. Lapithos Ware. Nicosia.

PAPADOPOULOS, G. 2010
'Δύο χρόνια σε σπηλιά για να γλιτώσει το Άουσβιτς', *Ta Nea On-line* 27.02.2010. http://www.tanea.gr [accessed 03.05.2010]

PAPADOPOULOU, B. & K. TSOURIS 1993
'Late Byzantine ceramics from Arta: Some examples', in *La ceramica nel mondo bizantino tra XI e XV secolo e i suoi rapporti con l'Italia*, S. Gelichi (ed.), Florence, 241-59.

PAPANIKOLA-BAKIRTZI, D. & N. ZEKOS (EDS.) 2010
Late Byzantine Glazed Pottery from Thrace. Reading the Archaeological Finds. Thessaloniki.

PAPATHANASIOU, A. 1998
Η Μαγνησία και το Πήλιο στον Ύστερο Μεσαίωνα (1204-1423). Volos.

PAPATHANASSOPOULOS, G. 1996
'Habitation in caves', in *Neolithic Culture in Greece*, G.A. Papathanassopoulos (ed.), Athens, 39-40.

PAPATHOMA, E. 1999
'Κατάλογος αντικειμένων', in Με αφορμή μία στάμνα, Athens, 93-130.

PAPATHOMA, E. 2001
Λεύκωμα. Κερμικών εμπόριο. Study Center of Modern Ceramics (Κέντρο Μελέτης Νεώτερης Κεραμεικής).

PAPAZACHOS, B. & K. PAPAZACHOU 1989
Οι σεισμοί της Ελλάδας. Thessaloniki.

PAPAZACHOS, B.C., T.M. PANAGIOTOPOULOS, D.M. TSAPANOS, D. MOUNTRAKIS & G. DIMOPOULOS 1983
'A study of the 1980 summer seismic sequence in Magnesia region of Central Greece', *Geophysical Journal of the Royal Astronomical Society* 75(1), 155-68.

PATRONIS, V. 2009
'Land ownership in Thessaly, from the Annexation to Kilerer (1881-1910)', in *Thessaly – History and Culture of Thessaly*, 467-9.

PAYNE, S. 1972
'Partial recovery and sample bias: the results of some sieving experiments', in *Papers in Economic Prehistory*, E. S. Higgs (ed.), London, 49-64.

PAYNE, S. 1973
'Kill-off patterns in sheep and goats: the mandibles from Aswan Kale', *Anatolian Studies* 23, 281-303.

PAYNE, S. 1985
'Morphological distinctions between the mandibular cheek teeth of sheep and goats', *Journal of Archaeological Science* 14, 609-14.

PAYNE, S. 1985
'Zooarchaeology in Greece: a reader's guide', in *Contributions to Aegean Archaeology: Studies in Honor of William A. McDonald*, C. Wilkie & W.D.E. Coulson (eds.), Minneapolis, 211-44.

PETROPOULOS, M. 1999
Τα Εργαστήρια των Ρωμαϊκών Λυχναριών της Πάτρας και το Λυχνομαντείο. Athens.

PIZZI, K. 1998
'Silentes Loquimur: Foibe and border anxiety in postwar literature from Trieste', *Journal of European Studies* 28, 217-229.

PLATAKIS, E.K. 1977
Ονόματα σπηλαίων τής Κρήτης σχετικά μέ θρύλους καί παραδόσεις. Iraklion.

POUQUEVILLE, F.C.L. 1820
Travels in Epirus, Albania, Macedonia and Thessaly. London.

POWELL, M.L. 1985
'The analysis of dental wear and caries for dietary reconstruction', in *The analysis of prehistoric diets*, R.I. Gilbert & J.H. Mielke (eds.), Orlando, 307-38.

PSILAKIS, N. 1994
Μοναστήρια και ερημητήρια της Κρήτης, Vol. 2. Iraklion.

PUECH, P.-F. & F. CIANFARANI 1988
'Reply to Formicola', *Current Anthropology* 29, 665-8.

RAGHOEBAR, G.M., G. BOERING, A. VISSING & B. STEGENGA 1991
'Eruption disturbances of permanent molars: a review', *Journal of Oral Pathology and Medicine* 20, 159-66.

ROBERTS, C.A. 2007
'A bioarchaeological study of maxillary sinusitis', *American Journal of Physical Anthropology* 133, 792-807.

ROBIN, C. & N.A. ROTHSCHILD 2002
'Archaeological ethnographies. Social dynamics of outdoor space', *Journal of Social Archaeology* 2(2), 159-72.

ROBINSON, H.S. 1959
Pottery of the Roman period: chronology (The Athenian Agora Vol. V). Princeton.

ROTROFF, S. 1997
Hellenistic pottery: Athenian and imported wheelmade table ware and related material (The Athenian Agora Vol. XXIX, parts 1-2). Princeton.

ROTROFF, S. 2006
Hellenistic pottery: the plain wares (The Athenian Agora Vol. XXXIII). Princeton.

RUTTER, J. 1975
'Ceramic evidence for northern intruders in southern Greece at the beginning of Late Helladic IIIC', *American Journal of Archaeology* 79, 17-32.

SAMPSON, A. 1992
'Late Neolithic remains at Tharrounia, Euboea: a model for the seasonal use of settlements and caves', *Annual of the British School at Athens* 87, 61-101.

SARAFIS, S.G. 1980
ELAS: Greek Resistance Army. London.

SCHEPARTZ, L.A., S. MILLER-ANTONIO & J.M.A. MURPHY 2009,
'Differential health among the Mycenaeans of Messenia: Status, sex, and dental health at Pylos', in *New directions in the skeletal biology of Greece*, L.A. Schepartz, S.C. Fox & C. Bourbou (eds.), Princeton, 155-76.

SCHEUER, L. & S. BLACK 2000
Developmental juvenile osteology. San Diego.

SCHMID, E. 1972
Atlas of Animal Bones. Amsterdam.

SCHULZ, P.D. 1977
'Task activity and anterior tooth grooving in prehistoric California Indians', *American Journal of Physical Anthropology* 46, 87-92.

FLORE, L. 2004
'Sea Silk', www.sardolog.com/bisso/english/quoi.htm [accessed 26.10.2010]

SEIZANIS, M. 1879.
Η πολιτική της Ελλάδος και η επανάστασις του 1878 εν Μακεδονία, Ήπειρω και Θεσσαλία. Athens.

SHACKLETON, N.J. 1968
'The Mollusca, the Crustacea, the Echinodermata (Appendix IX)', in *Excavations at Saliagos near Antiparos*, J.D. Evans & C. Renfrew (eds.), Annuals of the British School of Archaeology at Athens Supplement 5, London, 122-38.

SHACKLEY, M. 1981
Environmental archaeology, London.

SIFFRE, A. 1911
'Note sure une usure spéciale dels molaires du squelette de La Quina', *Bulletin de la Société Préhsitorique Française* 8, 741-3.

SILVER, I.A. 1969
'The ageing of domestic animals', in *Science in Archaeology: a survey of Proress and Research*, D. Brothwell & E. Higgs (eds.), New York, 283-302.

SIRMALI, M., H. TÜRÜT, S. TOPÇU, E. GÜLHAN, U. YAZICI, S. KAYA & I. TAŞTEPE 2003
'A comprehensive analysis of traumatic rib fractures: morbidity, mortality and management', *European Journal of Cardio-thoracic Surgery* 24, 133-8.

SIVIGNON, M. 1975
'La Thessalie, analyse geographique d'une province greque'. Institute des Etudes Phondaniennes des Universites de Lyon, *Memoires et documents* 17, Lyon.

SIVIGNON, M. 2009
'Population differentiation in Thessaly during 1881-1914', in *Thessalia – History and Culture of Thessaly*, Thessaloniki, 459-65.

SKALIDAKIS, G. & C. GIOVANOPOULOS 2009
'Greece, partisan resistance', *The International Encyclopedia of Revolution and Protest*, N. Immanuel (ed.), Blackwell Reference Online. http://www.revolutionprotestencyclopedia.com/public/book?id=g9781405184649_yr2011_9781405184649 [accessed 08.01.2012]

SKOURA, TH. 1993
Χριστιανικά λατρευτικά ιερά. Οι υπόσκαφες και σπηλαιώδεις στον ελλαδικό χώρο. *Η ΚΑΘΗΜΕΡΙΝΗ* 08.08.1993, 14.

SKOUVARAS, V. 1959
Το παλιότερο Αρματολίκι του Πηλίου κι οι Αρβανίτες στη Θεσσαλομαγνησία (1750-1790). Volos.

SKOUVARAS, V. 1962
Meteora. Volos.

SLAVIN, R.G., S.L. SPECTOR & L. BERNSTEIN
2005
'The diagnosis and management of sinusitis: A practice parameter update', *The Journal of Allergy and Clinical Immunology* 116, S13-S47.

SMALL, D.B. 1990
'Handmade burnished ware and prehistoric Aegean economics: an argument for indigenous appearance', *Journal of Mediterranean Archaeology* 3(1), 3-25.

SNODGRASS, A. 1971
The Dark Age of Greece: An Archaeological Survey of the Eleventh to the Eighth Centuries BC. Edinburgh.

SPYROPOULOS, P. 1997
Χρονικό των σεισμών της Ελλάδας – από την αρχαιότητα μέχρι σήμερα. Athens/Ioannina.

STÄHLIN, F. 1924
Das hellenische Thessalien. Stuttgart.

STÄHLIN, F. 1965
Studies in Ancient Greek Topography I. Berkeley.

STÄHLIN, F. 1969
Studies in Ancient Greek Topography II. Berkeley.

STANDRING, K. (ED.) 2009
Prespa walking guide. Society for the Protection of Prespa. Aghios Germanos.

STEWART, C. 1991
Demons and the Devil: Moral Imagination in Modern Greek Culture. Princeton.

STIROS S., P. TRIANTAFILLIDES & A. CHASAPIS
2004
'Geodetic evidence for active uplift of the Olympus Mt, Greece', *Bulletin of the Geological Society of Greece* 36(4), Proceedings of the 10th International Congress, Thessaloniki, 1697-705.

SUTTON, S.B., 2000
Contingent countryside. Stanford, California.

SYNODIOU, TH. 1995
Μοναστήρια και Εκκλησίες. *Η ΚΑΘΗΜΕΡΙΝΗ* 30.07.1995, 15-8.

TE RIELE, G.-J.-M.-J. 1977
'Les inscriptions trouvées à Goritsa', *Αρχαιολογίκον Δελτίον* 27(1972), B' 2 Chroniques, 408-11.

TELLER, F.J. 1879
Geologische Beschreibung des südöstlichen Thessalien. Denkschriften (Österreichische Akademie der Wissen-schaften. Mathematisch-Naturwissenschaftliche Klasse, Bd. 40), Vienna.

THE DRY STONE WALLING ASSOCIATION 2004
Dry Stone Walling, Techniques and Traditions. Doncaster.

THEOCHARIS, D.P. 1966A
'Η παλαιολιθική τέχνη στο Πήλιο', *Thessalika* 5, 76-82.

THEOCHARIS, D.P. 1966B
'Χρονικά. Σπήλαια Πηλίου', *Αρχαιολογικόν Δελτίον* 21, 255.

THEOCHARIS, D.P. 1967A
'Χρονικά', *Αρχαιολογικόν Δελτίον* 22, 297-8.

THEOCHARIS, D.P. 1967B
Η Αυγή της Θεσσαλικής Προϊστορίας (Εκδοση της Φιλάρχαιου Εταιρείας Βόλου, Θεσσαλικά μελετήματα, αριθμ. 1.). Volos.

THEOCHARIS, D.P. 1968
'Χρονικά. Ανασκαφή σπηλαίου Σαρακηνού', *Αρχαιολογικόν Δελτίον* 23, 262-3.

THEOCHARIS, D.P. 1969
'Χρονικά. Δοκιμαστική ἀνασκαφή σπηλαίων Πηλίου', *Ἀρχαιολογικόν Δελτίον* 24, 222-3.

THOMAS, J. 1998
Time, Culture, and Identity: An Interpretive Archaeology. New York.

TILLEY, C. 1994
A Phenomenology of Landscape. Oxford.

TOMKINS, P. 2009
'Domesticity by default. Ritual, ritualisation and cave use in the Neolithic Aegean', *Oxford Journal of Archaeology* 28, 125-53.

TOMKINS, P. 2010
'Neolithic Antecedents', in *The Oxford handbook of the Aegean Bronze Age*, E. Cline (ed.), Oxford, 31-49.

TOVI, F., D. BENHARROCH, A. GATOT & Y. HERTZANU 1992
'Osteoblastic osteitis of the maxillary sinus', *Laryngoscope* 102, 427-30.

TRATMAN, E.K. 1956
'Human teeth and archaeology', *The Advancement of Science* 12, 419-23.

TRATSA, M. 2010
Το Πήλιο γίνεται… Τουρκοβούνια. Κοινωνία (ΤΟ ΒΗΜΑ).http://www.tovima.gr/society/article/?aid=373499 [accessed 20.12.2010]

TREATY OF KÜÇÜK KAYNARCA 2011
Encyclopædia Britannica Online. http://www.britannica.com/EBchecked/topic/324324/Treaty-of-Kucuk-Kaynarca [accessed 01.12.2011]

TRIANTAPHYLLOU, S. 2001
A bioarchaeological approach to prehistoric cemetery populations from Central and Western Greek Macedonia. Oxford.

TSIVILAKOS, M.G., S.K. MANOLIS, O. VIKATOU & M.J. PAPAGRIGORAKIS 2002
'Periodontal disease in the Mycenaean (1450-1150 BC) population of Aghia Triada, W. Peloponnese, Greece', *International Journal of Anthropology* 17, 91-100.

TSOTSOROS, S.N. 1986
Οικονομικοί και κοινωνικοί μηχανισμοί στον ορεινό χώρο Γορτυνία 1715-1828. Athens.

TURNER II, C.G. 1988
'Reply to Formicola', *Current Anthropology* 29, 664-5.

TURNER, G. 2006
'Bahamian Ship Graffiti', *The International Journal of Nautical Archaeology* 35(2), 253-73.

UBELAKER, D.H., T.W. PHENICE & W.M. BASS 1969
'Artificial interproximal grooving of the teeth in American Indians', *American Journal of Physical Anthropology* 30, 145-50.

UNGAR, P.S., F.E. GRINE, M.F. TEAFORD & A. PÉREZ-PÉREZ 2001
'A review of interproximal wear grooves on fossil hominin teeth with new evidence from Olduvai Gorge', *Archives of Oral Biology* 46, 285-92.

VAN BOESCHOTEN, R. 2006
'Broken bonds and divided memories: Wartime massacres reconsidered in a comparative perspective', *Oral History* 35(1), 1-12.

VARPIO, M. & B. WELLFELT 1988
'Disturbed eruption of the lower second molar: Clinical appearance, prevalence, and etiology', *ASDC Journal of Dentistry for Children* 55, 114-8.

VAXEVANOPOULOS, M. 2006
Tectonic conditions in the speleogenetic process of Melissotripa cave in Kefalovriso of Elassona (Central Greece). MSc dissertation, Aristotle University of Thessaloniki.

VERDELIS, N. 1958
Ο Πρωτογεωμετρικός ρυθμός της Θεσσαλίας. Athens.

VERGELY, P., 1984
Tectoniques des ophiolites dans les Hellenides Internes (defonnation, metamorphisme et phenomenes sedimentaires). Consequences sur l' evolution des regions Tethysiennes Occidentales. PhD dissertation, University Paris-Sud, Orsay.

VERGOPOULOS K. 1975 (IN GREEK)
The rural issue in Greece. Athens.

VERNICOS N., S. DASCALOPOULOS, G. PAVLO-GEORGATOS & D. PAPADOPOULOS 2003
'Ξερολιθικά τοπία και κατασκευές: μία κοινή κληρονομιά', in *Proceedings of the Scientific Conference on Educational programmes in Monuments and Museums – Educational Material for the Environmental Education, Mytilini 17-19/10/2003*, Mytilini. Unpublished conference paper.

VETH, P., M. SPRIGGS & S. O'CONNOR 2005
'Continuity in tropical cave use: examples from East Timor and the Aru Islands, Maluku', *Asian Perspectives* 44(1), 180-92.

VICKERS, M. 1997
The Albanians: a modern history. London.

VLACHAKIS, V. 2003-05
Ο αιγυπτιακός ελληνισμός στον 19ο και 20ο αιώνα. Η περίπτωση των Πηλιορειτών. http://extras.ha.uth.gr/g-m/ln2/paper_02.asp [accessed 10.8.2010]

VON DEN DRIESCH, A. 1976
A Guide to the Measurement of animal bones from Archaeological Sites, Peabody Museum Bulletin 1. Cambridge MA.

VOYIATZIS, TH.K. 2008
Όταν τους ιδώμεν όλους υπό το πέλμα μας... Όταν τους πατήσωμεν όλους τελειωτικώς, θα ησυχάσωμεν...', Volos.

VROOM, J. 1998
'Early Modern archaeology in Central Greece: The contrast of artifact-rich and sherdless sites', *Journal of Mediterranean Archaeology* 11(2), 131-64.

VROOM, J. 2003
After Antiquity. Ceramics and Society in the Aegean from the 7th to the 20th century a.C. A Case Study from Boeotia, Central Greece. Leiden.

VROOM, J. 2005
Byzantine to Modern Pottery in the Aegean. An Introduction and Field Guide. Utrecht.

VROOM, J. 2007
'Pottery finds from a 'cess-pit' at the southern wall in Durrës, central Albania', in *Çanak. Late Antique and Medieval Pottery and Tiles in Mediterranean Archaeological Contexts* (Byzas 7), B. Böhlendorf-Arslan, A.O. Uysal & J. Witte-Orr (eds.), Istanbul, 319-34.

WACE, A.J.B. 1906
'The topography of Pelion and Magnesia', *The Journal of Hellenic Studies* 26, 143-68.

WALLACE, J.A. 1974
'Approximal grooving of teeth', *American Journal of Physical Anthropology* 40, 385-90.

WACE, A.J.B. & M.S. THOMPSON 1914
The Nomads of the Balkans: An Account of Life and Customs among the Vlachs of Northern Pindus. London (reprinted New York, 1971).

WATSON, P.J. 1979
Archaeological ethnography in Western Iran. Tucson, Arizona.

WATSON P.J. 1995
'Archaeology, Anthropology and the Culture Concept', *American Anthropologist* 97(4), 683-94.

WEDDE, M. 1990
'The "Ring of Minos" and Beyond: Thoughts on Directional Determination on Aegean Bronze Age Ship Iconography', *Hydra 7* (Working Papers in Middle Bronze Age Studies), 1-24.

WEIDENREICH, F. 1937
'The dentition of *Sinanthropus pekinensis*: A comparative odontography of the hominids', *Paleontologia Sinica* 101, 1-121.

WELLS, C. 1977
'Disease of the maxillary sinus in antiquity', *Medical and Biological Illustration* 27, 173-8.

WHITELAW, T.M. 1991
'The ethnoarchaeology of recent rural settlement and land use', in *Landscape Archaeology as Long-Term History. Nothern Keos in the Cycladic Islands from earliest Settlement until modern Times*, J.F. Cherry, J.C. Davis & E. Mantzourani (eds.), Los Angeles, 403-54.

WILLEY, P. & J.L. HOFMAN 1994
'Interproximal grooves, toothaches, and purple coneflowers', in *Skeletal biology in the Great Plains*, D. W. Owsley & R.L. Jantz (eds.), Washington, 147-57.

WILLIAMS, M. 2000
'Dark ages and dark areas: global deforestation in the deep past', *Journal of Historical Geography* 26, 28-46.

WILSON, J.M. 2005
'Recreational caves', in *Encyclopedia of Caves*, D.C. Culver & W.B. White (eds.), Burlington, San Diego and London, 469-75.

ZACHOS, K. 1999
'Zas cave on Naxos and the role of caves in the Aegean Late Neolithic', in *Neolithic society in Greece*, P. Halstead (ed.), Sheffield Studies in Aegean Archaeology 2, Sheffield, 153-63.

ZACHOU, K., G. MARATOU, K. ZACHOS & G. MARATOS 1965
Metallogenic map of Greece. Athens.

ZANOV, G. 2000 (IN BULGARIAN)
Ceramics during the Bulgarian Renaissance. Sofia.

ZERVAS, G. 1998
'Quantifying and optimizing grazing regimes in Greek mountain systems', *Journal of Applied Ecology* 35, 983-6.

LIAPIS, K. (ED.) 2002
Οι τρεις Μηλιώτες διδάσκαλοι του γένους. Cultural Organization of the Municipality of Milies. Volos.

ΣΠΗΛΑΙΟ ΜΕ ΜΥΚΗΝΑΪΚΑ ΣΤΗΝ ΚΑΡΛΑ
http://www.taxydromos.gr/perrisotereseidhseis/ tabid/152/articleType/articleView/articleId/35191/--. aspx [accessed 15.04.2011]

Appendix 1

Foreword

As far as we are aware, there has previously been no systematic attempt to map caves on Pelion for ethnographic or archaeological purposes. It was, therefore, a priority to produce basic plans of the surveyed caves that could be easily deciphered by the interested reader.

The plans presented here form only a sample of the surveyed caves, which constitute the most informative of the sites regarding architectural elements and artefact distributions. Thirty-one plans from different caves were made in varying scales 1:50 (19 plans), 1:100 (11 plans), and 1:200 (1 plan). The digitized plans were produced by tracing over the scanned, hand-recorded cave plans in AutoCAD. All drawings were subsequently processed to be visually enhanced and readable to the eye of the non-specialized reader. The layers' structure is identical to each drawing image and fully accessible to any further editing.

We have attempted stylistic consistency among the plans of the caves and conventions are retained throughout. The plans share an accompanying key to abbreviations and symbols. A range of symbols makes it simple to affiliate finds on the plans with major artifact material groups. In many cases, an identification of the artifact is included on the plan together with a symbol. The key also allows the reader to distinguish between intact/partly ruined and drystone/rubble walls and it describes the position of wooden structural elements. Most cave plans include height measurements from cave floor to roof and height above ground.

Fig. 23 of Appendix 1 shows the layout of the Chiron's Cave (MIL-1). It is based on but redrawn from a plan by Petrocheilou in 1970. It has been included for comparison with Fig. 28, which is the site plan recorded by the Pelion Cave Project.

A 'floating' grid coordinate system was found adequate in the context of the survey. Survey grids at each site used the cave or rockshelter as a guide, and parallel measuring tapes were laid across the site in the most convenient direction. The survey grids were not tied to a recognized mapping coordinate system, but a GPS provided a location point for each site.

Digitization was undertaken by Theodosios Touisouzoglou-Tousizoglou (CADU Architects), and the expenses for this work have been covered by part of a grant from the Danish Research Council for Culture and Communication (Humanities).

❊ ❊ ❊

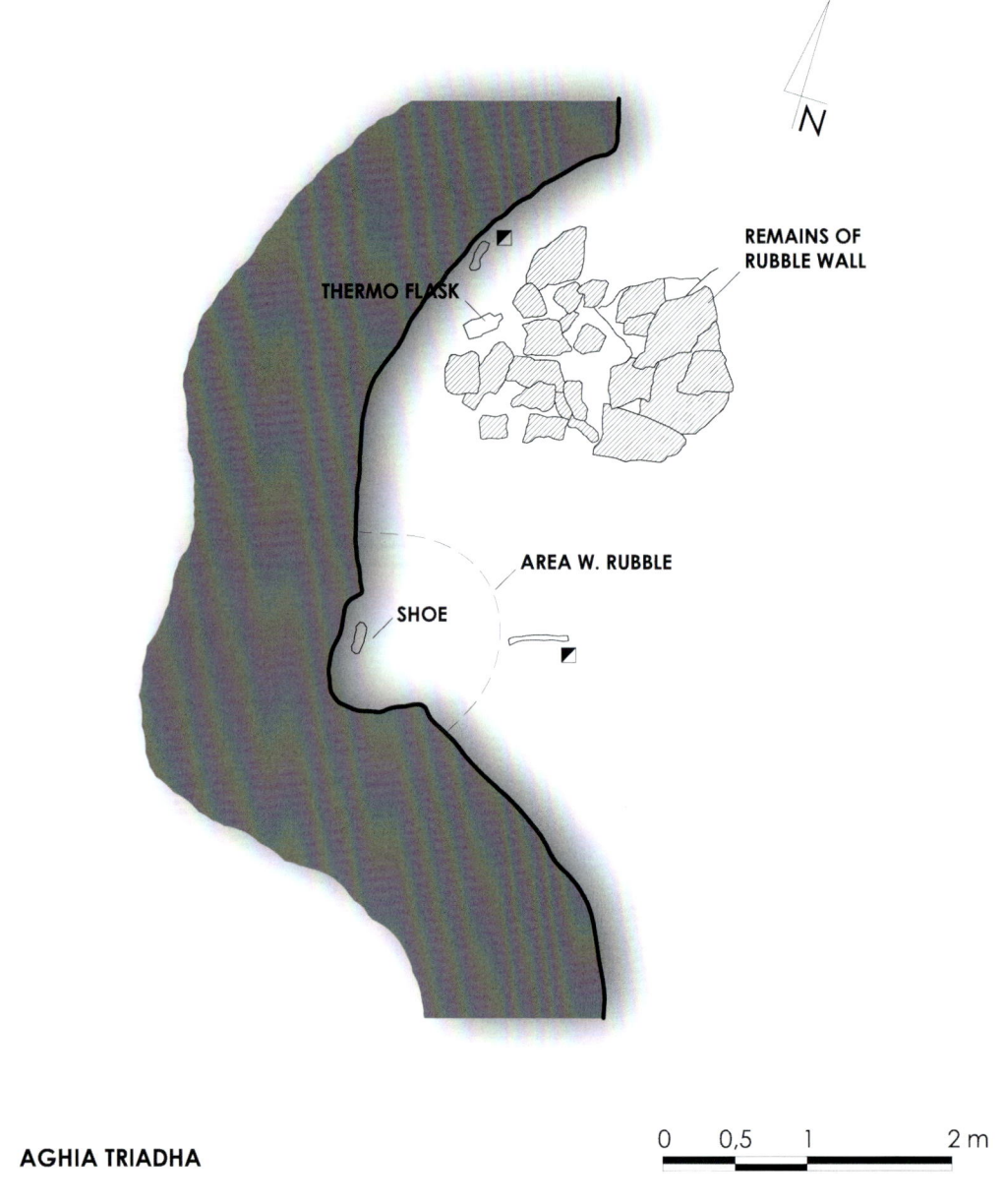

REMAINS OF
RUBBLE WALL

THERMO FLASK

AREA W. RUBBLE

SHOE

N

0   0,5   1   2 m

AGHIA TRIADHA

Fig. 1. Agia Triada I (MIL-19).

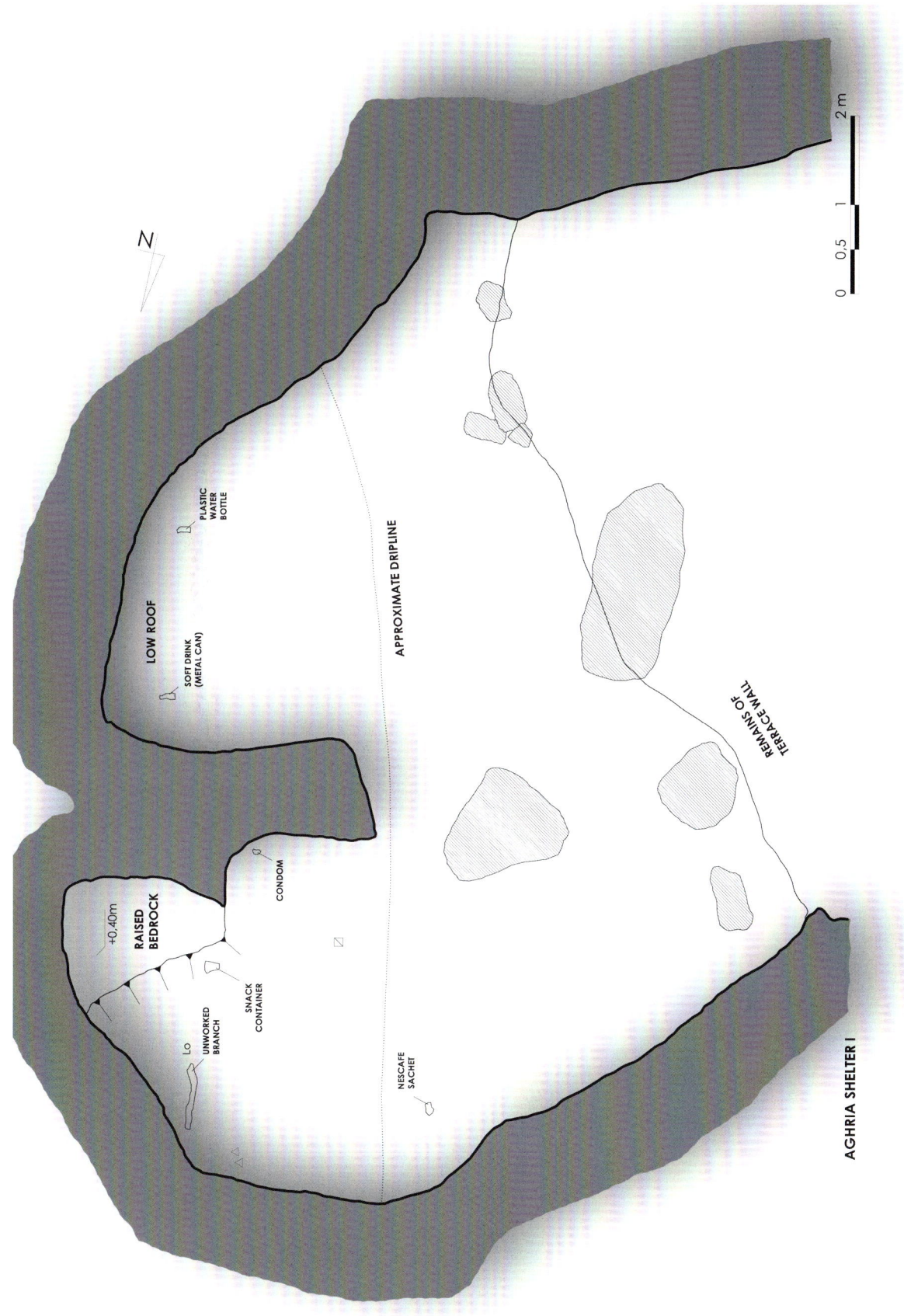

Fig. 2. Agria Rockshelter (AGR-3).

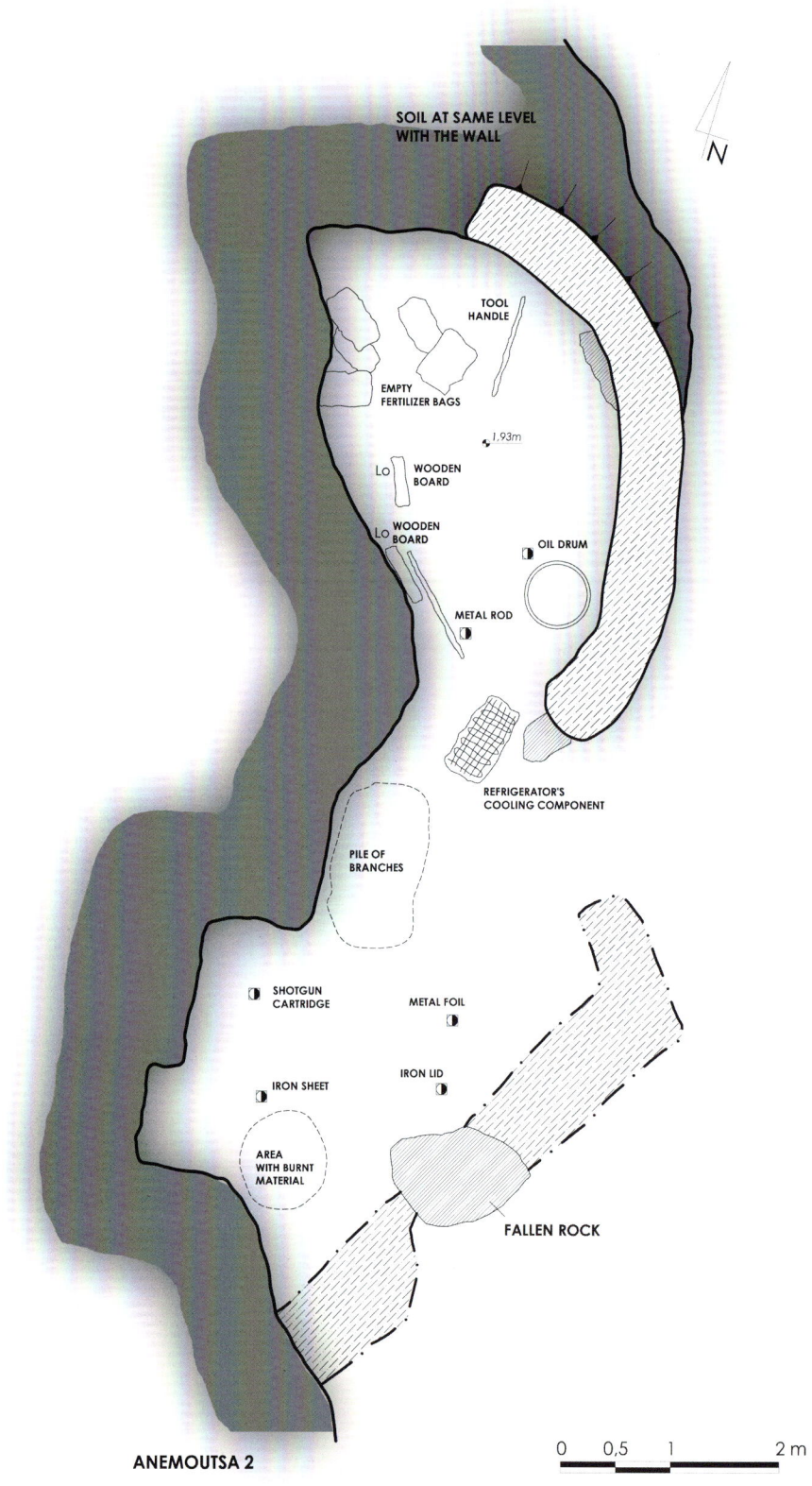

Fig. 3. Anemoutsa II (AGR-8).

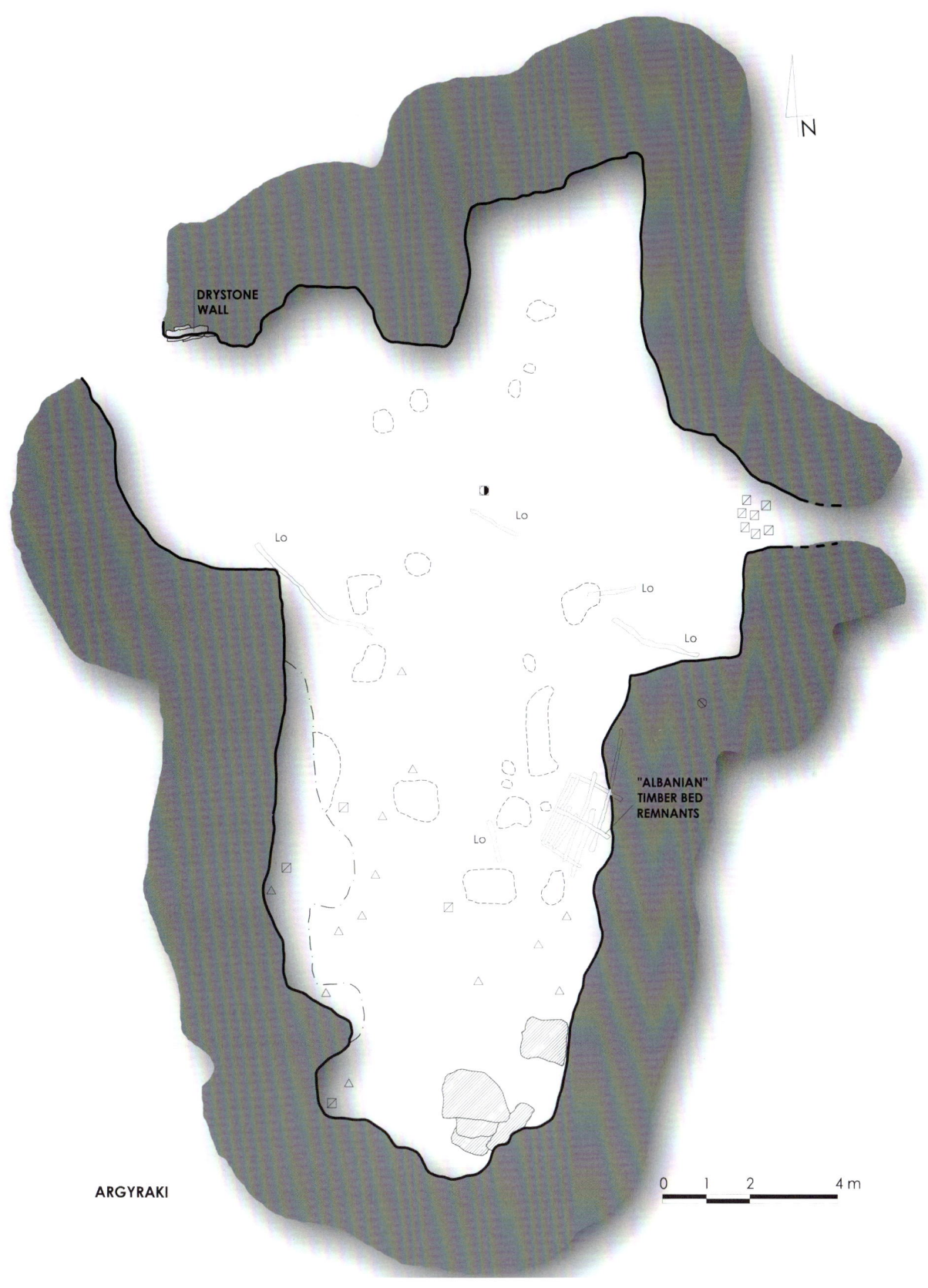

DRYSTONE
WALL

"ALBANIAN"
TIMBER BED
REMNANTS

ARGYRAKI

0   1   2   4 m

Fig. 4. Argyraki (MIL-3).

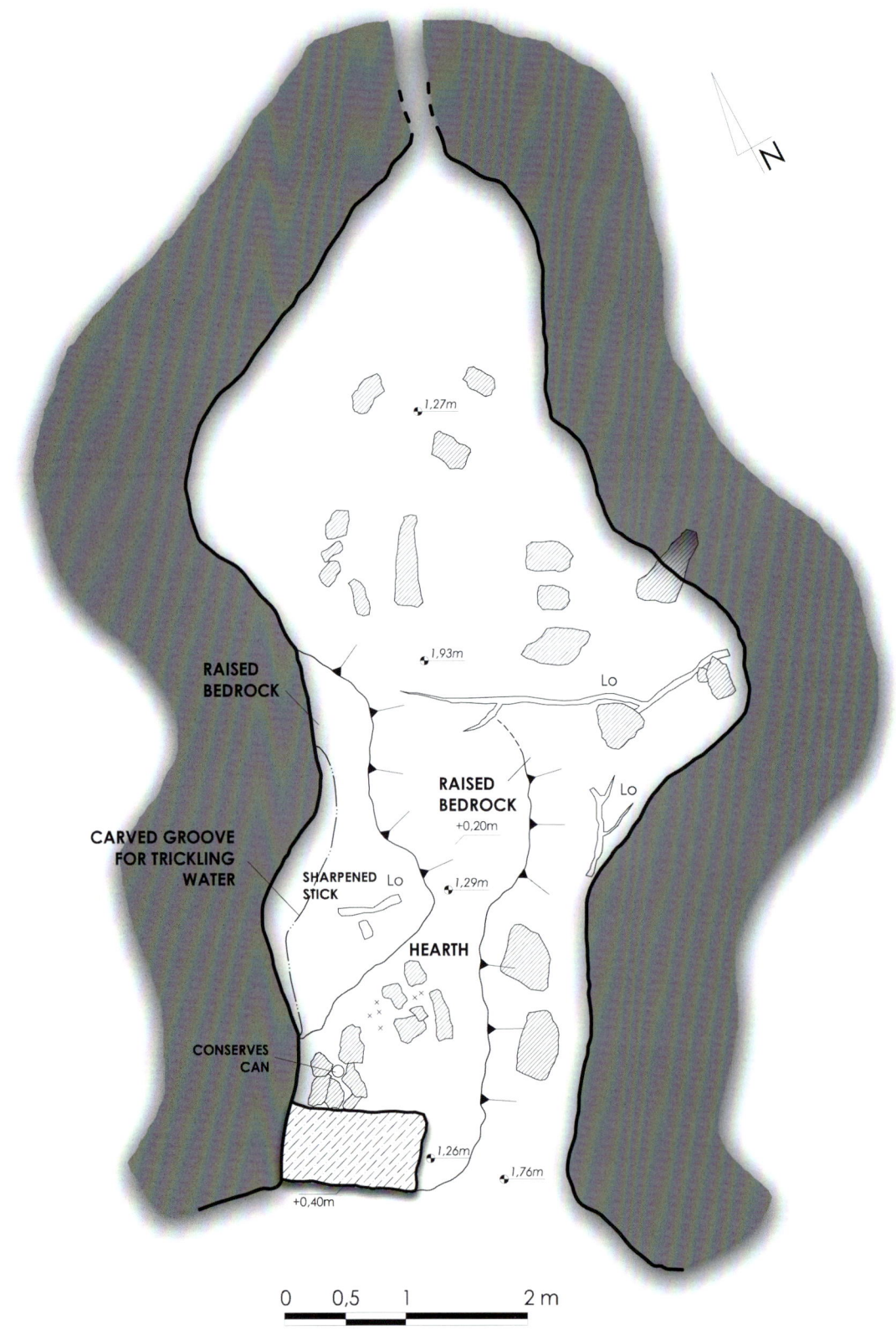

Fig. 5. Banikas Kastraki (ZAG-5).

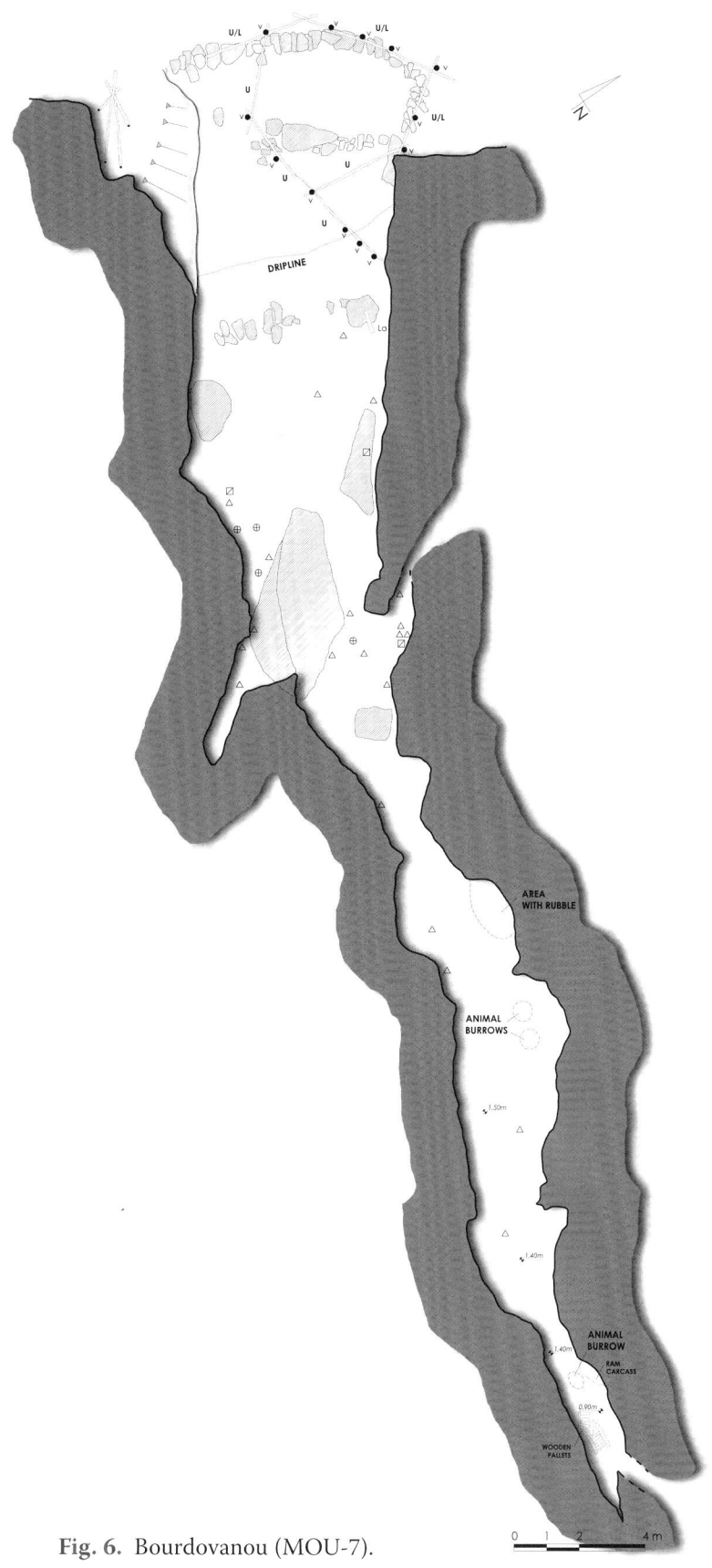

Fig. 6. Bourdovanou (MOU-7).

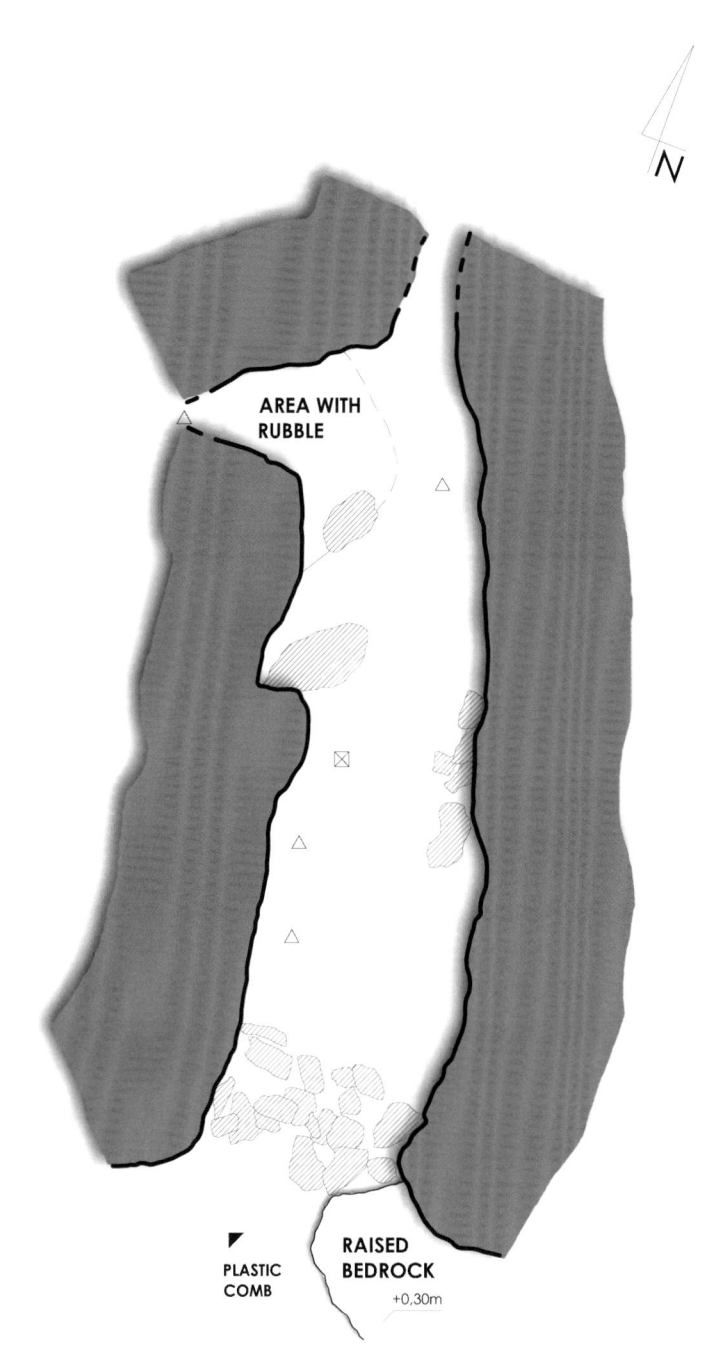

AREA WITH
RUBBLE

PLASTIC
COMB

RAISED
BEDROCK

+0,30m

N

0   0,5   1   2 m

Fig. 7. Goritsa II (POR-3).

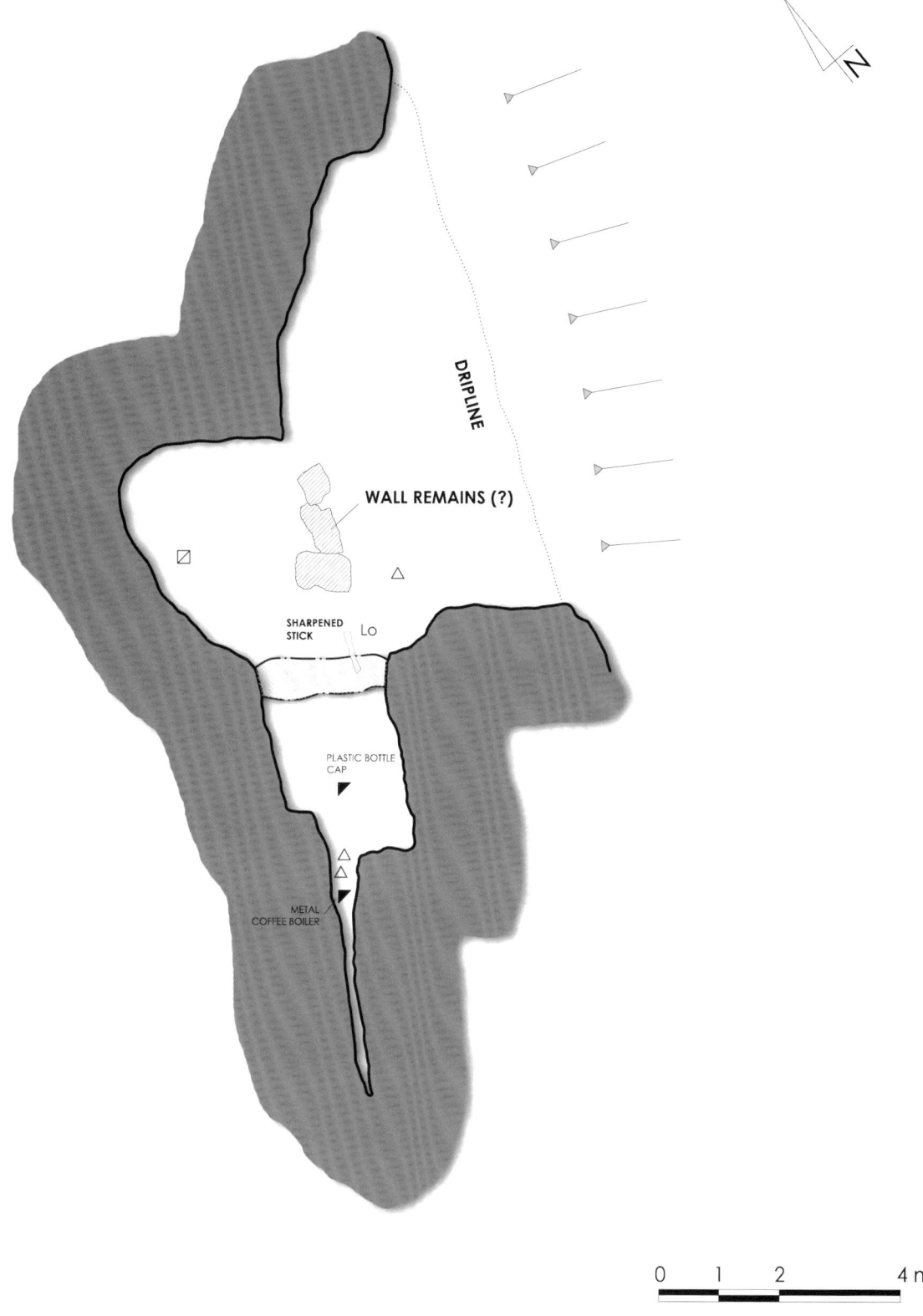

WALL REMAINS (?)

DRIPLINE

SHARPENED
STICK   Lo

PLASTIC BOTTLE
CAP

METAL
COFFEE BOILER

N

0   1   2   4 m

Fig. 8. Gidospilia (MAK-23).

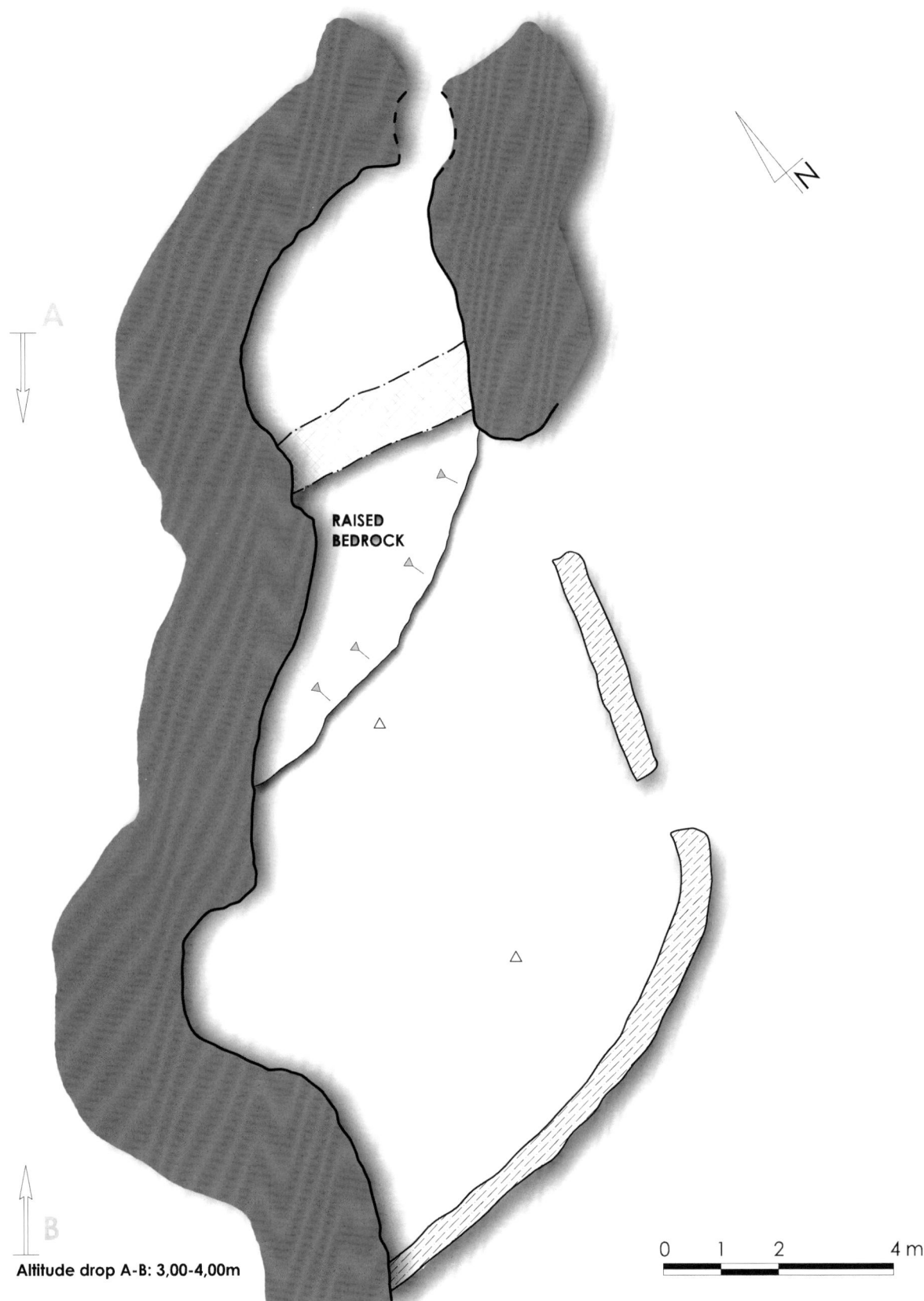

RAISED
BEDROCK

A

B

Altitude drop A-B: 3,00-4,00m

0   1   2   4 m

Fig. 9. Koukourava XII (MAK-16).

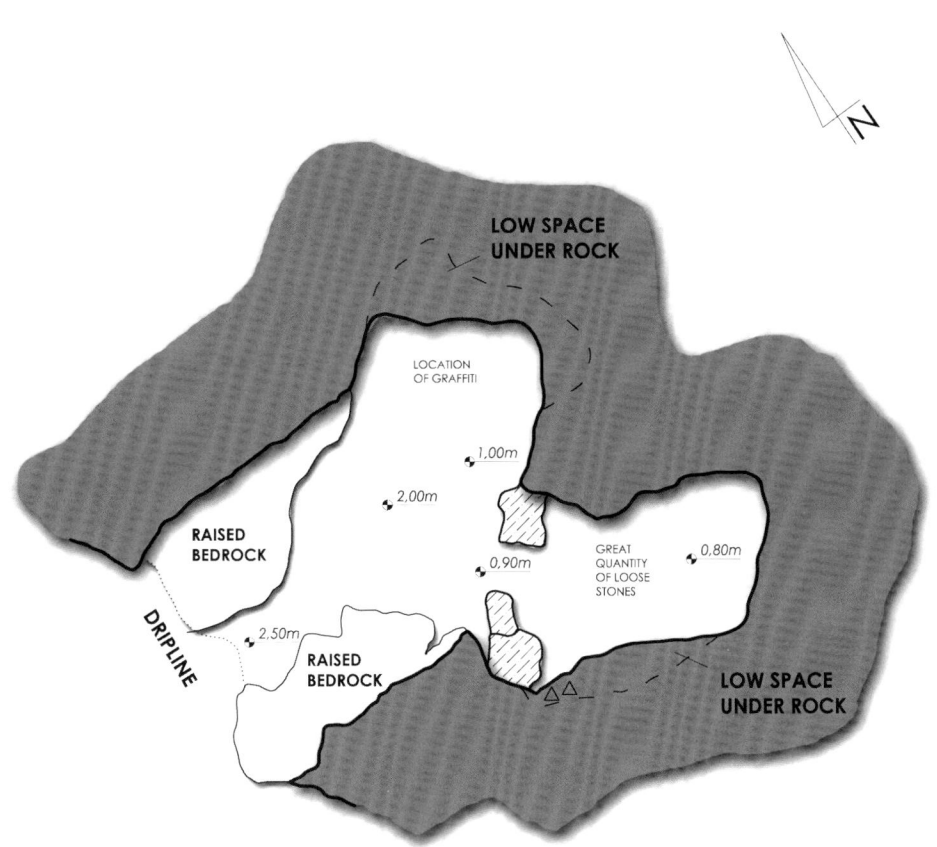

LOW SPACE
UNDER ROCK

LOCATION
OF GRAFFITI

1,00m

2,00m

RAISED
BEDROCK

GREAT
QUANTITY
OF LOOSE
STONES

0,80m

0,90m

DRIPLINE

2,50m

RAISED
BEDROCK

LOW SPACE
UNDER ROCK

N

0   1   2   4 m

I

Fig. 10. Koutra I (MOU-5).

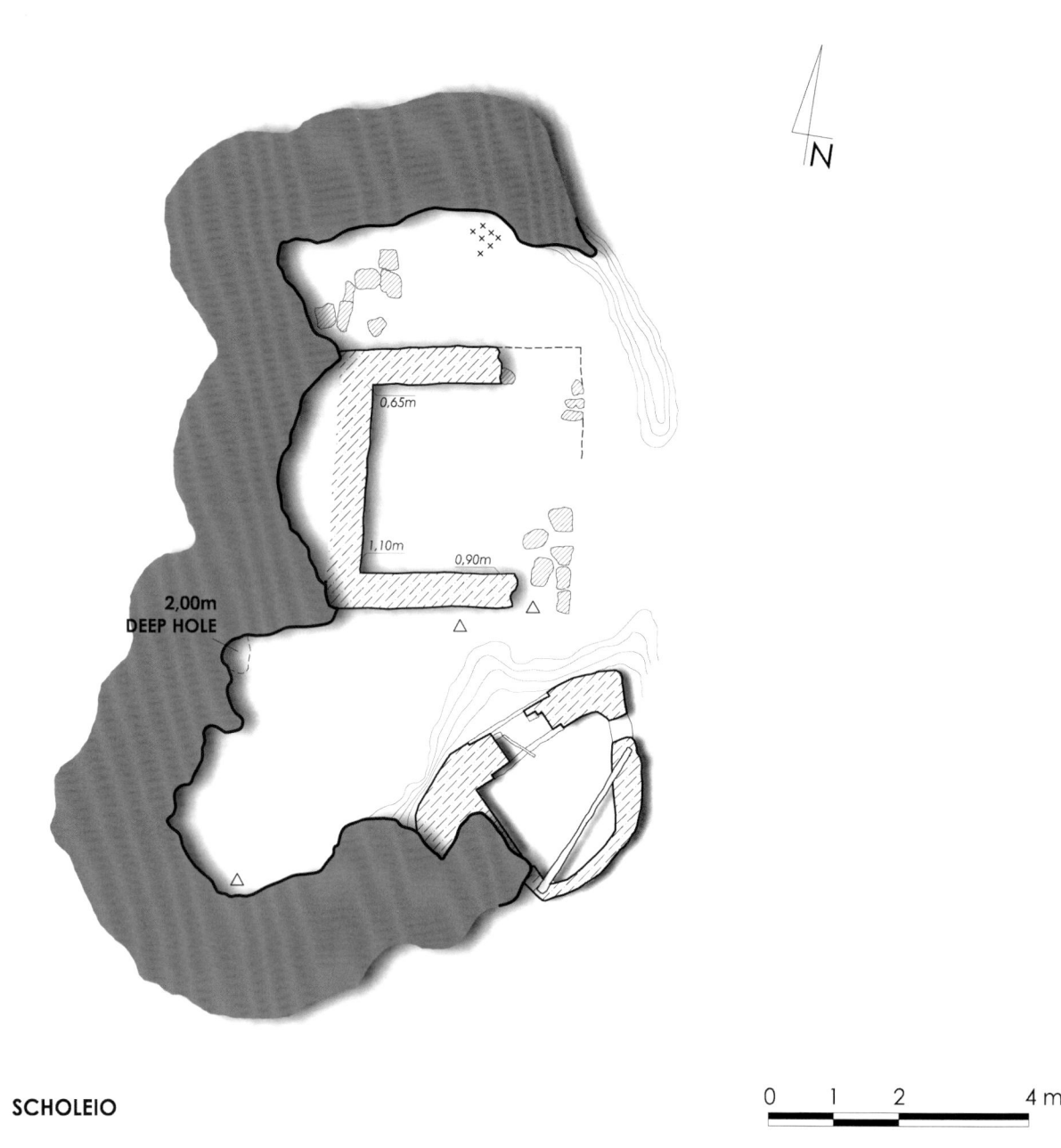

0,65m

1,10m

0,90m

2,00m
DEEP HOLE

SCHOLEIO

0 1 2 4 m

Fig. 11. Kryfo Scholeio (MOU-10).

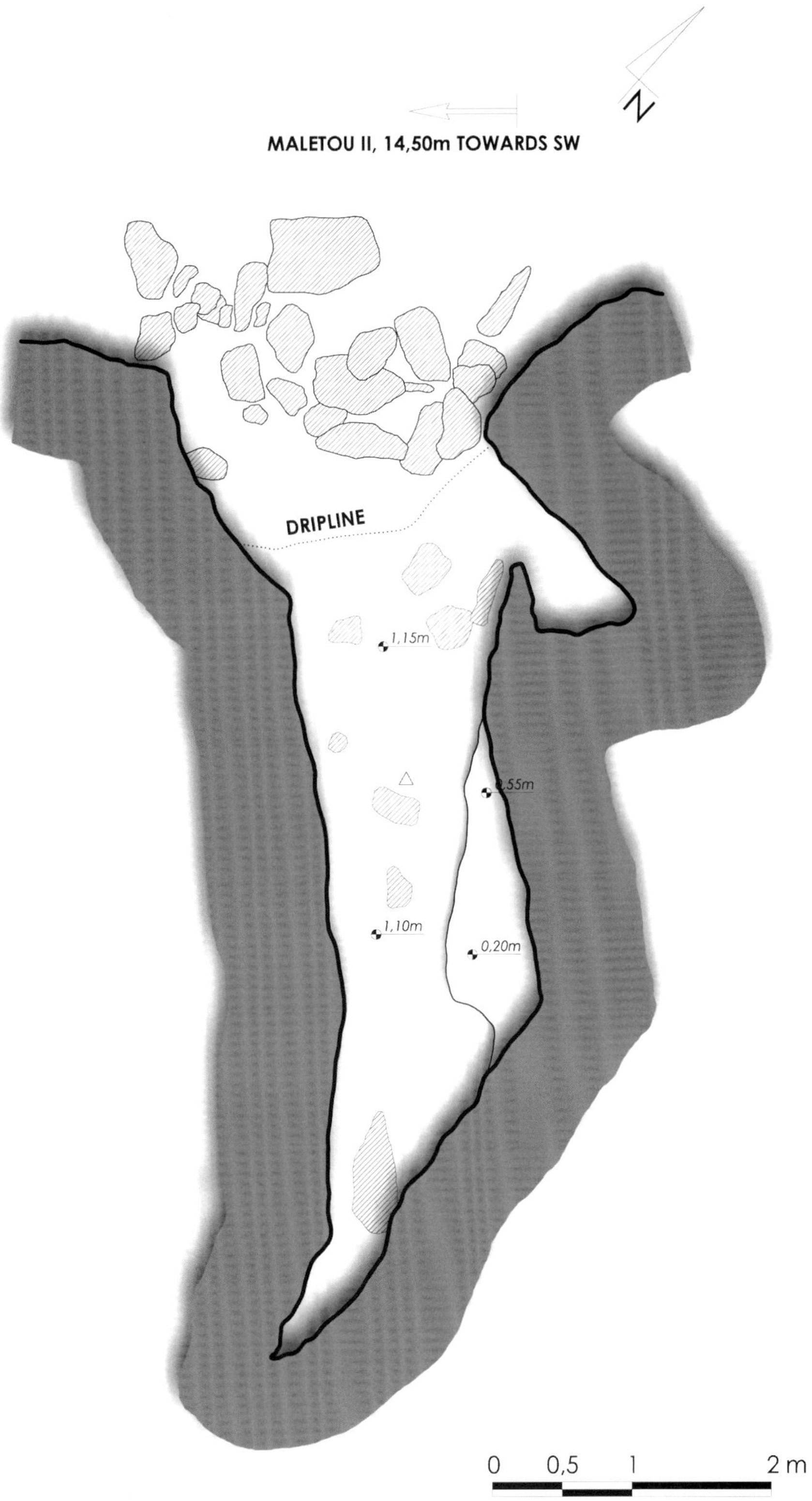

MALETOU II, 14,50m TOWARDS SW

DRIPLINE

1,15m

0,55m

1,10m

0,20m

0   0,5   1   2 m

Fig. 12. Maletou I (MOU-20).

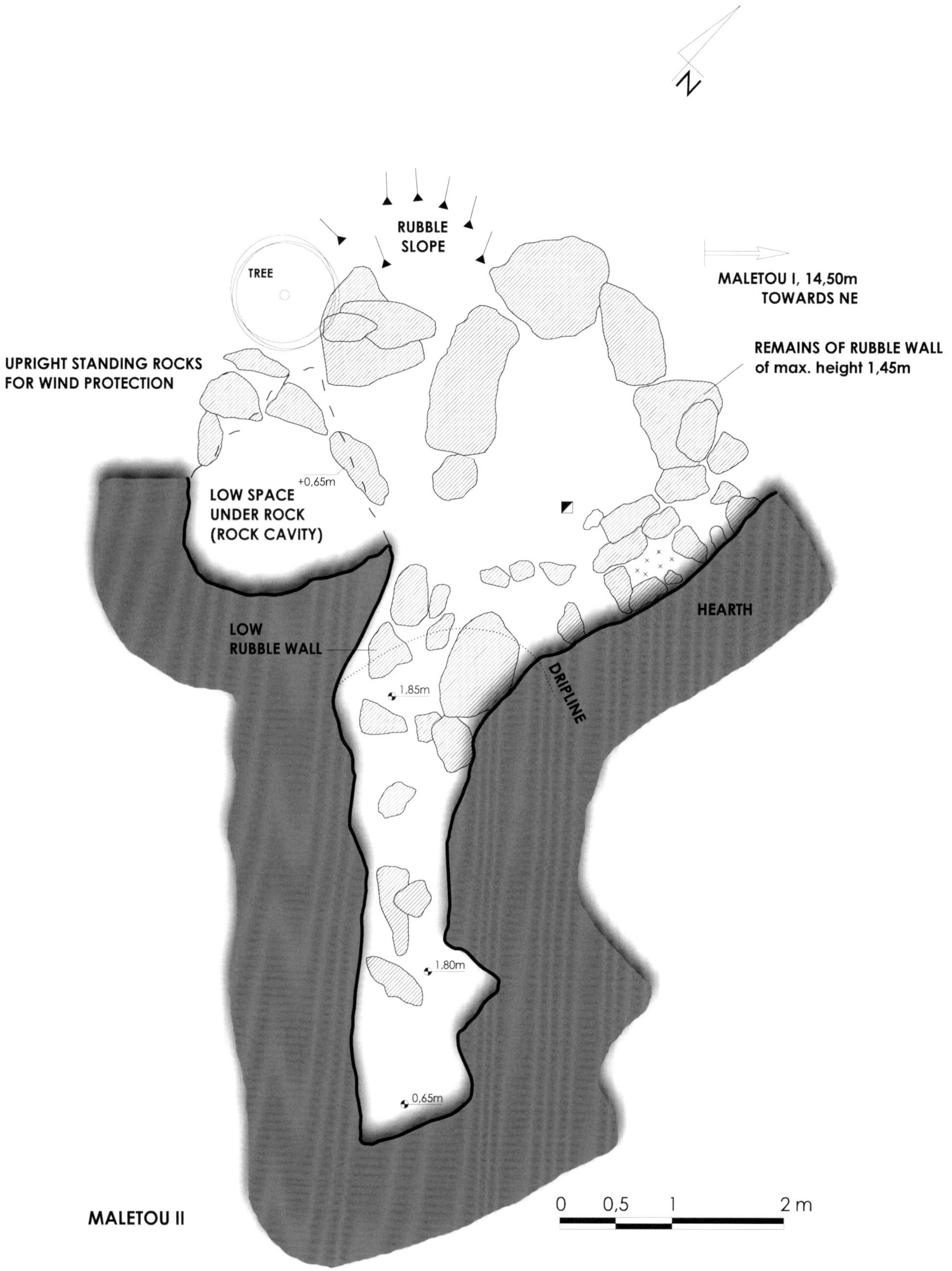

RUBBLE
SLOPE

TREE

MALETOU I, 14,50m
TOWARDS NE

UPRIGHT STANDING ROCKS
FOR WIND PROTECTION

REMAINS OF RUBBLE WALL
of max. height 1,45m

+0,65m

LOW SPACE
UNDER ROCK
(ROCK CAVITY)

HEARTH

LOW
RUBBLE WALL

DRIPLINE

1,85m

1,80m

0,65m

MALETOU II

0   0,5   1   2 m

Fig. 13. Maletou II (MOU-21).

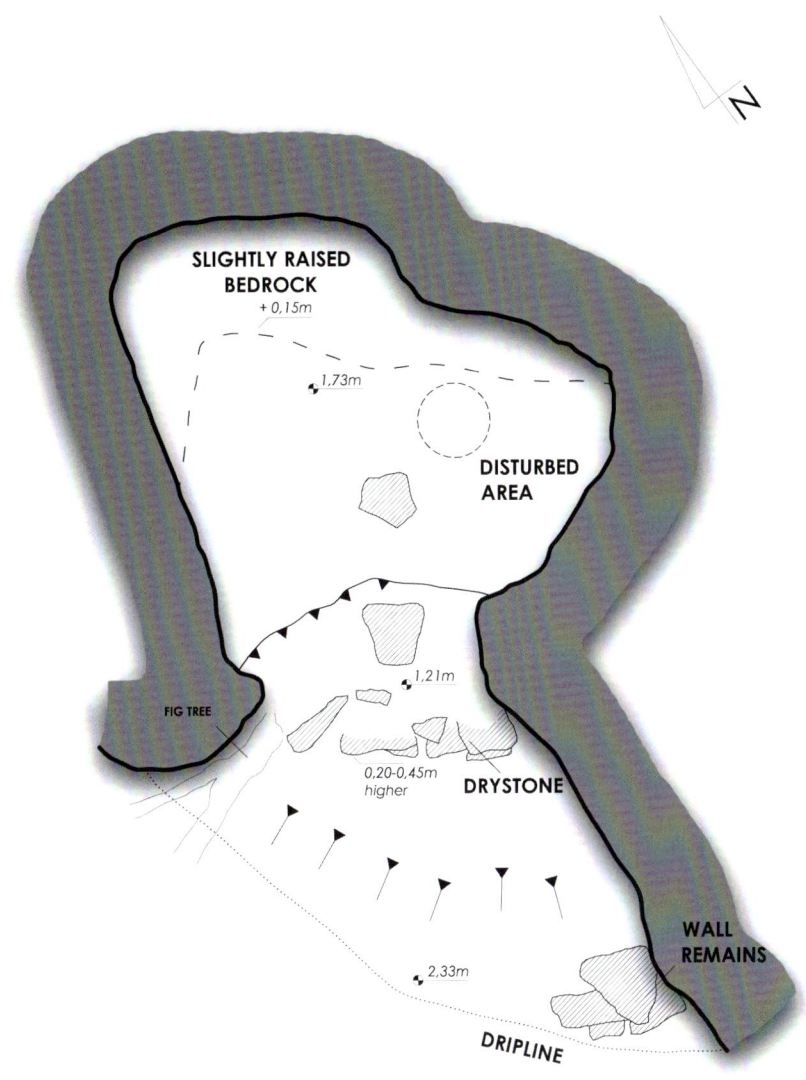

SLIGHTLY RAISED
BEDROCK
+ 0,15m

1,73m

DISTURBED
AREA

FIG TREE

1,21m

0,20-0,45m
higher

DRYSTONE

WALL
REMAINS

2,33m

DRIPLINE

MAROUKO'S CAVE

0   0,5   1   2 m

Fig. 14. Marouko's Cave (MIL-22).

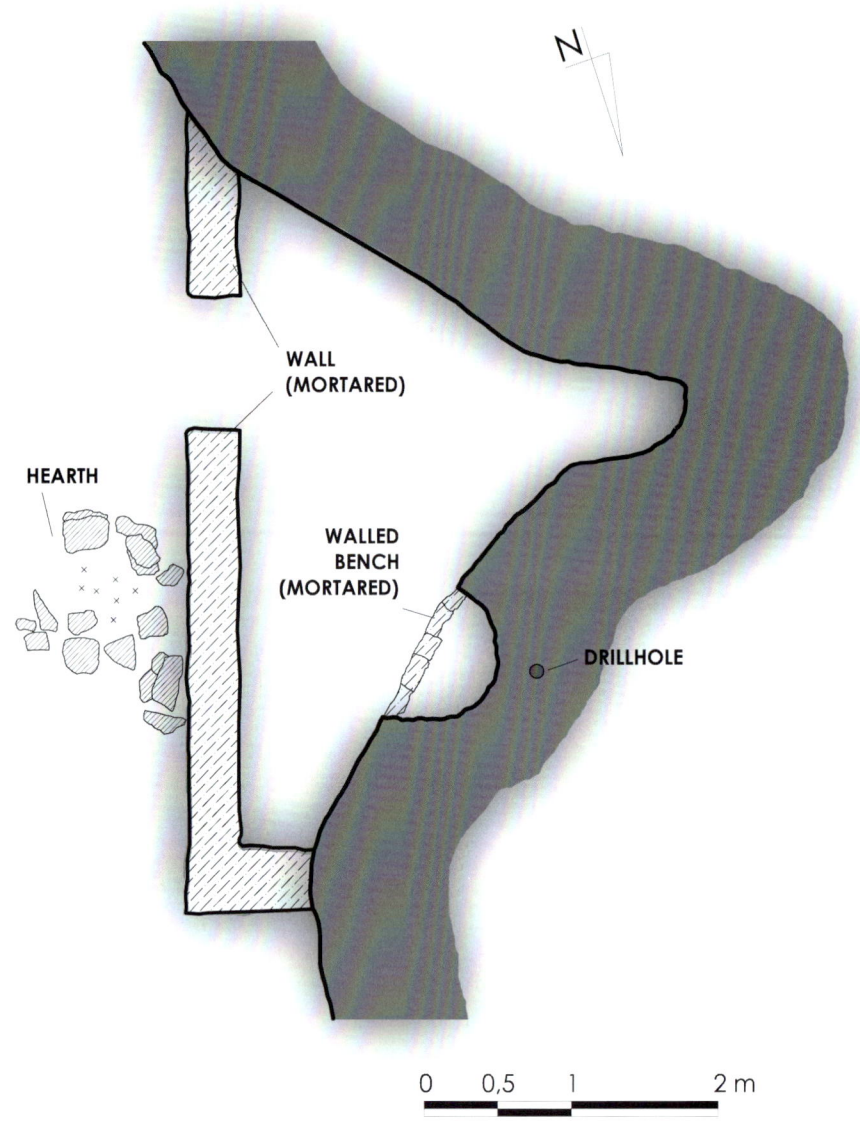

Fig. 15. Oghla I – Little Train Station (MIL-9).

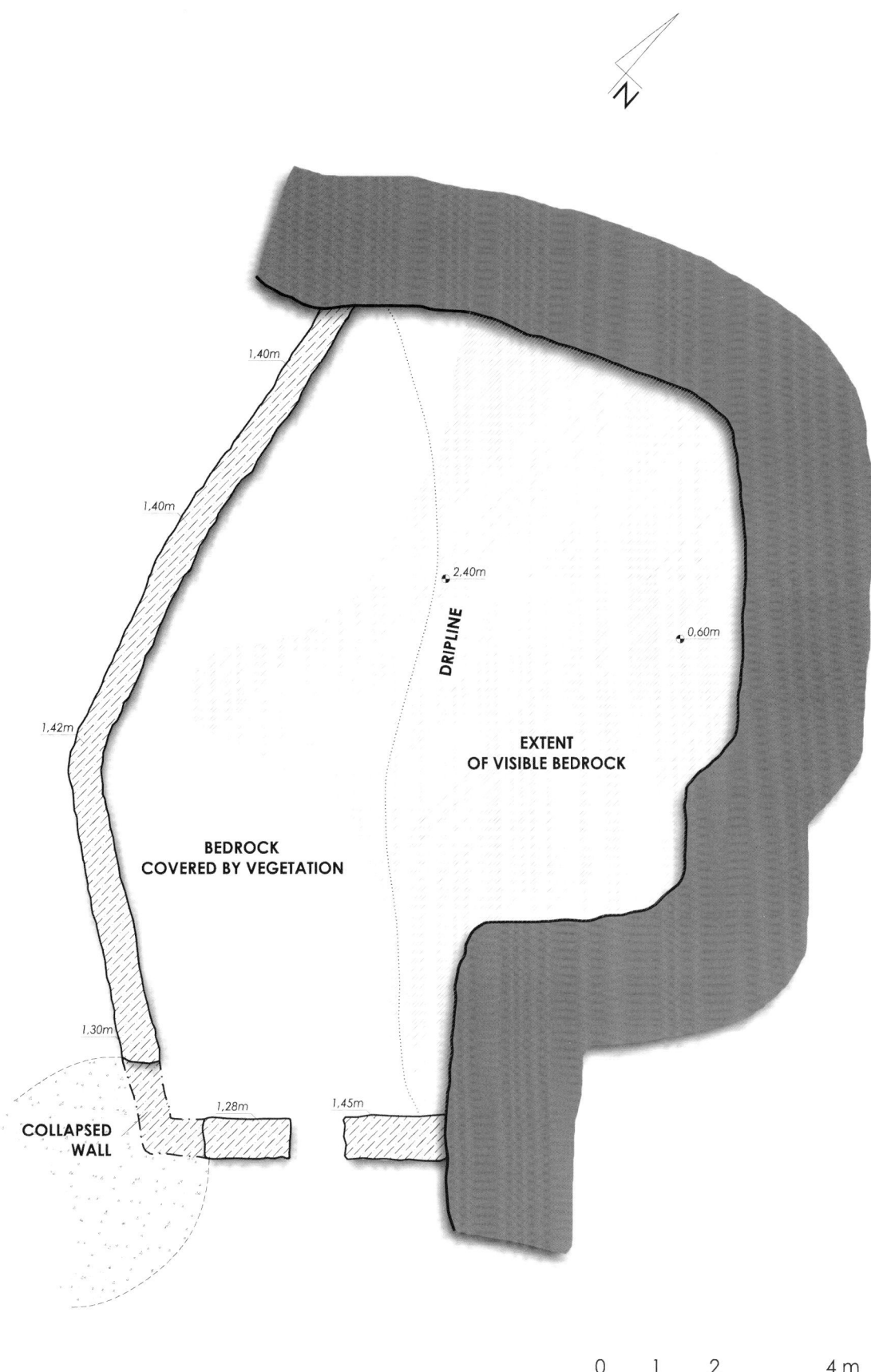

N

1,40m

1,40m

1,42m

DRIPLINE

2,40m

0,60m

EXTENT
OF VISIBLE BEDROCK

BEDROCK
COVERED BY VEGETATION

1,30m

1,28m

1,45m

COLLAPSED
WALL

0   1   2   4 m

Fig. 16. Pidima tis Grias (MIL-5).

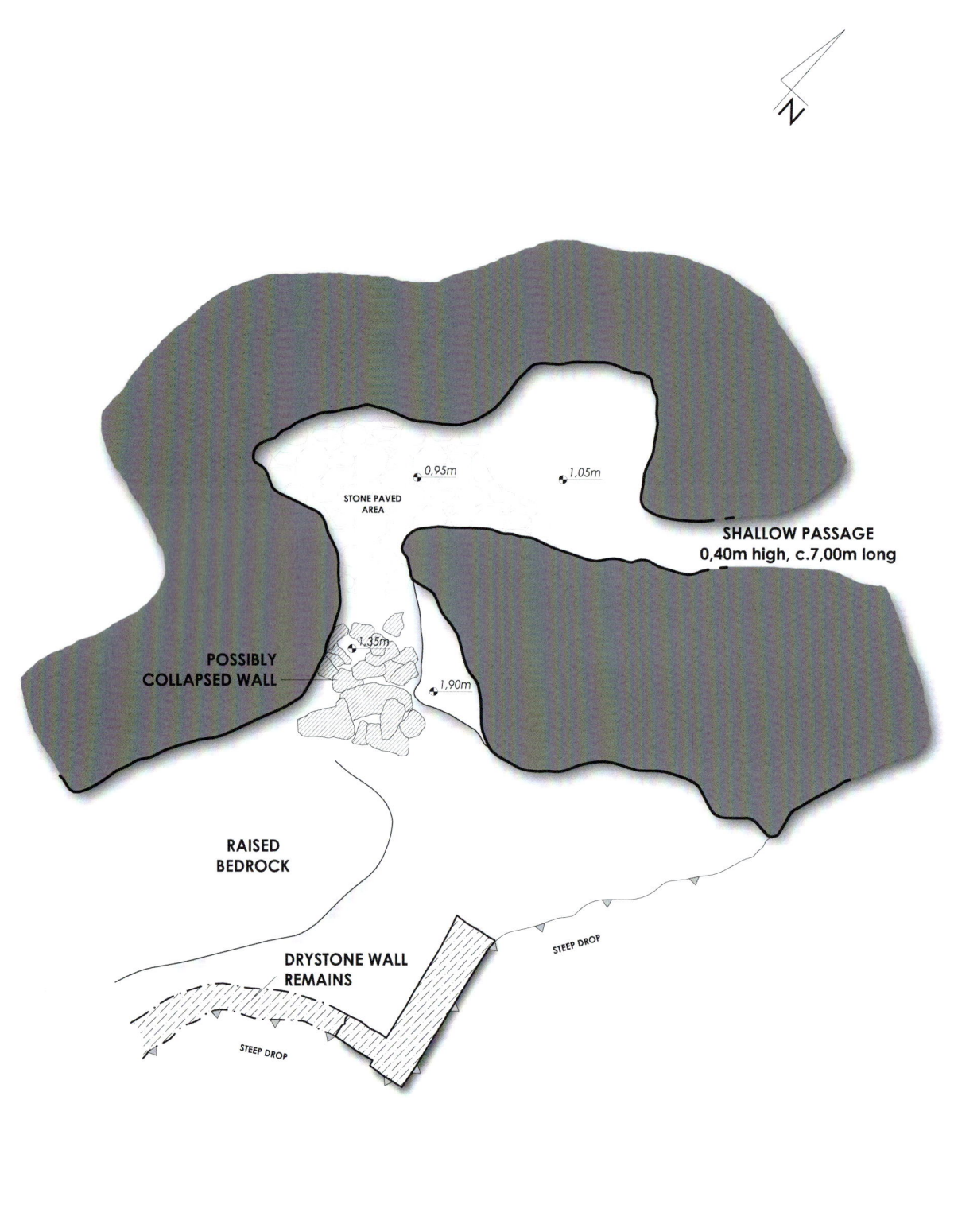

0,95m

1,05m

STONE PAVED
AREA

SHALLOW PASSAGE
0,40m high, c.7,00m long

1,35m

POSSIBLY
COLLAPSED WALL

1,90m

RAISED
BEDROCK

STEEP DROP

DRYSTONE WALL
REMAINS

STEEP DROP

STEEP DROP

0   1   2   4 m

Fig. 17. Spilia (KER-6).

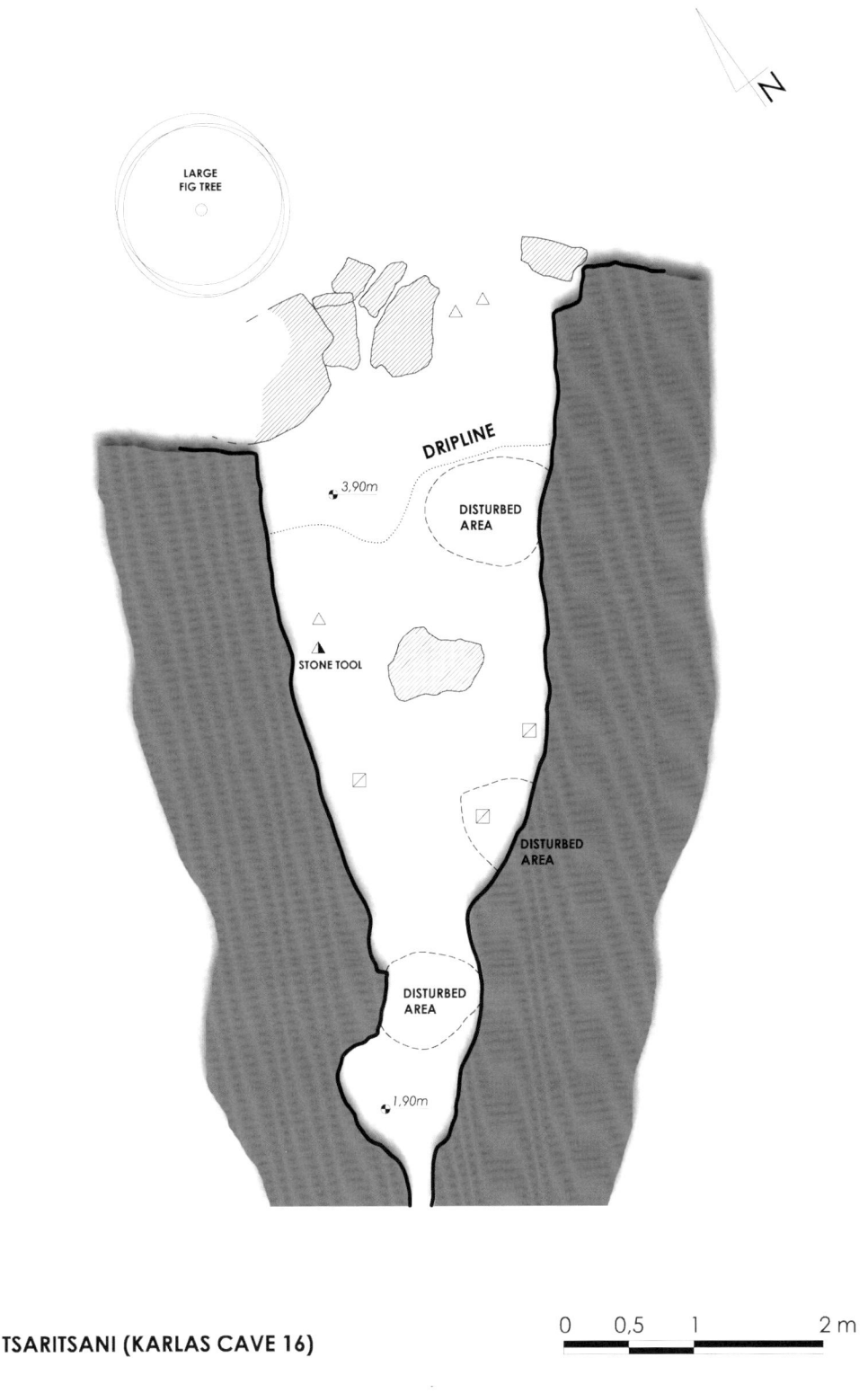

LARGE
FIG TREE

DRIPLINE

3,90m

DISTURBED
AREA

STONE TOOL

DISTURBED
AREA

DISTURBED
AREA

1,90m

TSARITSANI (KARLAS CAVE 16)

0   0,5   1   2 m

Fig. 18. Tsaritsani (KAR-16).

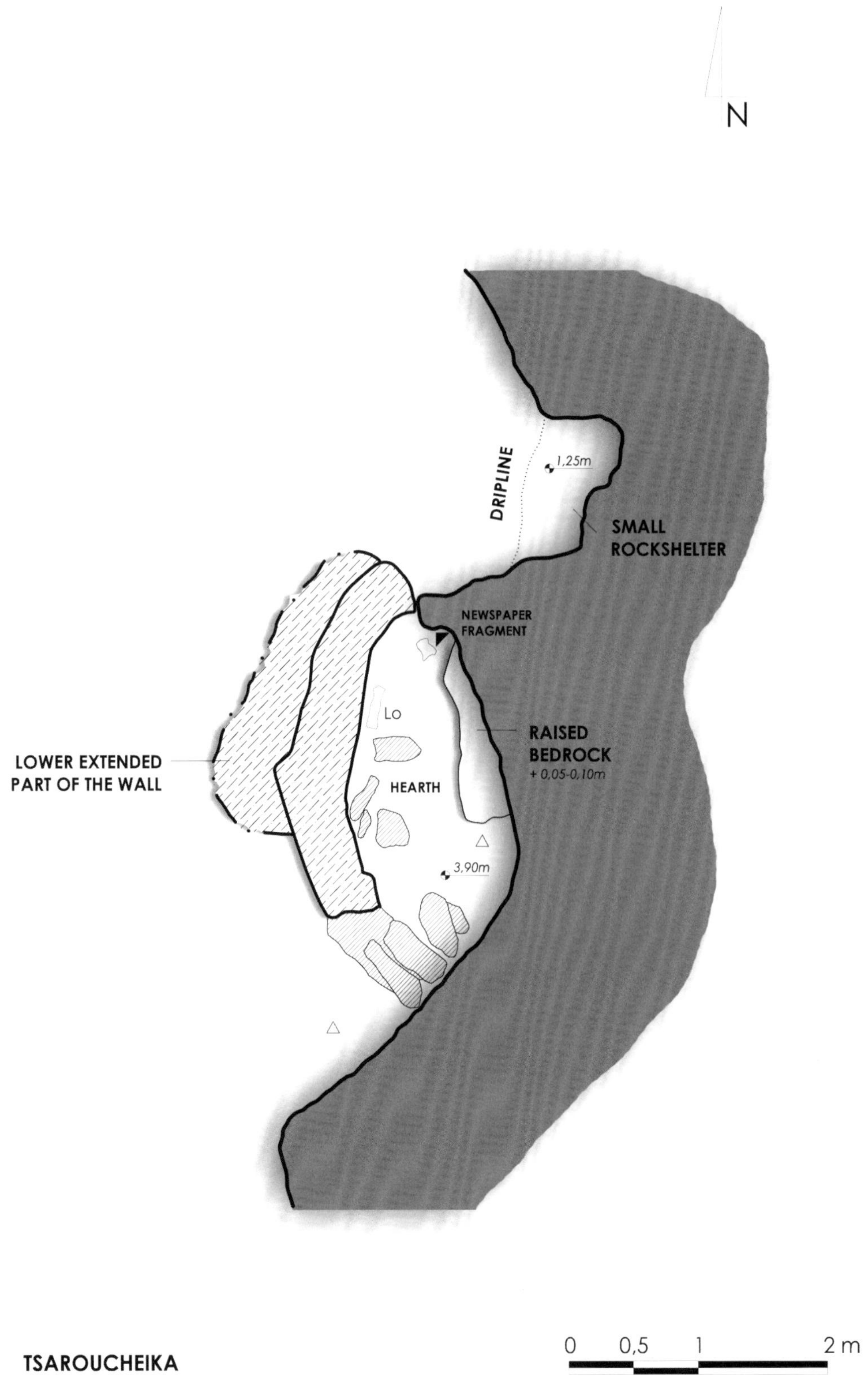

DRIPLINE

1,25m

SMALL
ROCKSHELTER

NEWSPAPER
FRAGMENT

Lo

RAISED
BEDROCK
+ 0,05-0,10m

HEARTH

LOWER EXTENDED
PART OF THE WALL

3,90m

TSAROUCHEIKA

0   0,5   1   2 m

Fig. 19. Tsaroucheika (MIL-21).

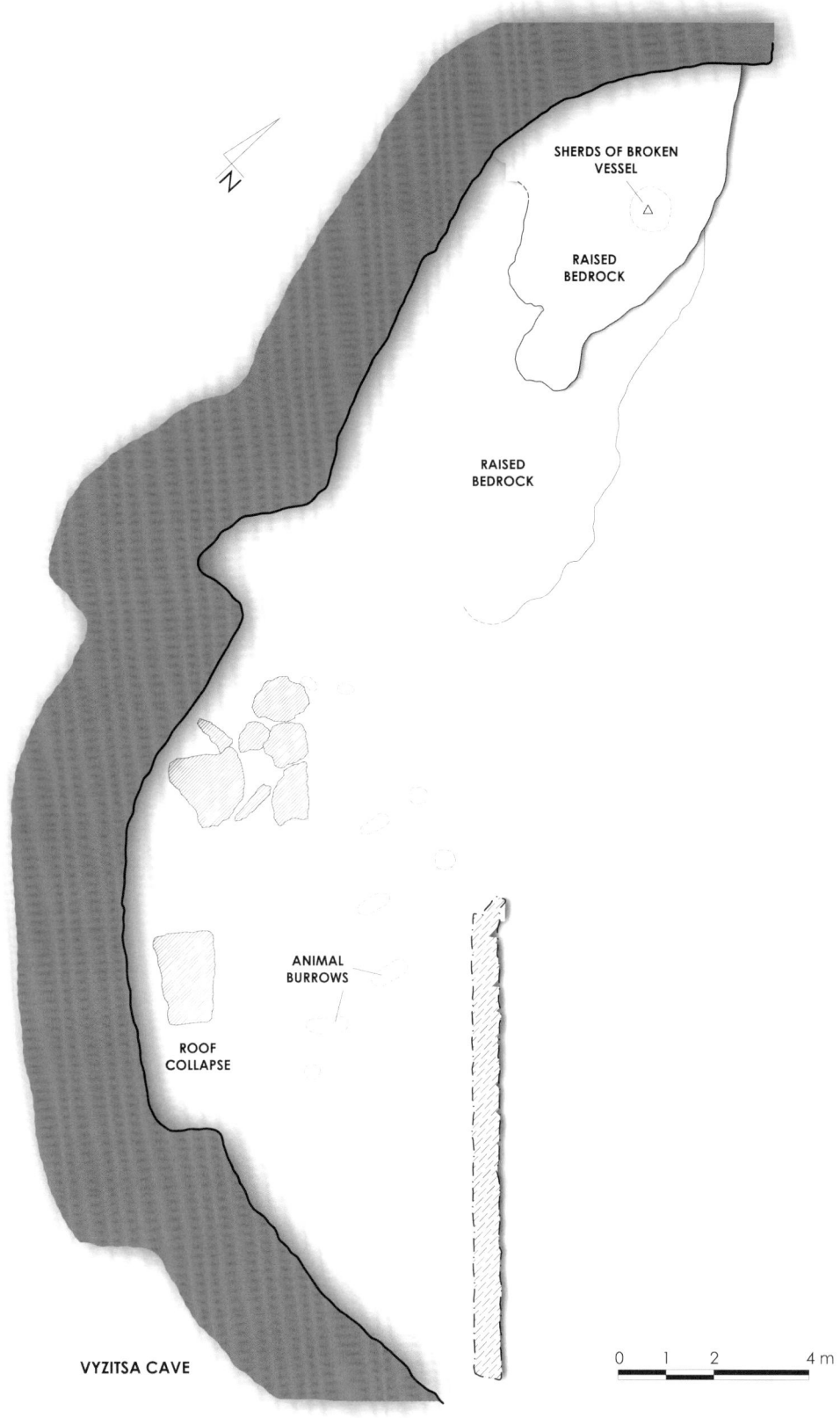

Fig. 20. Laios Cave (MIL-4).

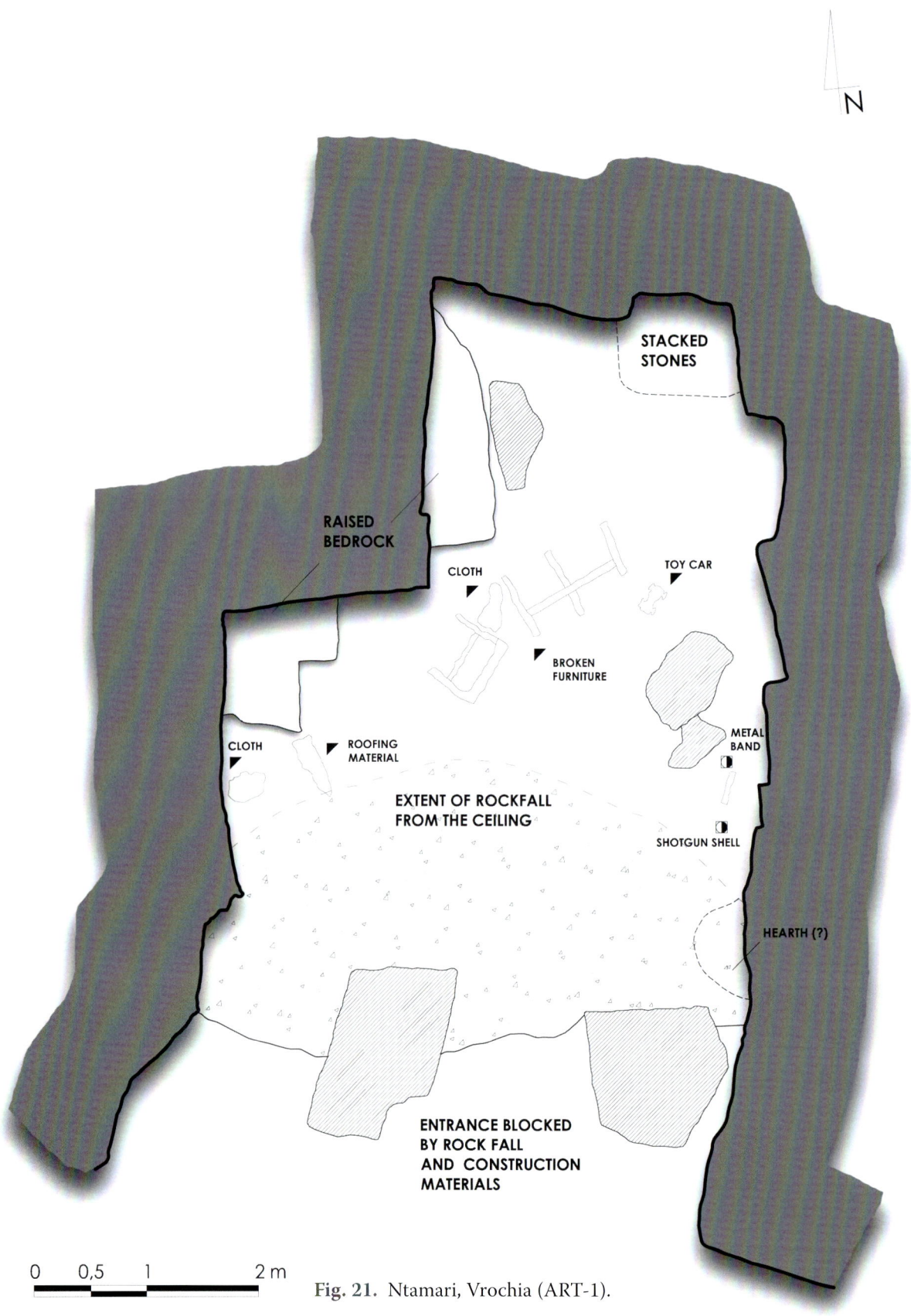

RAISED
BEDROCK

STACKED
STONES

CLOTH

TOY CAR

BROKEN
FURNITURE

METAL
BAND

CLOTH

ROOFING
MATERIAL

EXTENT OF ROCKFALL
FROM THE CEILING

SHOTGUN SHELL

HEARTH (?)

ENTRANCE BLOCKED
BY ROCK FALL
AND CONSTRUCTION
MATERIALS

0   0,5   1   2 m

Fig. 21. Ntamari, Vrochia (ART-1).

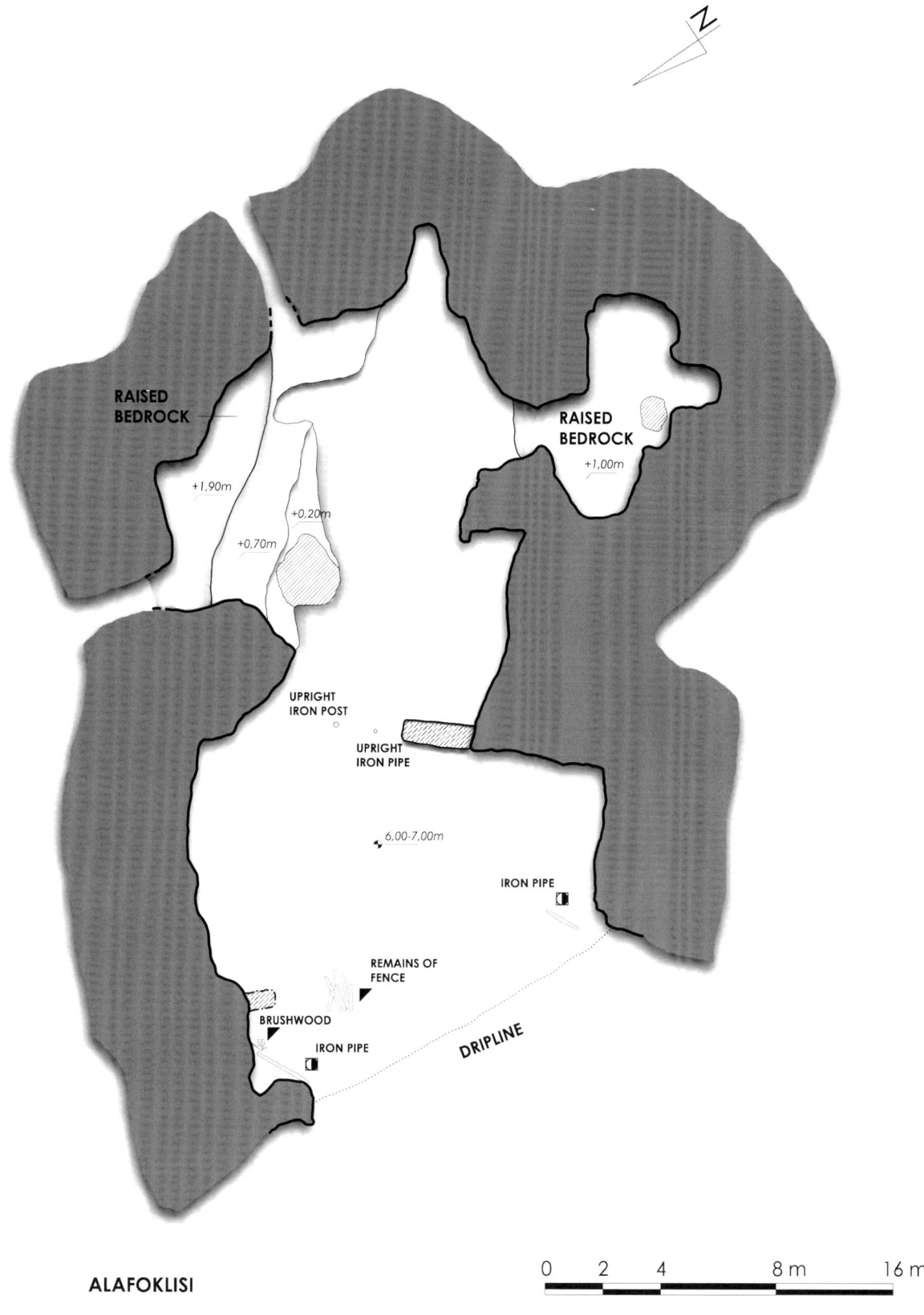

RAISED
BEDROCK

+1,90m

+0,20m

+0,70m

RAISED
BEDROCK

+1,00m

UPRIGHT
IRON POST

UPRIGHT
IRON PIPE

6,00-7,00m

IRON PIPE

REMAINS OF
FENCE

BRUSHWOOD

IRON PIPE

DRIPLINE

ALAFOKLISI

0   2   4   8 m   16 m

Fig. 22. Alafoklisi (MAK-3).

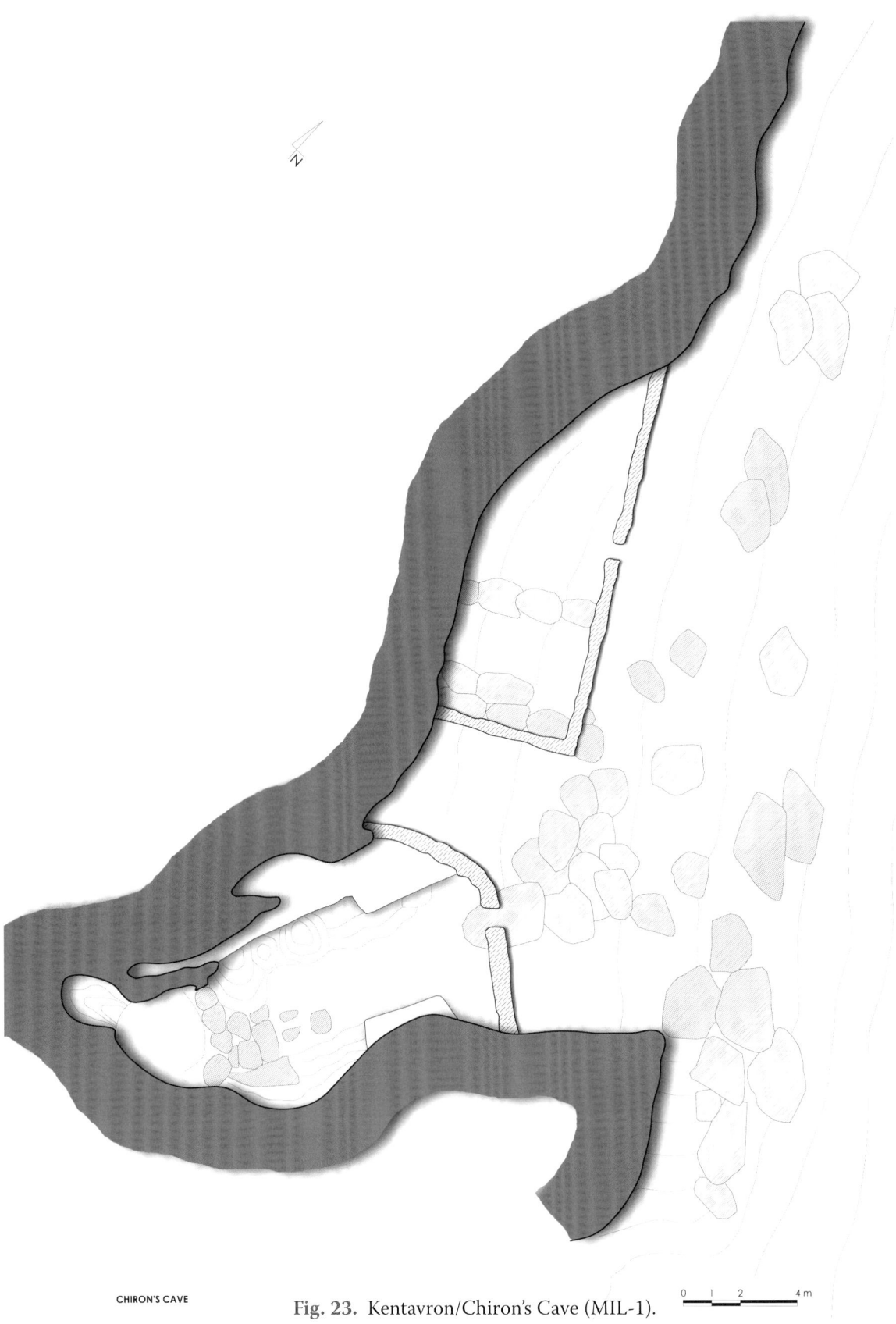

CHIRON'S CAVE

Fig. 23. Kentavron/Chiron's Cave (MIL-1).

0 1 2   4 m

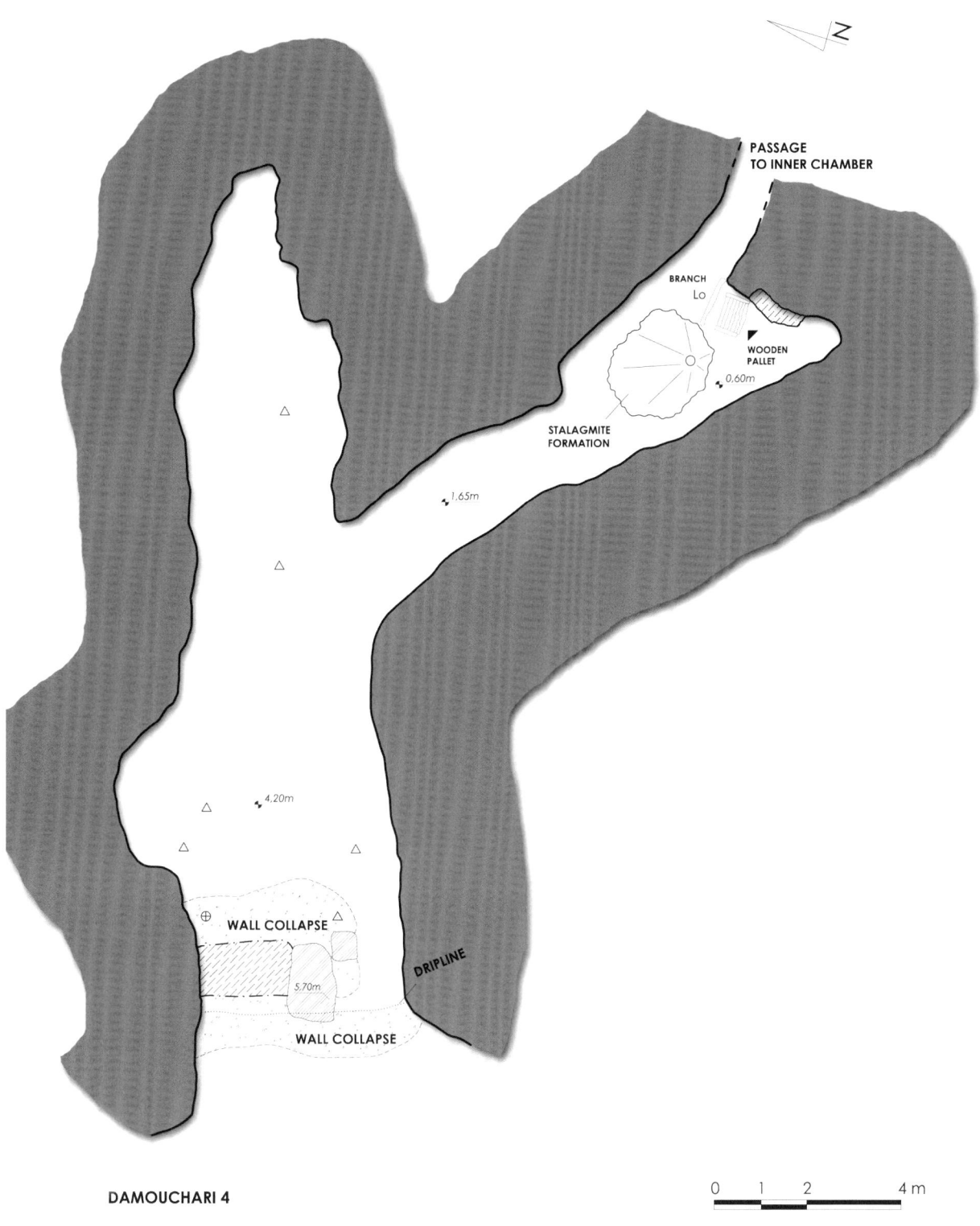

PASSAGE
TO INNER CHAMBER

BRANCH
Lo

WOODEN
PALLET

0.60m

STALAGMITE
FORMATION

1.65m

4.20m

WALL COLLAPSE

DRIPLINE

5.70m

WALL COLLAPSE

N

0 1 2 4 m

DAMOUCHARI 4

Fig. 24. Damouchari IV (MOU-16).

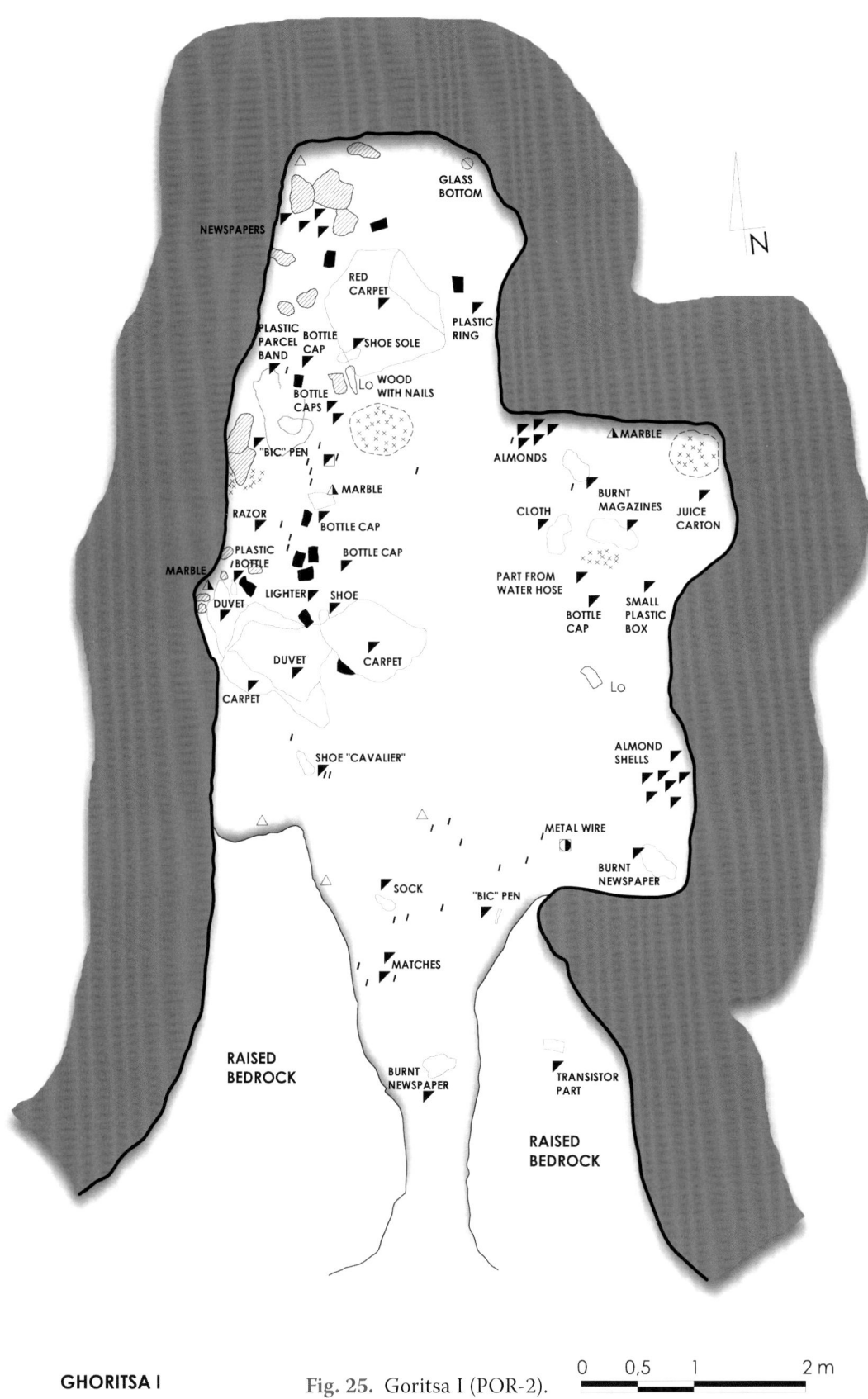

GHORITSA I   **Fig. 25.** Goritsa I (POR-2).   0   0,5   1   2 m

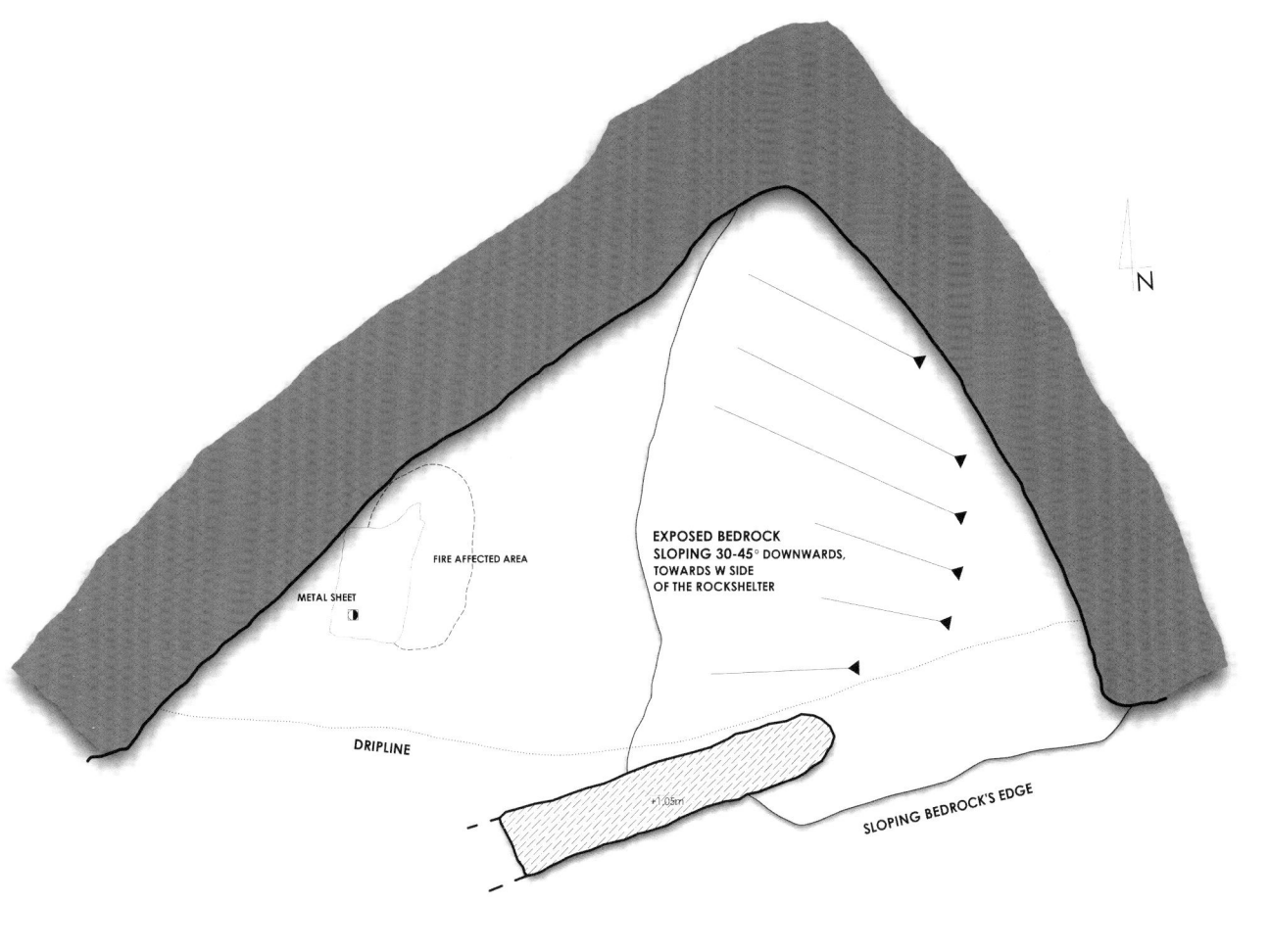

N

EXPOSED BEDROCK
SLOPING 30-45° DOWNWARDS,
TOWARDS W SIDE
OF THE ROCKSHELTER

FIRE AFFECTED AREA

METAL SHEET

DRIPLINE

+1.05m

SLOPING BEDROCK'S EDGE

KARLAS CAVE 5

0   0,5   1   2 m

Fig. 26. Lake Karla V (KAR-5).

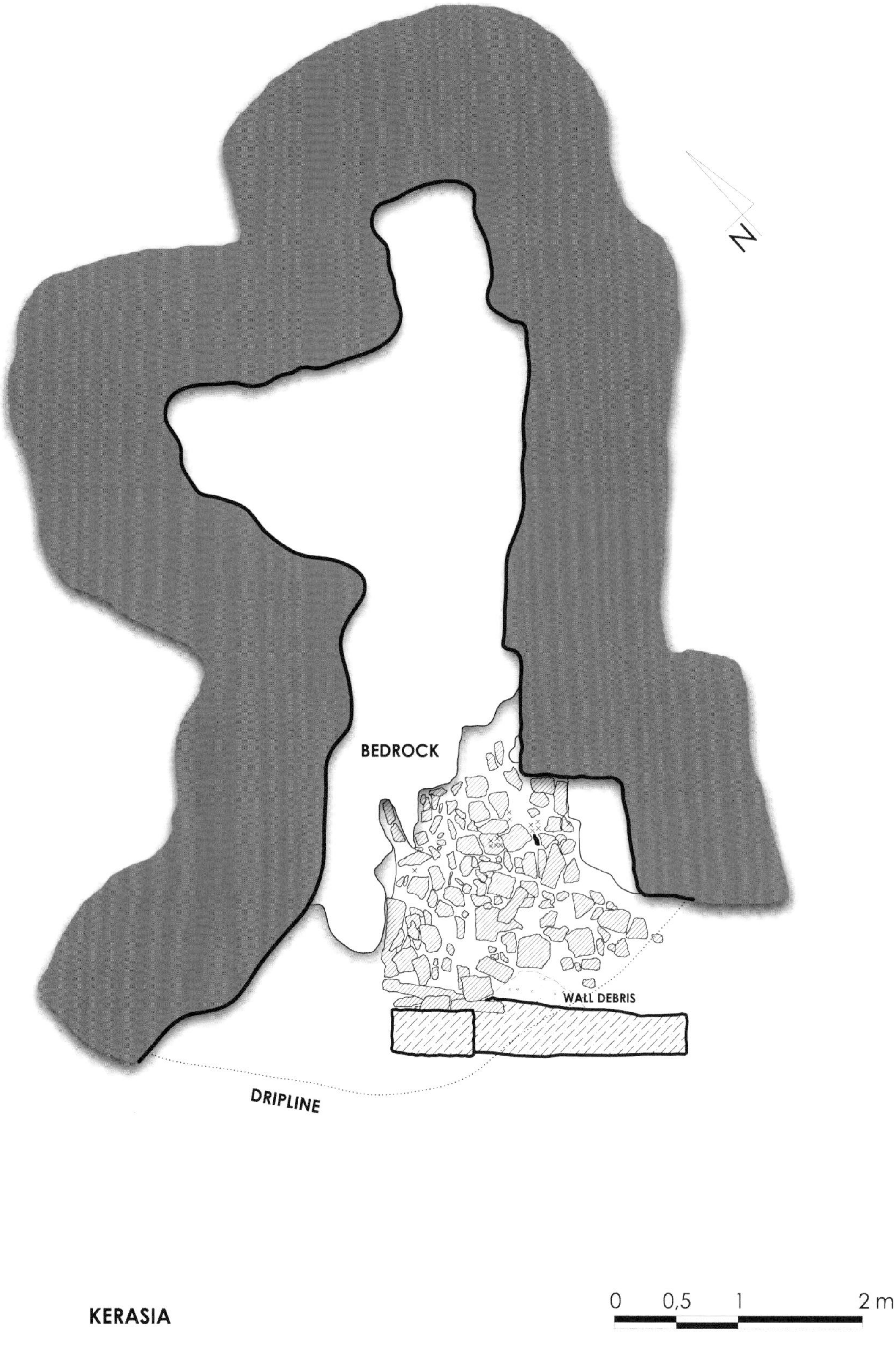

BEDROCK

WALL DEBRIS

DRIPLINE

KERASIA

0   0,5   1   2 m

Fig. 27. Spilia pros Panagia/Sogianni (KAR-22).

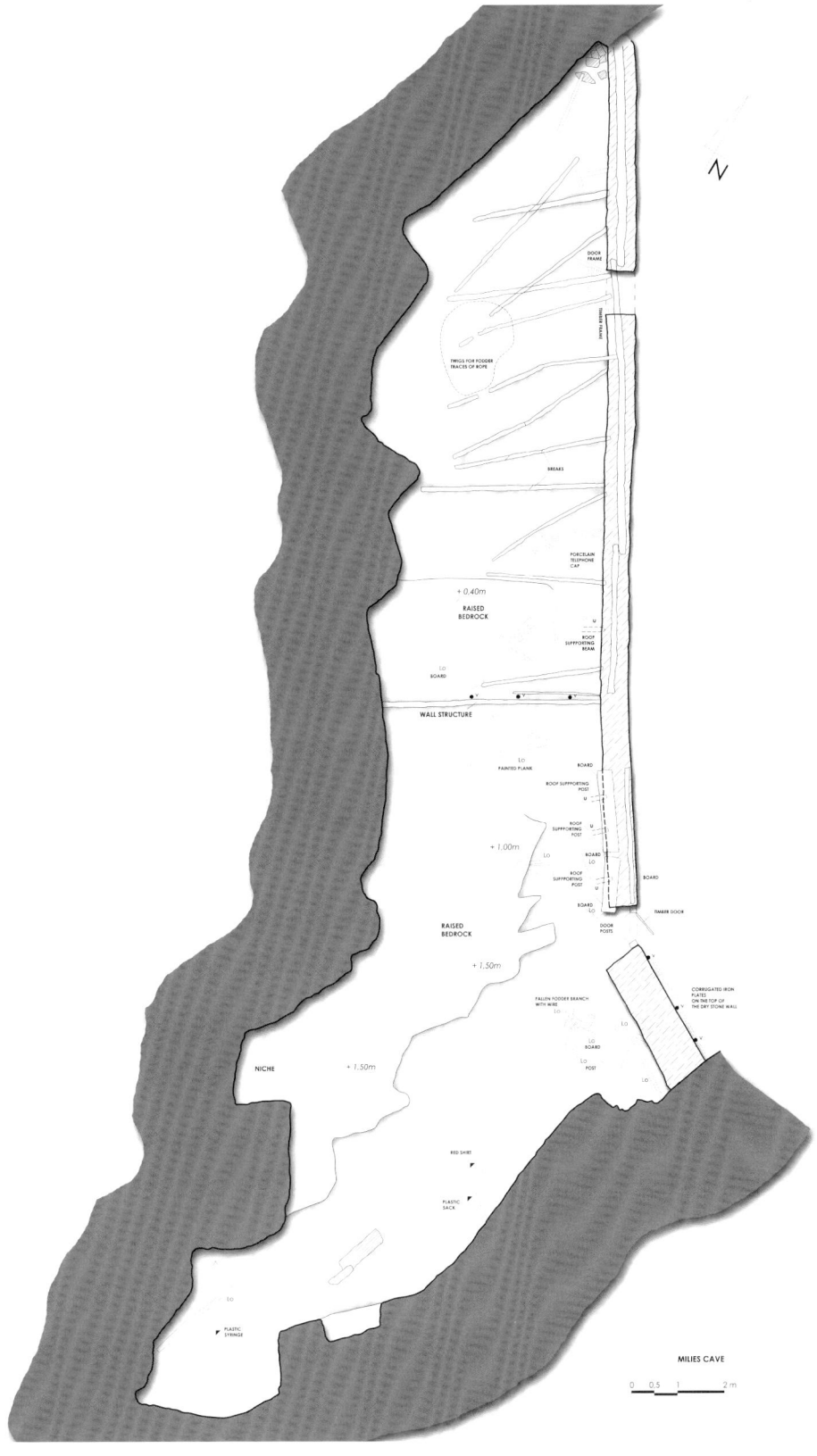

Fig. 28. Kentavron/Chironos (MIL-1).

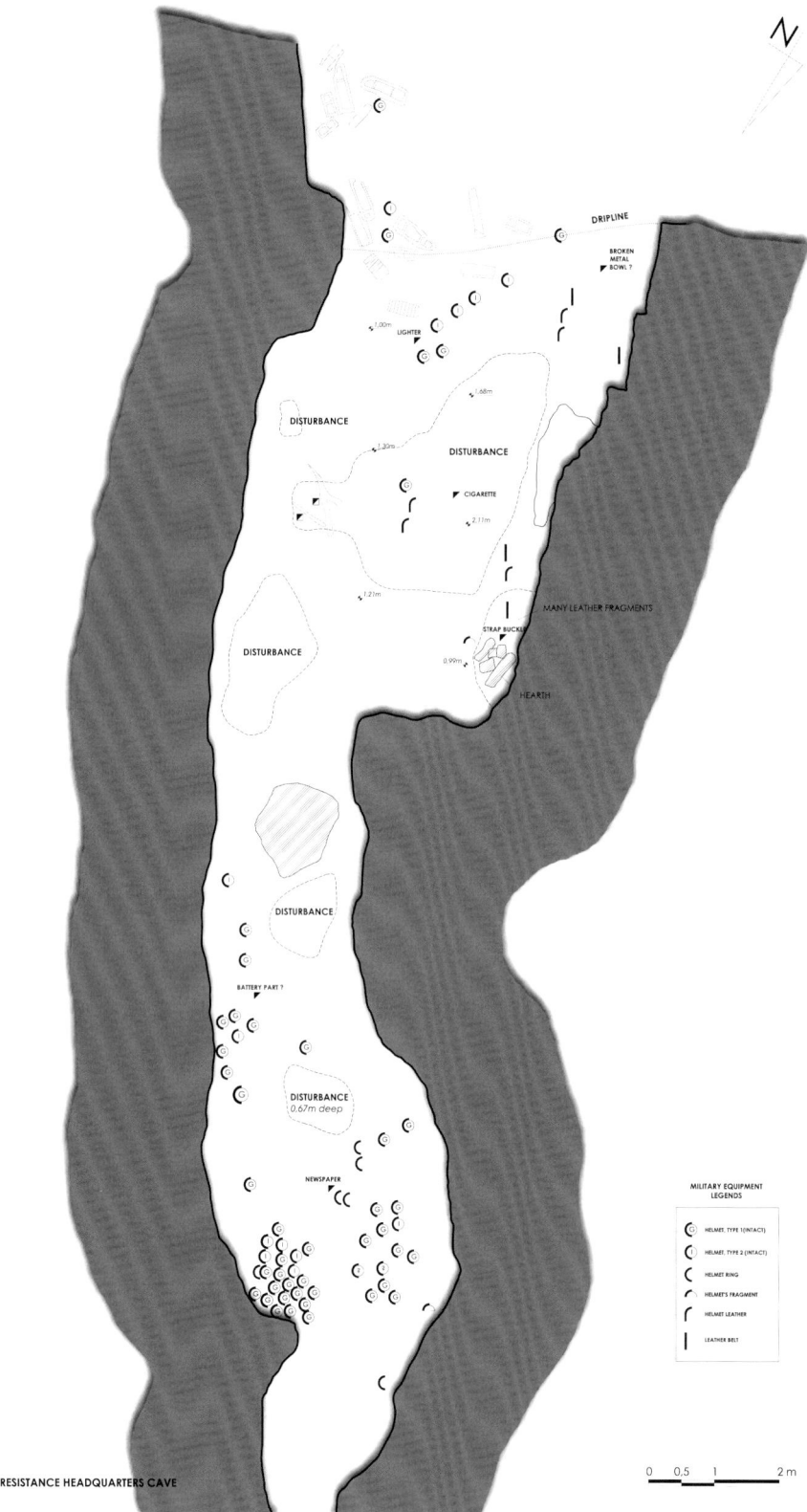

Fig. 29. Spilia me kranoi (KER-8).

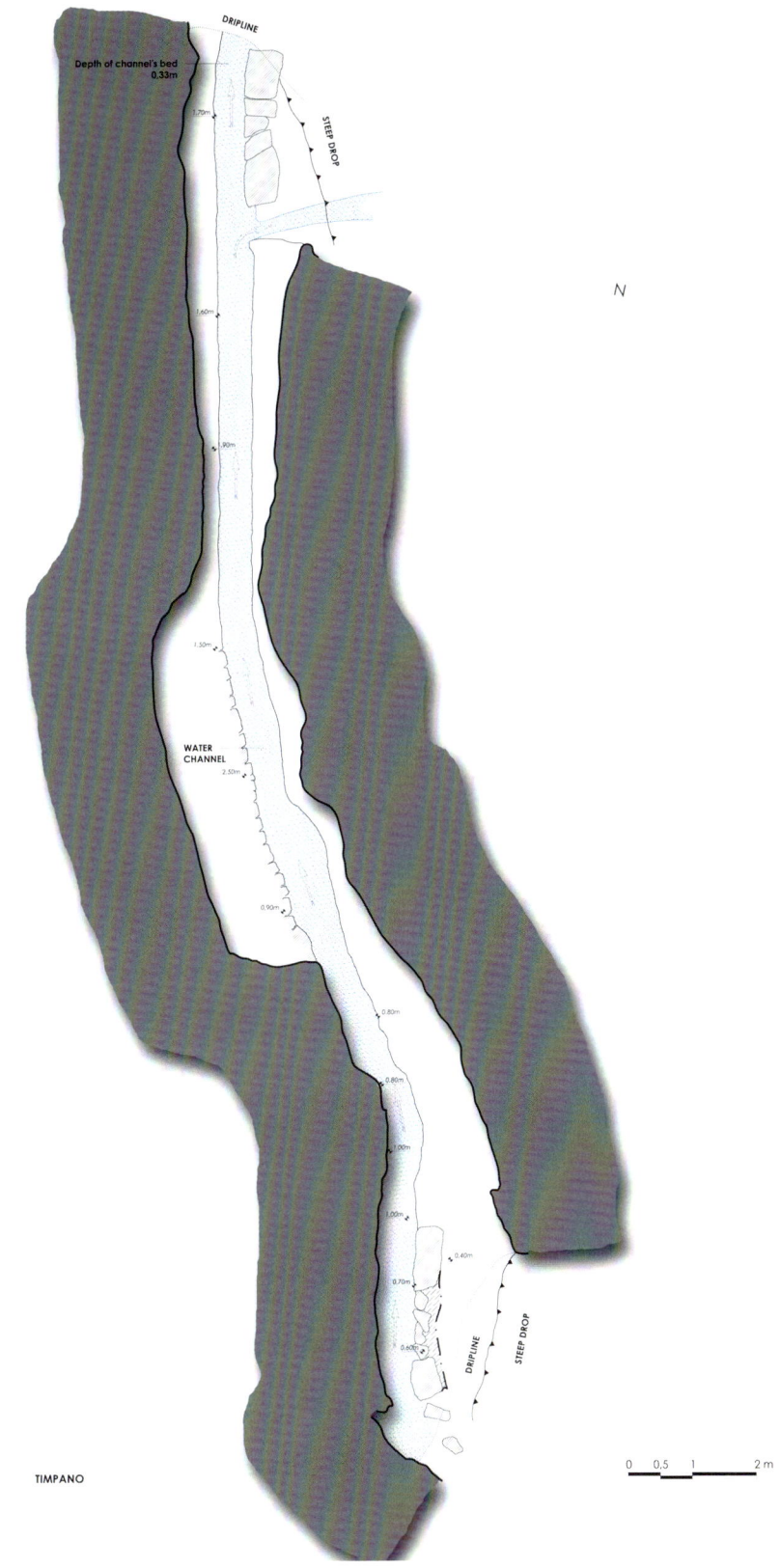

DRIPLINE

Depth of channel's bed
0.33m

1.70m

STEEP DROP

1.60m

N

1.90m

1.30m

WATER
CHANNEL

2.50m

0.90m

0.80m

0.60m

1.00m

1.00m

0.40m

0.70m

DRIPLINE

STEEP DROP

0.60m

TIMPANO

0   0,5   1   2 m

Fig. 30. Timpano (MOU-25).

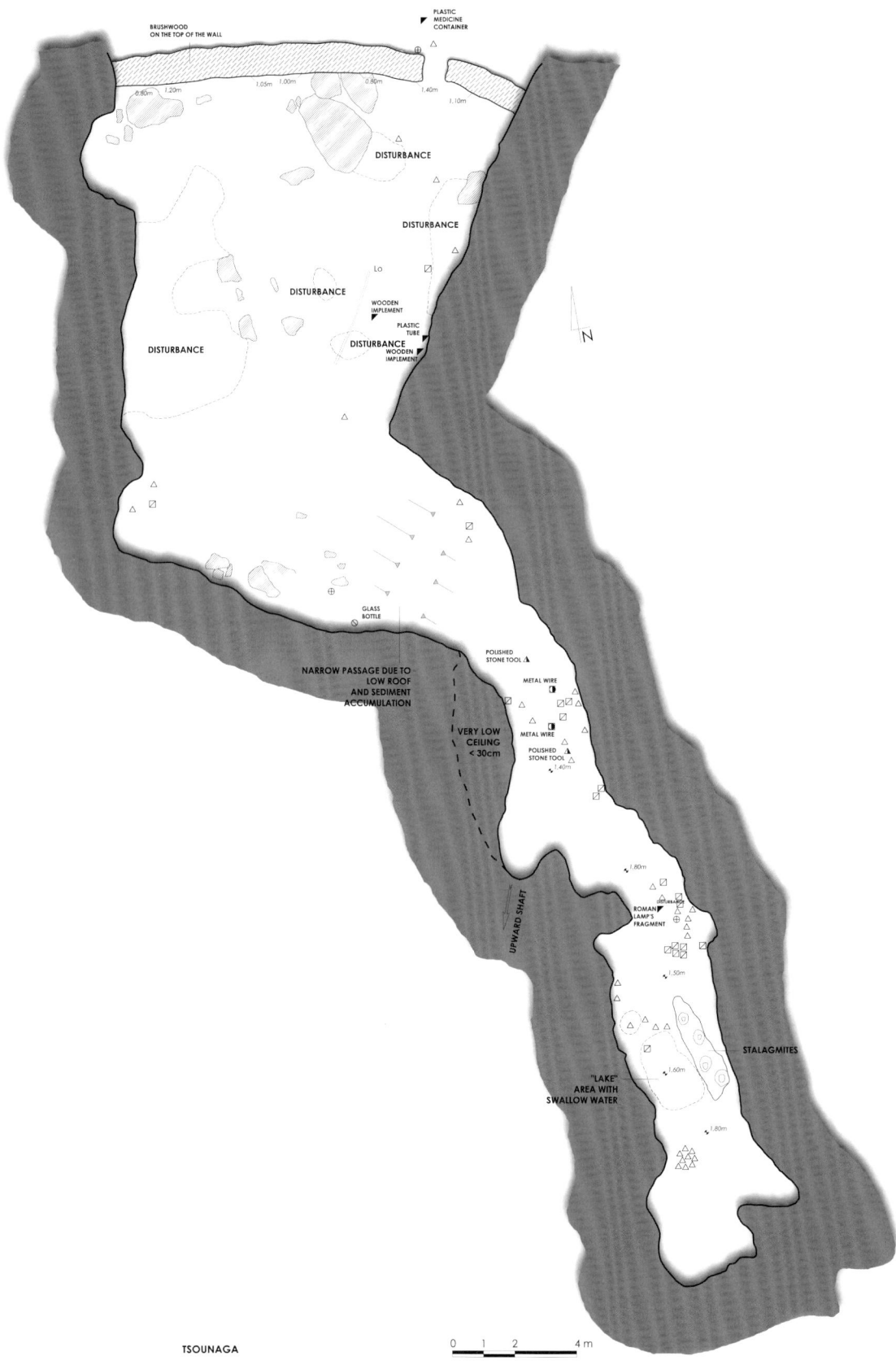

Fig. 31. Tsounaga (MOU-1).

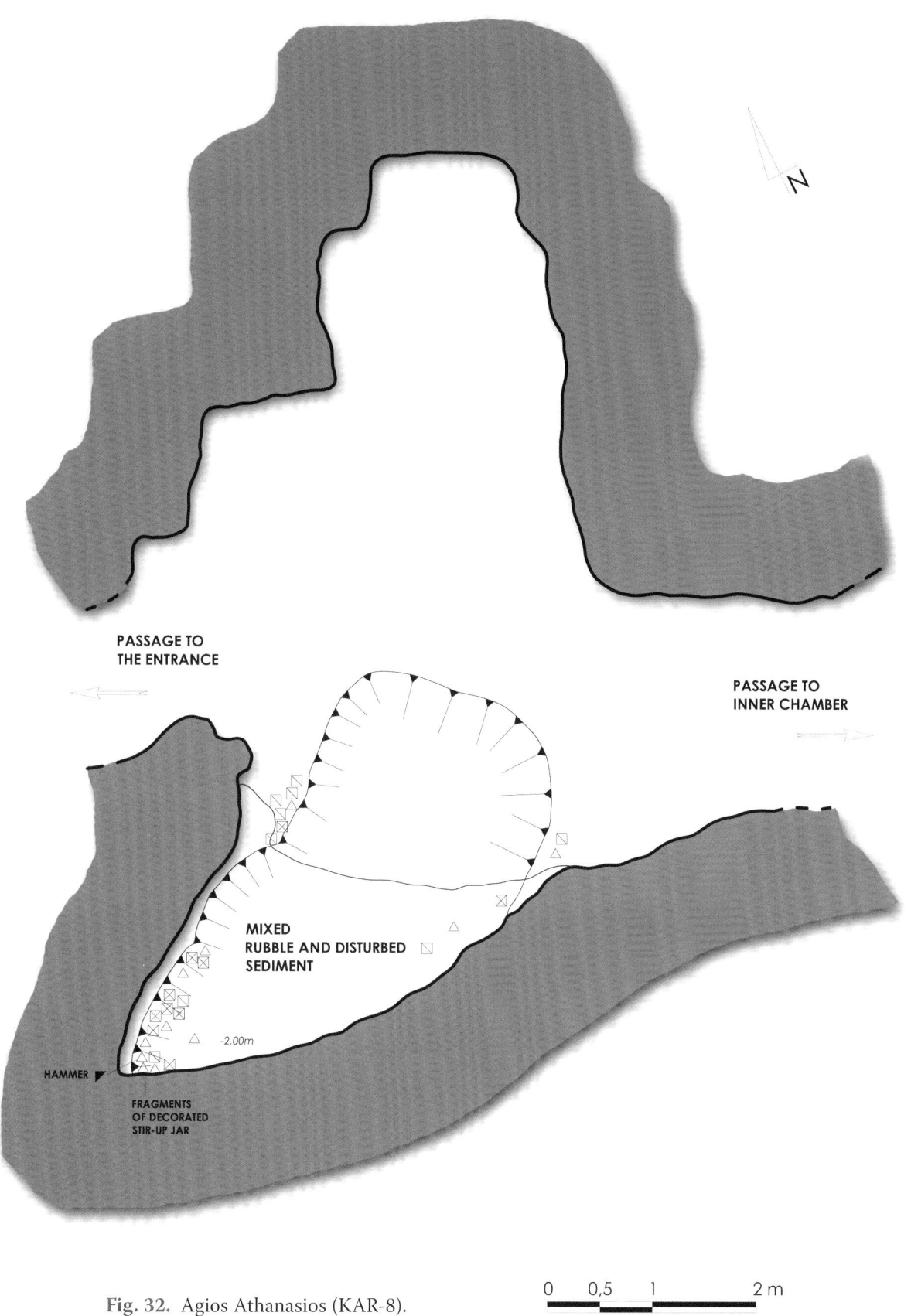

PASSAGE TO
THE ENTRANCE

PASSAGE TO
INNER CHAMBER

MIXED
RUBBLE AND DISTURBED
SEDIMENT

-2,00m

HAMMER

FRAGMENTS
OF DECORATED
STIR-UP JAR

N

0   0,5   1   2 m

Fig. 32. Agios Athanasios (KAR-8).

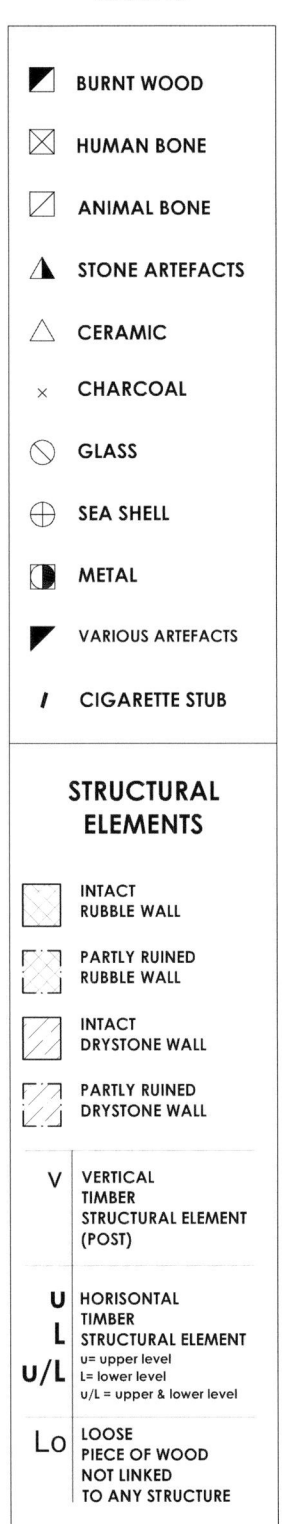

LEGEND

◤ BURNT WOOD

⊠ HUMAN BONE

◩ ANIMAL BONE

◣ STONE ARTEFACTS

△ CERAMIC

× CHARCOAL

⊘ GLASS

⊕ SEA SHELL

◖ METAL

◥ VARIOUS ARTEFACTS

/ CIGARETTE STUB

STRUCTURAL ELEMENTS

INTACT
RUBBLE WALL

PARTLY RUINED
RUBBLE WALL

INTACT
DRYSTONE WALL

PARTLY RUINED
DRYSTONE WALL

V VERTICAL
TIMBER
STRUCTURAL ELEMENT
(POST)

U HORISONTAL
L TIMBER
STRUCTURAL ELEMENT
u= upper level
U/L L= lower level
u/L = upper & lower level

Lo LOOSE
PIECE OF WOOD
NOT LINKED
TO ANY STRUCTURE

**Height
measurements**

+1,05m Height from Cave Floor to
Cave Roof

+1,05m Height above Ground
or Cave Floor

Appendix 2

Ioannis Voskos

Foreword

A full account of the Prehistoric to Roman period pottery collected by the survey is presented here. The catalogue functions as a supplement to the general discussions of fabrics, shapes, decoration and chronology found in chapter 2.6, where catalogue numbers denote specific references to items listed in the catalogue.

Rather than chronological, the catalogue shows the entries in site order to facilitate a comparison of the artefact assemblages from each cave. A presentation by site is better suited to the primary aim of this catalogue. A difficulty encountered in drawing up this catalogue was the abraded and fragmentary condition of most ceramic sherds that made dating and description of the vessels' original shape a challenging task. The aim was to create a comprehensive, brief and, to the extent possible, detailed catalogue,

appropriate for all types of potsherds, regardless of their date, class, and function.

All items have on their first line a catalogue number (Cat No.), an object description/shape, and occasionally a reference to an illustration. The site designation is found in parenthesis, followed by the portion of the object preserved and its dimensions, expressed in millimetres. Description of the fabric for the pottery includes the relative size, shape, colour, and type of inclusions, based on macroscopic inspection; and colour name, as determines by the Munsell Soil Colour Charts. Descriptions of the surface treatments are also given. Preservation is stated as 'very poor', 'poor', 'medium' or 'good'. A suggested chronological phase concludes each entry.

Besides the ceramics, the catalogue also includes two worked stones and four stone tools.

Catalogue of Prehistoric – Roman artefacts

Ceramics

3 COARSE COOKING POT

(ART-9) Rim (closed?). Max. pres. H. 35mm, Max. pres. W. 31mm, Th. 8mm. Very fine fabric, fine, mica and some white and brown inclusions (3-5%); evenly fired, exterior colour 7.5YR 7/4 (pink), interior colour 10YR 7/3 (very pale brown). Medium preservation.

Roman?

114 ROOF TILE

(POR-1) Pan tile fragment (painted). Max. pres. H. 84mm, Max. pres. W. 82mm, Th. 18mm. Fine-medium fabric, white and grey inclusions (3-5%); evenly fired, 5YR 5/6 (yellowish red). Medium-poor preservation.

Hellenistic?

115 ROOF TILE

(POR-1) Pan tile fragment (red painted). Max. pres. H. 81mm, Max. pres. W. 95mm, Th. 18mm. Fine-medium fabric, white and grey inclusions (3-5%); evenly fired, 10YR 7/4 (very pale brown), red paint 2.5YR 5/4 (reddish brown). Slightly curvilinear (Laconian type?); one surface covered with reddish slip. Medium preservation.

Hellenistic?

116 ROOF TILE

(POR-1) Pan tile fragment (black painted). Max. pres. H. 89mm, Max. pres. W. 59mm, Th. 19-20mm. Very fine fabric, whitish inclusions (1-2%); evenly fired, 2.5YR 3/2 (dusky red), black paint 2.5YR 2.5/1 (reddish black). Slightly curvilinear (Laconian type?); one surface has traces of black slip. Medium preservation.

Hellenistic?

117 ROOF TILE

(POR-1) Pan tile fragment (red painted). Max. pres. H. 92mm, Max. pres. W. 31mm, Th. 14-15mm. Very fine

to fine fabric, whitish and reddish inclusions (10%); evenly fired, 2.5YR 6/6 (light red), red paint 10R 4/6 (red). Flat surface; one surface has traces of red paint. Medium to poor preservation.

Hellenistic?

118 ROOF TILE

(POR-1) Pan tile fragment (not painted). Max. pres. H. 95mm, Max. pres. W. 41mm, Th. 20mm. Fine to medium fabric, schist, whitish and brownish inclusions (10-15%); evenly fired, 7.5YR 7/4 (pink). Slightly curvilinear. Medium preservation.

Hellenistic?

119 ROOF TILE

(POR-1) Cover tile fragment? Max. pres. H. 41mm, Max. pres. W. 59mm, Th. 15mm. Very fine fabric, whitish and reddish inclusions, grog? (2-3%); evenly fired, 2.5YR 6/8 (light red). Slightly curvilinear. Medium to poor preservation.

Hellenistic?

120 ROOF TILE

(POR-1) Pan tile fragment? Max. pres. H. 51mm, Max. pres. W. 45mm, Th. 18mm. Fine fabric, white, grey and black inclusions (15-20%); evenly fired, 2.5YR 6/6 (light red). Slightly curvilinear. Very poor preservation.

Hellenistic?

121 ROOF TILE

(POR-1) Pan tile fragment (not painted). Max. pres. H. 55mm, Max. pres. W. 32mm, Th. 17mm. Fine fabric, white and black inclusions (5-10%); evenly fired, 5YR 6/6 (reddish yellow). Poor preservation.

Hellenistic?

122 ROOF TILE

(POR-1) Pan or cover tile fragment. Max. pres. H. 34mm, Max. pres. W. 37mm, Th. 15mm. Fine to medium fabric, white, grey and brown inclusions (10-15%); evenly fired(?), colour unidentifiable. Poor preservation.

Hellenistic?

127 OPEN? VESSEL FIG. 2.6.1

(POR-3) Rim, handle. Max. pres. H. 76mm, Max. pres. W. 75mm (body), 48mm (handle), Th. 5-9mm (body). Medium to coarse fabric, quartz inclusions (10-15%); firing indeterminable, 10YR 6/3 (pale brown) and 10YR 5/3 (brown). Handle formed separately and attached to vessel; roughly smoothed, uneven exterior surface. Handmade; "Barbarian" or local ware. Medium preservation. Small 1990, 4, fig. 1, no 1.

Late Bronze Age

128 CARINATED OPEN VESSEL (LAMP?)

(POR-3) Body. Max. pres. H. 38mm, Max. pres. W. 20mm, Th. 3-5mm. Very fine fabric with no visible inclusions; evenly fired, 10YR 8/4 (very pale brown). Poor preservation.

Late Helladic or Roman

129 OPEN VESSEL

(POR-3) Body. Max. pres. H. 18mm, Max. pres. W. 41mm. Very fine fabric, some inclusions (2-3%); firing indeterminable, exterior surface: 2.5YR 4/2 (weak red), interior surface: 5YR 2.5/1 (black). Well burnished, traces of black paint on interior surface and reddish paint on the exterior. Medium to good preservation.

Prehistoric? Historic period?

202 SMALL OPEN VESSEL (PLATE?)

(KAR-8) Base and rim. Max. pres. H. 23mm, Max. pres. W. 61mm, Th. 5mm (base), 4-5mm (body). Very fine fabric with no inclusions; evenly fired, 5YR 6/6 (reddish yellow). Traces of the wheel on interior surface. Good preservation.

Roman

203 OPEN VESSEL FIG. 2.6.2

(KAR-8) Rim, body and handle. Max. pres. H. 88mm, Max. pres. W. 85mm (body), 32mm (handle), Th. 7-8mm (body), 9mm (handle). Medium to coarse fabric with mica, quartz and other greyish inclusions (15%); evenly fired, 10YR 3/1 (very dark grey). On top of both surfaces is a smoothed and burnished, thin dark wash; "Barbarian" or local ware. Good preservation. Small 1990, 4, fig. 1, no 1.

Late Helladic

204 ROOF TILE

(KAR-8) Cover tile fragment? Max. pres. H. 64mm, Max. pres. W. 62mm, Th. 20mm. Fine fabric with white, red and black inclusions (3-5%), evenly fired, 2.5YR 6/8 (light red), red paint 2.5YR 4/4 (reddish brown). Slightly curvilinear; traces of red paint on exterior surface. Medium preservation.

Hellenistic?

205 CLOSED VESSEL (JUG?)

(KAR-8) Body. Max. pres. H. 83mm, Max. pres. W. 56mm, Th. 9-10mm. Fine fabric with mica and grey and dark brown inclusions (5-7%), evenly fired, 7.5YR 6/6 (reddish yellow), red slip 5YR 5/3 (reddish brown). Traces of wheel ridges on interior surface and possible traces of red slip on exterior. Medium preservation.

Roman

206 CLOSED VESSEL (JAR?)

(KAR-8) Body. Max. pres. H. 77mm, Max. pres. W. 74mm, Th. 5mm. Medium fabric with inclusions (7-10%), evenly fired, exterior: 10YR 5/2 (greyish brown), interior: 5YR 5/6 (yellowish red). Interior surface is smoothed and wheel-ridged; exterior has a thin, dark wash and is burnished. Medium preservation.

Roman?

207, 248-249, 254 SMALL STIRRUP JAR FIG. 2.6.7

(KAR-8) Upper part of jar including neck, spout, a handle and part of the body. Max. pres. H. 123mm, Max. pres. W. 9mm (body), 10mm (handle), Th. 3-4mm (body), 5mm (handle). Very fine fabric with no visible inclusions, evenly fired, 10YR 8/4 (very pale brown). Exterior surface has black, red and light brown bands of paint around neck and body. Three parallel black bands around the body. More motives (red, brown) around the neck. The spout was attached to vessel when it was still leather-hard. Medium preservation.

Late Helladic III.

208 OPEN VESSEL

(KAR-8) Body. Max. pres. H. 6mm, Max. pres. W. 57mm, Th. 3-4mm. Very fine fabric with few inclusions (3%), evenly fired, 10YR 5/2 (greyish brown). Exterior surface has traces of dark paint. Medium preservation.

Late Bronze Age.

209 UNIDENTIFIABLE

(KAR-8)

210 CLOSED VESSEL? FIG. 2.6.13

(KAR-8) Handle, body. Max. pres. H. 111mm, Max. pres. W. 62mm, Th. 6mm. Fine fabric with mica and white-grey inclusions (10%), evenly fired (except handle), 2.5YR 6/6 (light red), red slip: 10R 4/3 (weak red). Traces of red burnished slip on exterior, smoothed. Handle is ridged and elliptical in section. Good preservation.

Hellenistic?

211 OPEN VESSEL FIG. 2.6.14

(KAR-8) Base, body. Max. pres. H. 89mm, Max. pres. W. 68mm, Th. 4-5mm (body) and 7mm (rim). Fine to medium fabric with white, grey and brown inclusions (10-15%), unevenly fired, core: 10YR 6/2 (light brownish grey), surfaces: 7.5YR 7/4 and 6/4 (pink and light brown). Wheel ridges and smoothed interior. Medium to good preservation. Rotroff 2006, fig. 46, nos 269-270.

Hellenistic?

212 DEEP BOWL (SKYPHOS) FIG. 2.6.9

(KAR-8) Body. Max. pres. H. 48mm, Max. pres. W. 59mm, Th. 4-5mm. Very fine fabric with few inclu-

sions (1-2%), evenly fired, 10YR 8/4 (very pale brown). Exterior: Dark brown paint (partly preserved), spiral or concentric circles motif; interior: covered with paint. Base of small horizontal handle is visible. Medium to good preservation.

Late Helladic

213 SMALL CLOSED VESSEL

(KAR-8) Body. Max. pres. H. 19mm, Max. pres. W. 28mm, Th. 3-4mm. Very fine fabric with few inclusions (1%), evenly fired, 10YR 7/4 (very pale brown). Painted red and brown parallel lines and bands. Good preservation.

Late Helladic

214 SMALL OPEN VESSEL? FIG. 2.6.15

(KAR-8) Base. Max. pres. H. 19mm, Max. pres. W. 61mm, D. 80mm (foot), Th. 5mm (foot). Very fine fabric with mica and white inclusions (3-5%), unevenly fired, core: 10YR 5/1 (grey), surfaces: 7.5YR 6/4 (light brown). Smoothed surface. Medium preservation. Rotroff 1997, fig. 1, nos 12-13.

Hellenistic-Roman?

215 OPEN VESSEL FIG. 2.6.16

(KAR-8) Rim and neck. Max. pres. H. 22mm, Max. pres. W. 44mm, Th. 6-7mm. Very fine fabric with mica and white-red inclusions (5%), evenly fired, 7.5YR 7/4 (pink). Wheel ridges. Medium preservation. Rotroff 1997, fig. 1, nos 12-13.

Hellenistic-Roman?

216-217 SMALL CLOSED VESSEL

(KAR-8) Body. Max. pres. H. 22mm, Max. pres. W. 19mm, Th. 6mm. Fine fabric with mica and white-dark brown inclusions (15%), unevenly fired, exterior: Gley 1 3/N (very dark grey), interior: 2.5YR 5/6 (red). Thin layer of slip above original clay. Medium preservation. Robinson 1959, pl. 13, K69-72 and also pl. 16, L4 and L6.

Roman

218-220 SMALL CLOSED VESSEL

(KAR-8) Body. Max. pres. H. 30mm, Max. pres. W. 20mm, Th. 4mm. Fine fabric with mica and white-red inclusions (15%), unevenly fired, exterior: 2.5YR 6/6 (light red), interior: 7.5YR 5/3 (brown). Possibly pattern-burnished? Three-four burnished lines on exterior surface. Medium preservation.

Undated

221-222 CLOSED VESSEL

(KAR-8) Body. Max. pres. H. 36mm, Max. pres. W. 57mm, Th. 4-5mm. Fine fabric with mica and white, grey and dark brown inclusions (15-20%), evenly fired, 5YR 5/6 (yellowish red). Wheel traces. Medium preservation.

Classical?

223 CLOSED VESSEL

(KAR-8) Body. Max. pres. H. 27mm, Max. pres. W. 25mm, Th. 5-6mm. Fine fabric with many white and grey inclusions (15%), unevenly fired, interior: 5YR 7/4 (pink), exterior: 10YR 7/4 (very pale brown), core: 7.5YR 7/2 (pinkish grey). Whitish slip above the original clay on exterior surface; wheel traces on interior surface. Poor preservation.

Classical?

224 – 231 CLOSED VESSEL (JAR OR FLASK?)

(KAR-8) Body. Max. pres. H. 54mm, Max. pres. W. 71mm, Th. 4-5mm. Fine to medium fabric with quartz(?) and grey inclusions (7-10%), unevenly fired, surfaces: 5YR 6/6 (reddish yellow), core: 5Y 5/1 (grey). Red polished slip on exterior; wheel traces on interior surface. Medium to good preservation.

Classical-Hellenistic?

232 – 234 CLOSED VESSEL (PITHOS?)

(KAR-8) Body. Max. pres. H. 68mm, Max. pres. W. 47mm, Th. 11mm. Coarse fabric with many inclusions (30-40%), unevenly fired, exterior: 5YR 5/6 (yellowish red), interior: 7.5YR 5/3 (brown), core: 10YR 4/1 (dark grey). Both surfaces burnished, thin reddish wash on exterior. Medium preservation.

Bronze Age

235 SMALL DRINKING CUP

(KAR-8) Body. Max. pres. H. 17mm, Max. pres. W. 22mm, Th. 3mm. Very fine fabric with few inclusions (less than 1%), evenly fired, 10YR 6/3 (pale brown). Burnish interior with wheel marks, polished exterior surface. Yellow Minyan type. Good preservation.

Middle to Late Helladic

236 CLOSED VESSEL

(KAR-8) Body. Max. pres. H. 23mm, Max. pres. W. 55mm, Th. 3-4mm. Very fine fabric with few inclusions (2-3%), evenly fired, Gley 1 4/N (dark grey). Traces of black paint on exterior. Medium preservation.

Late Bronze Age

237, 238 CLOSED VESSEL

(KAR-8) Body. Max. pres. H. 20mm, Max. pres. W. 26mm, Th. 5mm. Very fine to fine fabric with inclusions (10%), unevenly fired, exterior: 7.5YR 4/1 (dark grey), interior: 7.5YR 6/4 (light brown). Both surfaces smoothed and smoothed; thin, black wash on exterior. Medium preservation

Prehistoric?

239 CLOSED VESSEL

(KAR-8) Body. Max. pres. H. 23mm, Max. pres. W. 32mm, Th. 4mm. Fine to medum fabric with inclusions (15-20%), firing and colur unidentifiable. Interior smoothed and exterior burnished. Medium preservation.

Roman or post-Roman?

240 – 243 UNIDENTIFIABLE

(KAR-8)

244 CLOSED(?) VESSEL

(KAR-8) Body. Max. pres. H. 31mm, Max. pres. W. 38mm, Th. 2-3mm. Fine fabric with inclusions (3-5%), evenly fired, 10YR 3/2 (very dark greyish brown). Traces of paint. Medium preservation.

Prehistoric?

245 CLOSED VESSEL

(KAR-8) Body. Max. pres. H. 26mm, Max. pres. W. 32mm, Th. 3-4mm. Fine fabric with inclusions (3-5%), evenly fired, 10YR 6/3 (pale brown). Burnished; four (accidental?) parallel grooves. Medium preservation.

Prehistoric?

246 SMALL DRINKING VESSEL? FIG. 2.6.12

(KAR-8) Handle. Max. pres. H. 46mm, Max. pres. W. 8-12mm. Very fine fabric with no inclusions, evenly fired, Gley 1 2.5/N (black). Handle covered by black polished paint. Good preservation.

Classical or Hellenistic?

247 CLOSED VESSEL (AMPHORA?)

(KAR-8) Body. Max. pres. H. 70mm, Max. pres. W. 76mm. Th. 6-7mm. Very fine fabric with a few mica and very few mid-large grits (1-2%), evenly fired, 7.5YR 6/6 (reddish yellow). Wheel traces on interior. Medium preservation.

Roman?

250 CLOSED VESSEL

(KAR-8) Body. Max. pres. H. 48mm, Max. pres. W. 57mm. Th. 5mm. Fine to medium fabric with some inclusions (7-10%), evenly fired, 10YR 6/4 (light yellowish brown). Burnished. Medium preservation.

Prehistoric?

251 CLOSED VESSEL

(KAR-8) Body. Max. pres. H. 35mm, Max. pres. W. 29mm. Th. 3-4mm. Fine fabric with inclusions (5%), firing and colour indeterminable. Poor preservation.

Prehistoric?

252 OPEN BOWL WITH S-PROFILE (DIPPER?) FIG. 2.6.8

(KAR-8) Body. Max. pres. H. 87mm, Max. pres. W. 106mm. Fine to medium fabric with inclusions (10-15%), unevenly fired, surfaces: 7.5YR 6/4 (light brown), core: 10YR 5/1 (grey). Smoothed surface. Rim inclined slightly outwards. Medium to good preservation. Catling 2009, Vol. I, 221, 415-416 and Catling 2009, Vol. II, 266, fig. 270.

Late Helladic?

253 CLOSED VESSEL

(KAR-8) Body. Max. pres. H. 58mm, Max. pres. W. 29mm, Th. 7mm. Fabric indeterminable, clay: 5YR 6/8 (reddish yellow). Very poor preservation.

Roman?

255 OPEN VESSEL FIG. 2.6.3

(MIL-1) Rim. Max. pres. H. 49mm, Max. pres. W. 72mm, Th. 9mm. Medium fabric with traces of mica and inclusions (10%), unevenly fired, exterior: 5YR 5/4 (reddish brown), interior: 7.5YR 5/4 (brown), core: 2.5Y 5/1 (grey). Both surfaces smoothed and burnished. Rim slightly outward leaning. Good preservation. Small 1990, 4, figs. 1, 2 and 6.

Late Helladic?

324, 325 CLOSED VESSEL (AMPHORA?)

(MAK-12) Body (three re-joined sherds). Max. pres. H. 112mm, Max. pres. W. 57mm. Th. 6-7mm. Very fine fabric with mica and white inclusions (5-7%), unevenly fired, exterior: 5YR 5/6 (yellowish red), interior: 2.5YR 5/8 (red). Six parallel incised lines on one sherd. Medium to poor preservation.

Roman (2nd century AD)

326 CLOSED VESSEL (JUG/JUGLET OR LAGYNOS)

(MAK-12) Body. Max. pres. H. 51mm, Max. pres. W. 46mm, Th. 5-7mm. Very fine fabric with few inclusions (1%), evenly fired, 5YR 6/6 (reddish yellow), paint: 5YR 4/6 (yellowish red). Horizontal band (22mm wide) of red gloss. Medium preservation. Rotroff 1997, fig. 20, nos 125-126.

Late Hellenistic

327 OPEN VESSEL
(MAK-12) Body. Max. pres. H. 30mm, Max. pres. W. 54mm, Th. 4-7mm. Very fine fabric with no visible inclusions, evenly fired, 5YR 6/8 (reddish yellow). Medium preservation.

Classical-Hellenistic?

328 – 330 CLOSED VESSEL
(MAK-15) Body. Max. pres. H. 23mm, Max. pres. W. 29mm, Th. 5-7mm. Very fine fabric with mica and white-grey inclusions (7-10%), evenly fired, clay: 5YR 6/6 (reddish yellow). Possible traces of black slip. Very poor preservation.

Roman?

331 MEDIUM-SIZED CLOSED VESSEL
(MAK-16) Base, lower part. Max. pres. H. 22mm, Max. pres. W. 86mm, D. 90mm, Th. 5mm. Very fine fabric with few inclusions (1%), evenly fired, 5YR 6/6 (reddish yellow). Wheel ridges on interior. Medium to good preservation.

Roman?

332-333 CLOSED VESSEL
(MAK-16) Base. Max. pres. H. 10mm, Max. pres. W. 54mm, Th. 4-5mm (base). Very fine to fine fabric with mostly white-grey inclusions (15-20%), evenly fired, exterior: 7.5YR 6/4 (light brown), interior: 5YR 5/6 (yellowish red). Traces of wheel ridges. Medium preservation.

Late Hellenistic – Early Roman?

334-338 MEDIUM-LARGE CLOSED VESSEL
(MIL-13) Body and rim. Max. pres. H. 28mm, Max. pres. W. 155mm, Th. 10-11mm. Coarse fabric with quartz, mica, schist and white, grey and dark inclusions (15-20%), evenly fired, 7.5YR 5/4 (brown). Both surfaces smoothed and exterior burnished. Vessel likely for storage or transportation of liquids. Medium to poor preservation.

Prehistoric?

341-342 CLOSED VESSEL
(POR-3) Body. Max. pres. H. 55mm, H. 67mm, Th. 12-14mm. Medium fabric with mainly schist inclusions (20-25%), evenly fired, 2.5YR 4/4 (reddish brown) and 2.5YR 4/6 (red). Interior smoothed. Very poor preservation.

Roman?

343 THIN-WALLED OPEN VESSEL (DRINKING?)
(POR-3) Body. Max. pres. H. 49mm, Max. pres. W. 37mm, Th. 1-3mm. Very fine fabric with few inclusions (1%), evenly fired, 10YR 8/4 (very pale brown). Possible traces of red paint on exterior. Carinated? Very poor preservation.

Archaic?

344 ROOF TILE
(POR-3) Pan tile fragment (painted). Max. pres. H. 107mm, Max. pres. W. 95mm, Th. 17-19mm. Medium to coarse fabric with quartz, grog and dark brown-reddish inclusions (10%), evenly fired, 5YR 6/6 (reddish yellow). Poor preservation.

Roman?

350 BOWL?
(KAR-15) Body. Max. pres. H. 45mm, Max. pres. W. 62mm, Th. 4-9mm. Fine to medium fabric with mica, quartz and red-grey inclusions (10-15%), unevenly fired, 7.5YR 6/8 (reddish yellow), core: 10YR 6/2 (light brownish grey). Thin reddish crust on interior; smoothed. Poor to medium preservation.

Roman?

351 CLOSED VESSEL
(KAR-15) Body. Max. pres. H. 34mm, Max. pres. W. 22mm, Th. 5-6mm. Very fine to fine fabric with some mica and larger amounts of white-grey inclusions

(7-10%), evenly fired, 5YR 6/6 (reddish yellow). Exterior burnished, interior smoothed. Medium preservation.

Hellenistic-Roman?

352 CLOSED VESSEL

(KAR-15) Body. Max. pres. H. 30mm, Max. pres. W. 25mm, Th. 5-6mm. Very fine fabric with mica and some white and dark brown inclusions (2-3%), unevenly fired, interior: 7.5YR 6/6 (reddish yellow), exterior: 10YR 4/2 (greyish brown). Well burnished; exterior with thin crust of dark brown slip. Medium to good preservation.

Classical-Hellenistic?

353 OPEN VESSEL

(KAR-15) Body. Max. pres. H. 10mm, Max. pres. W. 32mm. Fine fabric with mainly schist and other grey and black inclusions (25-30%), evenly fired, 5YR 5/6 (yellowish red). Poor preservation.

Hellenistic-Roman?

354 MEDIUM TO LARGE OPEN VESSEL

(KAR-15) Body. Max. pres. H. 82mm, Max. pres. W. 65mm. Fine to medium fabric with quartz and black and grey inclusions (5-10%), evenly fired, clay: 10YR 6/2 (light brownish grey), slip: 2.5Y 7/2 (light grey). Light-coloured slip on exterior. Handmade vessel. Medium to good preservation.

Bronze Age?

355 CLOSED VESSEL

(KAR-15) Body. Max. pres. H. 77mm, Max. pres. W. 27mm, Th. 8-10mm. Fine fabric with white and grey inclusions (2-5%), evenly fired, clay: 10YR 7/3 (very pale brown), slip: 2.5Y 7/3 (pale yellow). Light-coloured slip on exterior. Handmade vessel. Medium to good preservation.

Bronze Age?

356 OPEN VESSEL?

(KAR-15) Body. Max. pres. H. 56mm, Max. pres. W. 32mm, Th. 7mm. Fine fabric with micra and few gray and brown inclusions (1-2%), unevenly fired, slip: 2.5Y 7/3 (pale yellow), clay: 5YR 6/6 (reddish yellow), core: 10YR 5/3 (brown). Polished slip on exterior. Thin yellowish coating on interior. Handmade. Medium preservation.

Bronze Age?

357 CLOSED VESSEL

(KAR-15) Body. Max. pres. H. 67mm, Max. pres. W. 65mm, Th. 7mm. Very fine to fine fabric with white and grey inclusions (10%), evenly fired, slip: 2.5Y 7/3 (pale yellow), clay: 10YR 7/4 (very pale brown). Handmade. Medium preservation.

Bronze Age?

358 OPEN VESSEL?

(KAR-15) Body. Max. pres. H. 66mm, Max. pres. W. 36mm, Th. 8-9mm. Very fine to fine fabric with reddish inclusions (3-5%), unevenly fired, core: 2.5Y 5/2 (greyish brown), clay: 7.5YR 7/6 (reddish yellow). Exterior polished. Handmade. Good preservation.

Bronze Age?

359 OPEN VESSEL?

(KAR-15) Body. Max. pres. H. 26mm, Max. pres. W. 36mm, Th. 5mm. Very fine fabric with white and red inclusions (3-5%), evenly fired, clay: 5YR 6/6 (reddish yellow), slip: 10YR 7/4 (very pale brown). Thin whitish slip and traces of red paint on exterior. Medium preservation.

Bronze Age?

360 CLOSED VESSEL (PITHOS)

(KAR-15) Base. Max. pres. H. 55mm, Max. pres. W. 117mm, D. 120-130mm, Th. 9-10mm (body) 19mm (base). Medium to coarse fabric with many white and grey inclusions (20-25%), unevenly fired, interior: 7.5YR 4/2 & 5/1 (brown & grey), slip: 2.5YR 4/4 (reddish brown). Slap building? Exterior thin layer of reddish slip. Poor preservation.

Roman?

361 CLOSED VESSEL (PITHOS)

(KAR-15) Body. Max. pres. H. 53mm, Max. pres. W. 39mm, Th. 16-17mm. Medium fabric with mica, quartz and other grey inclusions (20-25%), unevenly fired, clay: 2.5YR 6/6 (light red), core: 2.5Y 5/1 (grey). Poor preservation.

Early Historical period

362 CLOSED VESSEL

(KAR-15) Body. Max. pres. H. 35mm, Max. pres. W. 30mm, Th. 3mm. Fine fabric with white-grey inclusions (3-5%), unevenly fired, interior: 7.5YR 5/2 (brown), exterior: 5YR 5/4 (reddish brown). Wheel ridges? Medium preservation.

Hellenistic-Roman

363-376 UNIDENTIFIABLE

(KAR-15)

377 SMALL CLOSED VESSEL   FIG. 2.6.11

(KAR-15) Rim and body. Body. Max. pres. H. 32mm, Max. pres. W. 35mm, Th. 4mm. Very fine fabric with few white-grey inclusions (1-2%), evenly fired, 10YR 7/4 (very pale brown), red paint: 5YR 5/6 (yellowish red). Both surfaces well burnished. Reddish paint on exterior parallel to rim. Good preservation.

Archaic-Classical

378 BOWL

(KAR-15) Rim. Max. pres. H. 59mm, Max. pres. W. 28mm, Th. 6mm. Fine fabric with few red-brown inclusions (3-5%), evenly fired, slip: 5Y 8/2 (pale yellow), clay: 7.5YR 7/4 (pink). Good preservation.

Bronze Age

379 CLOSED VESSEL?

(KAR-15) Handle. Max. pres. H. 77mm, Max. pres. W. 60mm, Th. 20mm (handle). Fine fabric with mica and very few other inclusions (1%), unevenly fired, clay: 7.5YR 7/4 (pink), core: 2.5Y 6/2 (light brownish grey),

slip: 2.5Y 8/3 (pale yellow). Medium to good preservation.

Prehistoric?

380 OPEN VESSEL?

(KAR-15) Body. Max. pres. H. 97mm, Max. pres. W. 72mm, Th. 10-12mm. Very fine fabric with mica and white-grey inclusions (5-7%), evenly fired, 5YR 6/6 (reddish yellow). Both surfaces burnished. Medium preservation.

Roman?

381 OPEN VESSEL?

(KAR-15) Body. Max. pres. H. 35mm, Max. pres. W. 21mm, Th. 4-5mm. Very fine fabric with white-grey inclusions (5-7%), evenly fired, 7.5YR 6/6 (reddish yellow). Both surfaces well burnished. Medium preservation.

Roman?

382 CLOSED VESSEL (PITHOS)

(KAR-15) Body. Max. pres. H. 93mm, Max. pres. W. 55mm, Th. 14-16mm. Medium to coarse fabric with mica, schist, quartz and other white-grey inclusions (30-40%), unevenly fired, surfaces: 2.5YR 6/6 (light red), core: 10YR 6/4 (light yellowish brown). Poor preservation.

Early Historical?

383 OPEN VESSEL?

(KAR-15) Body. Max. pres. H. 26mm, Max. pres. W. 29mm, Th. 8-9mm. Medium to coarse fabric with mica, schist, quartz and other white-grey inclusions (30%), unevenly fired, surfaces: 5YR 5/6 (yellowish red), core: 10YR 5/1 (grey). Thin layer of reddish slip (or paint?) on both surfaces. Poor preservation.

Early Historical?

384-385 CLOSED VESSEL

(KAR-15) Body. Max. pres. H. 28mm, Max. pres. W. 31mm, Th. 11-13mm. Fine to medium fabric with mica and reddish and grey inclusions (7-10%), unevenly fired, surfaces: 10YR 6/4 (yellowish brown), core: 10YR 6/2 (light brownish grey). Exterior well burnished. Thick-walled sherd, possibly pithoid. Very poor to medium preservation.

Roman?

386 OPEN VESSEL?

(KAR-15) Body and part of handle. Max. pres. H. 41mm, Max. pres. W. 34mm, Th. 7-8mm. Very fine fabric with mica and white-grey inclusions (3-5%), unevenly fired, surfaces: 10YR 7/4 (very pale brown) & 10YR 6/4 (light yellowish brown), core: Gley 1 4/1 (very dark grey), red paint: 2.5YR 4/8 (red). Exterior red painted and traces of red paint; also on interior. Possibly for serving food or drink. Poor preservation.

Archaic or Classical

387 CLOSED VESSEL

(KAR-15) Body. Max. pres. H. 69mm, Max. pres. W. 46mm, Th. 7-8mm. Very fine fabric with mica and white inclusions (2-3%), evenly fired, 7.5YR 6/3 (light brown), red paint: 2.5YR 5/6 (red). Exterior red painted. Possibly vessel for transportation or storage of liquids. Medium preservation.

Archaic or Classical?

388 OPEN VESSEL

(KAR-15) Body. Max. pres. H. 90mm, Max. pres. W. 85mm, Th. 20mm. Fine to medium fabric with few inclusions (5%), unevenly fired, slip: 2.5Y 8/3 (pale yellow), clay: 7.5YR 6/4 (light brown), core: 2.5Y 6/2 (light brownish grey). Light-coloured slip. Handmade vessel. Medium preservation.

Bronze Age?

389 CLOSED VESSEL

(KAR-15) Body. Max. pres. H. 46mm, Max. pres. W. 59mm, Th. 4-5mm. Fine to medium fabric with mica and few white-grey inclusions (5-7%), evenly fired, slip: 2.5Y 8/3 (pale yellow), clay: 5YR 5/2 (reddish grey).

Light-coloured slip on exterior. Handmade vessel. Medium preservation.

Bronze Age?

390 OPEN VESSEL

(KAR-15) Body. Max. pres. H. 55mm, Max. pres. W. 48mm, Th. 13mm. Very fine fabric with few inclusions (1%), unevenly fired, slip: 2.5YR 6/6 (red), clay: 2.5YR 6/3 (light reddish brown). Reddish slip on interior. Poor to medium preservation.

Early Historical?

391 CLOSED VESSEL

(KAR-15) Body. Max. pres. H. 24mm, Max. pres. W. 30mm, Th. 8mm. Fine fabric with mica and other red-brown inclusions (5-10%), evenly fired, 10YR 7/4 (very pale brown). Exterior well burnished. Medium preservation.

Roman?

392 CLOSED VESSEL

(KAR-15) Body. Max. pres. H. 67mm, Max. pres. W. 51mm, Th. 10-11mm. Very fine fabric with mica and white-grey inclusions (10-15%), evenly fired?, 10YR 7/3 (very pale brown). Surfaces well burnished. Wheelmade. Medium preservation.

Early Historical

393 CLOSED VESSEL? (PITHOS?)

(KAR-15) Base.

Date?

394 CLOSED VESSEL (PITHOS?)

(KAR-16) Body. Max. pres. H. 106mm, Max. pres. W. 65mm, Th. 13-16mm. Medium fabric with mica, quartz and other grey-white inclusions (20-25%), evenly fired, 5YR 5/3 (reddish brown). Exterior burnished, interior smoothed. Poor preservation.

Early Historical

395 CLOSED VESSEL

(KAR-16) Body. Max. pres. H. 21mm, Max. pres. W. 19mm, Th. 8mm. Fine fabric with mica, schist and grey-white and red inclusions (10-15%), firing and colour indeterminable. Very poor preservation.

Hellenistic-Roman?

396 CLOSED VESSEL (PITHOS?)

(KAR-16) Body. Max. pres. H. 82mm, Max. pres. W. 72mm, Th. 16-18mm. Fine to medium fabric with mica and other white-grey inclusions (20%), unevenly fired, clay: 5YR 5/4 (reddish brown), core: unidentifiable. Exterior burnished, interior smoothed. Poor preservation.

Hellenistic-Roman?

397 CLOSED VESSEL

(KAR-16) Body. Max. pres. H. 26mm, Max. pres. W. 33mm, Th. 8-9mm. Fine to medium fabric with mica and other grey-white inclusions (10%), evenly fired, 5YR 6/6 (reddish yellow). Exterior polished, interior burnished. Handmade. Good preservation.

Date?

398 CLOSED VESSEL

(KAR-16) Body. Max. pres. H. 24mm, Max. pres. W. 27mm, Th. 8mm. Fine to medium fabric with white-grey inclusions (5-10%), evenly fired, clay: 10YR 7/2 (light grey), slip: 2.5 7/2 (light grey). Thin layer of polished slip. Medium preservation.

Date?

399 CLOSED VESSEL

(KAR-16) Body. Max. pres. H. 39mm, Max. pres. W. 28mm, Th. 7-8mm. Fine fabric with white-grey inclusions (3-5%), evenly fired, clay: 10YR 7/4 (very pale brown), slip: 2.5 7/3 (pale yellow). Exterior polished, interior burnished. Medium preservation.

Date?

400 CLOSED VESSEL

(KAR-16) Body. Max. pres. H. 28mm, Max. pres. W. 43mm, Th. 9mm. Fine to medium fabric with mica and grey and red inclusions (5-10%), 10YR 6/3 (pale brown). Exterior polished. Medium preservation.

Date?

401 CLOSED VESSEL

(KAR-16) Body. Max. pres. H. 38mm, Max. pres. W. 32mm, Th. 8-9mm. Fine to medium fabric with white-grey inclusions (5%), 7.5YR 7/4 (pink), slip: 2.5Y 7/3 (pale yellow). Exterior polished. Medium preservation.

Date?

402 CLOSED VESSEL

(KAR-16) Body. Max. pres. H. 55mm, Max. pres. W. 50mm, Th. 7-11mm. Fine fabric with few red inclusions (1-2%), 7.5YR 7/4 (pink), slip: 2.5Y 8/3 (pale yellow). Exterior polished, interior smoothed. Medium preservation.

Date?

403 CLOSED VESSEL

(KAR-16) Body. Max. pres. H. 42mm, Max. pres. W. 40mm, Th. 10-11mm. Fine to medium fabric with few red inclusions (3-5%), 7.5YR 6/3 (light brown), slip: 2.5Y 8/3 (pale yellow). Exterior polished, interior smoothed. Medium preservation.

Date?

404 UNIDENTIFIABLE

(KAR-16)

405 OPEN VESSEL?   FIG. 2.6.6

(KAR-16) Body and part of handle. Max. pres. H. 85mm, Max. pres. W. 113mm, Th. 8mm. Very fine fabric with mica, quartz and white-grey inclusions (3-5%), unevenly fired, interior: 7.5YR 5/4 (brown), exterior: 7.5YR 5/4-6/4 (brown-light brown), core: 2.5Y 4/1 (dark grey). Well smoothed and burnished. Thin layer of buff clay on both surfaces. Handle is slightly curvilinear with ovoid cross-section. Handmade. Medium preservation.

Bronze Age?

409 CLOSED VESSEL

(KAR-22) Body. Max. pres. H. 23mm, Max. pres. W. 31mm, Th. 4mm. Very fine fabric with mica and few whitish inclusions (1-2%), unevenly fired, interior: 5YR 6/6 (reddish yellow), exterior: 10YR 7/4 (very pale brown). Thin crust of whitish slip on exterior. Incised-impressed decoration of three curvilinear motifs (Grooved Ware). Good preservation.

Late Roman?

410 CLOSED VESSEL

(KAR-23) Body. Max. pres. H. 29mm, Max. pres. W. 38mm, Th. 3-4mm. Fine fabric with schist, quartz and other grey and white inclusions (25-30%), unevenly fired, interior: 5YR 5/6 (yellowish red), exterior: 5YR 5/4 (reddish brown). Very poor preservation.

Roman?

411 SMALL OPEN DRINKING VESSEL

(KAR-23) Rim. Max. pres. H. 18mm, Max. pres. W. 20mm, Th. 2mm. Very fine fabric with inclusions of mica, unevenly fired, exterior: 10YR 6/4 (light yellowish brown), interior: 7.5YR 6/6 (reddish yellow). Wheelmade. Medium preservation.

Roman?

418 JUG WITH TREFOIL MOUTH   FIG. 2.6.26

(KER-3) Upper part of jug. Max. pres. H. 100mm (body) 77mm (handle), Max. pres. W. 120mm (body) 27mm (handle), D. 120-130mm, Th. 4-6mm. Fine fabric with mica, grey and reddish inclusions (10-15%), evenly fired, 5YR 5/6 (yellowish red). Traces of wheel-ridges on both surfaces. Medium preservation. Robinson 1959, pl. 7, G188 and pl. 23, M101.

Middle Roman (2nd-3rd centuries AD

419-420 SMALL BOWL   FIG. 2.6.19 & 2.6.20

(MOU-1) Body and rim. Max. pres. H. 57mm, Max. pres. W. 130mm, D. 100mm (base) 150-160mm (rim), Th. 2-3. Fine fabric with much mica and many white-grey inclusions (10-15%), evenly fired, 5YR 5/6 (yellowish red). Wheel traces on both surfaces, possible

traces of burning on exterior. Rim is outturned. Good preservation.

Hellenistic-Roman?

421 OPEN VESSEL

(MOU-1) Body. Max. pres. H. 30mm, Max. pres. W. 49mm, Th. 10-11mm. Fine to medium fabric with mica, schist, quartz and other white, grey and reddish inclusions (7-10%), unevenly fired, exterior: 2.5YR 4/4 (reddish brown), interior: 5YR 5/4 (reddish brown), core: 2.5Y 5/1 (grey). Thin layer of fine reddish clay on both surfaces; well burnished. Good preservation.

Prehistoric?

422 CLOSED VESSEL (PITHOID)

(MOU-1) Body. Max. pres. H. 187mm, Max. pres. W. 163mm, Th. 30mm. Medium to coarse fabric with mica, schist, quartz and other white and grey inclusions (30-40%), evenly fired, 5YR 6/6 (reddish yellow). Exterior burnished, interior smoothed. Poor to medium preservation.

Hellenistic-Roman?

423 LARGE DEEP BOWL?

(MOU-1) Rim. Max. pres. H. 134mm, Max. pres. W. 70m, Th. 9-10mm. Fine fabric with mostly white inclusions (10-15%), evenly fired, 2.5YR 3/1 (very dark grey). Both surfaces well burnished; firing clouds. Good preservation.

Date?

424 OPEN VESSEL   FIG. 2.6.10

(MOU-1) Rim. Max. pres. H. 53mm, Max. pres. W. 45mm, Th. 10-12mm. Medium fabric with mica, quartz and other white and reddish inclusions (10%), unevenly fired, exterior: 5YR 4/4 (reddish brown), interior: 7.5YR 5/4 (brown), core: 10R 5/1 (reddish grey). Thin layer of fine reddish clay on both surfaces; well burnished. Good preservation. Possibly from same vessel as 421.

Prehistoric?

425 CLOSED VESSEL

(MOU-1) Body. Max. pres. H. 55mm, Max. pres. W. 47mm, Th. 10-17mm. Medium to coarse fabric with mostly schist, mica, quartz and other white-grey inclusions (30-40%), evenly fired, 5YR 5/6 (yellowish red) & 2.5YR 5/8 (red). Very poor preservation.

Early Historical?

426 CLOSED VESSEL (FOR TRANSPORT OF LIQUIDS?)

(MOU-1) Body. Max. pres. H. 104mm, Max. pres. W. 72mm, Th. 10-12mm. Fine fabric with much mica and white-grey inclusions (7-10%), unevenly fired, surfaces: 5YR 6/6 (reddish yellow), core: 10YR 5/1 (grey), slip: 10YR 3/2 (very dark greyish brown). Wheelmade; traces of polished slip on exterior. Very poor preservation.

Hellenistic-Roman?

427 CLOSED VESSEL?

(MOU-1) Base. Max. pres. H. 28mm, Max. pres. W. 76mm, Th. 6-7mm (body) 9-10mm (base). Fine fabric with inclusions (7-10%), evenly fired, 2.5YR 5/4 (reddish brown). Black paint on exterior part of base. One diagonal incised line. Good preservation.

Hellenistic-Roman?

428 CLOSED VESSEL

(MOU-1) Body. Max. pres. H. 25mm, Max. pres. W. 47mm, Th. 4mm. Fine fabric with mica and few inclusions (1%), unevenly fired, exterior: 10YR 7/4 (very pale brown), interior: 5YR 7/6 (reddish yellow). Wheel ridges on interior; curvilinear incisions on exterior. Medium preservation.

Late Roman

429 CLOSED VESSEL

(MOU-1) Body. Max. pres. H. 59mm, Max. pres. W. 46mm, Th. 3-4mm. Fine fabric with much mica and white and dark brown inclusions (15-20%), unevenly fired, surfaces: 5YR 6/6 (reddish yellow), core:

10YR 6/2 (light brownish grey). Rough exterior and smoothed interior. Poor to medium preservation.

Hellenistic-Roman?

430 CLOSED VESSEL (RODIAN AMPHORA?)

(MOU-1) Body. Max. pres. H. 54mm, Max. pres. W. 49mm, Th. 9-12mm. Very fine fabric with much mica and few white and dark brown inclusions (3-5%), evenly fired, 7.5YR 6/6 (reddish yellow). Wheel ridge on interior. Medium preservation. Joncheray 1976, 22, pl. V, nos 52-53.

Hellenistic-Roman?

431 CLOSED VESSEL

(MOU-1) Body. Max. pres. H. 40mm, Max. pres. W. 34mm, Th. 5mm. Fine fabric with mica, evenly fired, 5YR 6/6 (reddish yellow). Poor preservation.

Hellenistic-Roman?

432 CLOSED VESSEL

(MOU-1) Base. Max. pres. H. 25mm, D. 104mm, Th. 15mm. Very fine to fine fabric with few mica and other white inclusions, evenly fired, 5YR 6/6 (reddish yellow). Good preservation.

Hellenistic

433 CLOSED VESSEL

(MOU-1) Body. Max. pres. H. 59mm, Max. pres. W. 43mm, Th. 4-5mm. Very fine fabric with no inclusions except mica, evenly fired, 5YR 6/6 (reddish yellow). Two incised lines, slightly curvilinear. Poor to medium preservation.

Hellenistic-Roman

434 LAMP FIG. 2.6.17

(MOU-1) Fragment. Max. pres. H. 50mm (spout), Max. pres. W. 20mm (spout), Th. 3mm (body) 5mm (spout). Very fine fabric with no visible inclusions, Gley 1 3/N (very dark grey). Traces of burnishing. Black painted. Medium preservation. Howland 1958, 121, pl.45, no 504. BE 8617 in Doulgeri-Intzesiloglou 1994, 371-372, pl.278a-b; also BE 16027 in Nikolaou 2004, pl. 12a and BE 9335 in Malakasioti 2004, 95-97, pl. 19.

Hellenistic?

435 CLOSED VESSEL

(MOU-1) Body. Max. pres. H. 144mm, Max. pres. W. 94mm, Th. 7-9mm. Fine fabric with much mica and many grey and dark brown inclusions (3-5%), evenly fired, 7.5YR 6/6 (reddish yellow). Rough exterior; wheel ridges on interior. Medium preservation.

Late Hellenistic?

436, 440 CLOSED VESSEL   FIG. 2.6.28

(MOU-1) Base. Max. pres. H. 42mm, D. 90mm (body) 30mm (foot), Th. 2-4mm. Very fine fabric with mica and white-grey inclusions (5-7%), unevenly fired, surfaces: 2.5YR 4/4 (reddish brown), core: 2.5YR? (dark reddish grey). Possible traces of red paint. Same vessel as sherd 440? Good preservation. Robinson 1959, 101-102, M190-M194.

Roman

437 JUG OR AMPHORA?

(MOU-1) Neck and body. Max. pres. H. 45mm, Max. pres. W. 59mm, Th. 5mm (body) 3mm (neck). Very fine fabric with a few white inclusions (3%), unevenly fired, exterior: 10YR 7/4 (very pale brown), clay: 5YR 6/6 (reddish yellow). Shallow incisions parallel to the neck. Possible wheel ridge on interior. Medium preservation.

Late Hellenistic – Early Roman

438 CLOSED VESSEL

(MOU-1) Body. Max. pres. H. 62mm, Max. pres. W. 65mm, Th. 6-7mm. Very fine to fine fabric with quartz, mica and other white-grey and dark brown inclusions (20-25%), unevenly fired, exterior: 5YR 6/6 (reddish yellow), interior: 5YR 6/4 (light reddish brown), core: 10YR 6/1 (grey). Wheel ridges on interior. Poor to medium preservation.

Hellenistic-Roman?

439 OPEN VESSEL   FIG. 2.6.27

(MOU-1) Base. Max. pres. H. 14mm, Max. pres. W. 59mm, D. 59mm (base), Th. 2-3mm (body), 5mm (base). Very fine fabric with few dark brown inclusions (1-2%), evenly fired, clay: 5YR 6/6 (reddish yellow), red slip: 2.5YR 5/6 (red). Red slip on both surfaces. Easter sigilata C – Çandarli (L19). Very good preservation. Hayes 1985, 76-77, pl. XVII, nos 5-7.

Early Roman (1st-2nd century AD)

441 CLOSED VESSEL (AMPHORA)

(MOU-1) Body. Max. pres. H. 44mm, Max. pres. W. 63mm, Th. 10mm. Very fine fabric with few dark brown inclusions (2-3%), unevenly fired, exterior: 7.5YR 7/6 (reddish yellow), interior: 5YR 7/6 (reddish yellow), core: 7.5YR 7/4 (pink). One incised line on interior. Medium preservation. Joncheray 1976, 22, pl. V, nos 52-53.

Late Hellenistic

442 CLOSED VESSEL

(MOU-1) Body. Max. pres. H. 24mm, Max. pres. W. 37mm, Th. 8-9mm. Medium to coarse fabric with mica, pieces of schist and white-grey inclusions (10-15%), unevenly fired, exterior: 2.5YR 5/6 (red), interior: 2.5YR 3/2 (dusky red). Very poor preservation.

Roman?

443 SMALL OPEN DRINKING VESSEL

(MOU-1) Body. Max. pres. H. 20mm, Max. pres. W. 26mm, Th. 3-4mm. Fine fabric with few white inclusions (5%), unevenly fired, both surfaces: 2.5YR 4/2 (weak red), clay: 2.5YR 5/6 (red). Thin layer of brownish clay covers interior. Poor preservation.

Hellenistic-Roman?

444 CLOSED VESSEL (PITHOS?)

(MOU-1) Body. Max. pres. H. 128mm, Max. pres. W. 111mm, Th. 10mm. Fine fabric with mica and few dark brown inclusions (1-2%), exterior: 7.5YR 6/6 (reddish yellow). Shallow, horizontal incision parallel with the body of the vessel. Very poor to poor preservation.

Hellenistic?

PLATE?

(MOU-1) Body. Max. pres. H. 89mm, Max. pres. W. 113mm. Very fine fabric with mica and other grey and white inclusions (3-5%), evenly fired?, 5YR 6/6 (reddish yellow). Wheel traces on interior. Poor to medium preservation.

Hellenistic?

450 PLATE?

(MOU-1) Body. Max. pres. H. 53mm, Max. pres. W. 102mm, Th. 4-8mm. Fine fabric with mica and white-grey and dark brown inclusions (15-20%), evenly fired, 5YR 6/6 (reddish yellow). Curvilinear incised parallel lines on exterior (accidental?). Medium preservation. Possibly from same vessel as 451.

Hellenistic

451 PLATE?

(MOU-1) Body and part of ring foot. Max. pres. H. 55mm, Max. pres. W. 68mm, Th. 4-6mm. Fine fabric with white-grey inclusions (5-7%), unevenly fired, surfaces: 2.5YR 6/6 (light red), core: 7.5YR 5/3 (brown), paint: 7.5YR 4/1 (dark grey). Black paint covers interior, band of black paint on exterior (width 3cm). Medium to good preservation. Possibly from same vessel as 450. Papaconstantinou 2000, 211-212, plates 11-12; Rotroff 1997, 142-150.

Hellenistic

454 OPEN VESSEL?   FIG. 2.6.18

(MOU-1) Base (ring foot). Max. pres. H. 28mm, Max. pres. W. 82mm, Th. 6mm (body and base). Fine fabric with mostly mica inclusions (15%), evenly fired, 2.5YR 4/2 (weak red). Thin layer of black slip above the original reddish clay on both surfaces. Medium preservation. Rotroff 1997, 117-119.

Hellenistic

455 PLATE?

(MOU-1) Rim and neck. Max. pres. H. 42mm, Max. pres. W. 38mm, D. 50-55mm, Th. 3mm. Very fine fabric with few inclusions (1-2%), evenly fired, clay: 5YR 6/6 (reddish yellow), surfaces: 10YR 7/4 (very pale brown). Thin pale coating on both surfaces. Rilled rim and neck of vessel for serving food. Good preservation.

Hellenistic

456 OPEN VESSEL?   FIG. 2.6.29

(MOU-1) Base. Max. pres. H. 27mm, Max. pres. W. 116mm, D. 82mm (foot), Th. 5mm (body). Fine fabric with mica and white inclusions (5%), evenly fired, 5YR 5/6 (yellowish red), Wheel ridges on both surfaces. Base has projecting foot or low ring base. Medium preservation.

Roman?

457 MEDIUM-SIZED CLOSED VESSEL (AMPHORA?)

(MOU-1) Body. Max. pres. H. 132mm, Max. pres. W. 129mm, Th. 7-8mm. Very fine fabric with white-grey inclusions (3-5%), evenly fired, coating: 5YR 6/4 (light reddish brown), interior and clay: 2.5YR 6/6 (light red). Both surfaces smoothed, thin coating on exterior. Rodian amphora? Medium preservation.

Hellenistic

460-520 AMPHORA

(MOU-11) Body. Max. pres. H. 70mm, Max. Pres. W. 97mm, Th. 4-7mm. Fine to medium fabric with mica and a few white-grey inclusions (10-15%), evenly fired, clay: 7.5YR 6/4 (light brown). Grooved decoration on most sherds. Medium preservation. Possibly from same vessel as 521. Ntina 2010, 564 and pl. 3.

Late Roman

521 AMPHORA   FIG. 2.6.30

(MOU-11) Handle. Max. pres. H. 58mm, Max. Pres. W. 21mm, Th. 32mm (handle). Fine to medium fabric with mica and many white-grey inclusions (10-15%), evenly fired, clay: 7.5YR 6/4 (light brown). Longitudinal ridge on handle. Medium preservation. Possibly from same vessel as 460-520. Robinson 1959, 122, N7.

Late Roman

522 AMPHORA?   FIG. 2.6.31

(MOU-11) Handle. Max. pres. H. 48mm, Max. Pres. W. 18mm, Th. 26mm (handle) 6-8mm (body). Fine fabric with mica and white inclusions (5-7%), evenly fired, 5YR 6/6 (reddish yellow). Well smoothed. Handle is ovoid in cross-section. Poor to medium preservation.

Roman

523 OPEN VESSEL?

(AFE-5) Body. Max. pres. H. 36mm, Max. Pres. W. 46mm, Th. 9-10mm. Fine fabric with much mica and a few white-grey inclusions (10-15%), unevenly fired, core: 10YR 6/2 (light brownish grey), surfaces: 5YR 5/6 (yellowish red). Poor preservation.

Roman?

524 ??

(MOU-2) Body or rim? Max. pres. H. 26mm, Max. Pres. W. 43mm, Th. 6mm. Fine fabric with few mica and schist and some grey-dark inclusions (7-10%), evenly fired?, 10YR 5/1 (grey). Very poor preservation.

Undatable

525-531 CLOSED VESSEL?

(MOU-2) Body. Max. pres. H. 39mm, Max. Pres. W. 35mm, Th. 5mm. Very fine fabric with much mica and few white inclusions (7-10%), evenly fired, 5YR 6/6 (reddish yellow). Very poor preservation.

Roman?

532 CLOSED VESSEL?   FIG. 2.6.21

(MOU-7) Base. Max. pres. H. 23mm, Max. Pres. W. 88mm, D. 70mm, Th. 9mm (body) 3mm (base). Fine fabric with much mica and some white-grey inclusions (10-15%), unevenly fired, core: 10YR 5/1 (grey), clay: 7.5YR 5/4 (brown). Lower part of base (ring foot). Medium preservation.

Hellenistic?

534 AMPHORA ("DRESSEL TYPE")

(MOU-7) Body. Max. pres. H. 80mm, Max. Pres. W. 61mm, Th. 11-17mm. Fine fabric with mica, white-grey and dark inclusions, including one sea-shell (10-15%), unevenly fired, core: 10YR 7/3 (very pale brown), clay: 7.5YR 6/6 (reddish yellow). Possible traces of red paint on exterior, above thin layer of whitish clay (10YR 7/4, very pale brown). Poor to medium preservation. Joncheray 1976, pl. IV, nos 41-45.

Early Roman

535 AMPHORA

(MOU-7) Body. Max. pres. H. 71mm, Max. Pres. W. 51mm, Th. 7mm. Unevenly fired, core: 10YR 6/2 (light brownish grey), clay: 2.5YR 5/6 (red). Wheel ridges on exterior. Medium preservation.

Middle to Late Roman

536 SMALL CLOSED? VESSEL

(MOU-7) Body. Max. pres. H. 44mm, Max. Pres. W. 38mm, Th. 4-5mm. Fine fabric with few mica and white and dark inclusion (5-7%), evenly fired, exterior: 5YR 5/4 (reddish brown), thin layer: 10R 5/6 (red). Thin layer of reddish clay covers the original clay on interior. Good preservation.

Middle Roman

537 CLOSED VESSEL

(MOU-7) Body. Max. pres. H. 54mm, Max. Pres. W. 60mm, Th. 3-4mm. Fine fabric with mostly quartz and other white-grey inclusions (20%), evenly fired, 2.5YR 5/6 (red). 2-3 grooved lines run parallel to rim; possible wheel ridges on interior. Poor preservation.

Hellenistic?

538 AMPHORA

(MOU-7) Body. Max. pres. H. 154mm, Max. Pres. W. 166mm, Th. 10-12mm. Very fine fabric with no inclusions, evenly fired, 2.5YR 6/6 (light red). Poor to medium preservation.

Roman?

539 JUG? FIG. 2.6.32

(MOU-7) Base. Max. pres. H. 18mm, Max. Pres. W. 79mm, Th. 9mm (body) 4mm (base). Coarse fabric with mostly mica, schist and quartz (30-40%), unevenly fired, exterior: 5YR 5/6 (yellowish red), core: 2.5Y 4/1 (dark grey). Poor preservation.

Roman?

540 BOWL?

(MOU-7) Rim. Max. pres. H. 69mm, Max. Pres. W. 79mm, Th. 8mm. Medium fabric with mostly mica and some inclusions (7-10%), evenly fired with traces of firing clouds, 7.5YR 4/2 (brown). Both surfaces are smoothed and burnished. Good preservation.

Prehistoric?

541 OPEN VESSEL

(MOU-7) Body. Max. pres. H. 49mm, Max. Pres. W. 52mm, Th. 10mm. Medium fabric with mostly mica and some inclusions (10-15%), unevenly fired, core: 10YR 5/2 (greyish brown), surfaces: 2.5YR 5/6 (red). Traces of burnishing on exterior. Medium preservation.

Roman?

542 PITHOS?

(MOU-7) Body. Max. pres. H. 47mm, Max. Pres. W. 71mm, Th. 12-14mm. Fine to medium fabric with some inclusions (7-10%), evenly fired, 2.5YR 5/6 (red). Exterior burnished and thin reddish wash on interior above original clay. Poor to medium preservation.

Roman?

543 MEDIUM-SIZED CLOSED VES-SEL FIG. 2.6.24

(MOU-7) Body. Max. pres. H. 83mm, Max. Pres. W. 78mm, Th. 9-10mm. Very fine fabric, evenly fired, 10YR 6/2 (light brownish grey). Wheel traces on interior, exterior painted with two dark brown parallel, slightly curvilinear bands. Good preservation.

Hellenistic?

544 CLOSED VESSEL?

(MOU-7) Body. Max. pres. H. 56mm, Max. Pres. W. 53mm, Th. 7-8mm. Fine fabric with mica, quartz and white-grey inclusions (15-20%), unevenly fired, core: 5YR 5/3 (reddish brown), clay: 2.5YR 6/6 (light red). Exterior burnished, wheel traces on interior. Medium preservation.

Hellenistic?

545 CLOSED VESSEL? FIG. 2.6.22

(MOU-7) Handle. Max. pres. H. 106mm, D. 22mm. Fine fabric with mica, schist and many white-grey and dark inclusions (20-25%), unevenly fired, core: 10YR 5/3 (brown), exterior: 5YR 6/6 (reddish yellow). Handle has round cross-sections and is covered by 9-10 diagonal grooves. Medium preservation.

Hellenistic?

546 PITHOS?

(MOU-7) Body (and neck?). Max. pres. H. 89mm, Max. Pres. W. 104mm, Th. 10-13mm. Very fine to fine fabric with mostly white inclusions, surfaces: 2.5YR 5/6 (red). Wheel ridges. Poor preservation.

Roman?

548 CLOSED VESSEL?

(MOU-7) Base and foot. Max. pres. H. 42mm, Max. Pres. W. 76mm, D. 55mm, Th. 6mm. Fine to medium fabric with white-grey inclusions, evenly fired, 2.5YR 6/6 (light red). Poor preservation.

Hellenistic?

549 CLOSED VESSEL? FIG. 2.6.23

(MOU-7) Base. Max. pres. H. 18mm, D. 110mm, Th. 6-8mm (body) 8-10m (base). Very fine fabric with inclusions (10-15%), unevenly fired, interior: 5YR 5/6 (yellowish red), exterior: 7.5YR 6/4 (light brown), core: 10YR 6/2 (light brownish grey). Wheel ridges. Medium preservation.

Hellenistic?

550 AMPHORA?

(MOU-7) Body. Max. pres. H. 71mm, Max. Pres. W. 92mm, Th. 9-11mm. Fine fabric with few mica and white-grey inclusions (10-15%), unevenly fired, exterior: 2.5YR 6/6 (light red), interior: 5YR 6/4 (light reddish brown). Wheel ridges on interior, some burnishing on exterior. Medium preservation.

Roman?

551 CLOSED VESSEL?

(MOU-7) Body. Max. pres. H. 50mm, Max. Pres. W. 61mm, Th. 11-14mm. Very fine fabric with much mica and white-grey inclusions (15-20%). Very poor preservation.

Roman?

552 LARGE-SIZED VESSEL (STORAGE?)

(MOU-7) Body. Max. pres. H. 96mm, Max. Pres. W. 113mm, Th. 9-12mm. Very coarse to coarse fabric with mostly mica and schist (c. 10%), unevenly fired, core: 10YR 5/1 (grey), surfaces: 5YR 5/6 (yellowish red). Interior partly smoothed, thin reddish coating on exterior. Medium preservation.

Prehistoric?

556 CLOSED VESSEL

(MOU-20) Body. Max. pres. H. 29mm, Max. Pres. W. 21mm, Th. 4-5mm. Very fine fabric with no mica, evenly fired, clay: 5YR 7/6 (reddish yellow), crust: 10YR 7/4 (very pale brown). Thin pale crust on exterior covers original clay. Medium preservation.

Roman?

559 UNIDENTIFIABLE

(KER-7)

560 OPEN VESSEL

(MOU-16) Part of base and body. Max. pres. H. 41mm, Max. Pres. W. 47mm, Th. 6-8mm (body) 8mm (base). Fine to medium fabric with mica and mainly white inclusions (15-20%), evenly fired, 5YR 5/4 (reddish

brown), black slip: Gley 1 3/N (very dark grey). Traces of black slip on interior. Cooking vessel? Poor preservation.

Roman?

562 CLOSED VESSEL   FIG. 2.6.25

(MOU-16) Body. Max. pres. H. 77mm, Max. Pres. W. 75mm, Th. 5-6mm. Fine fabric with mainly mica and many greyish and reddish inclusions (10-15%), evenly fired, clay: 5YR 6/6 (reddish yellow), exterior: 10YR 5/3 (brown), first level of decoration: 7.5YR 4/2 (brown), second level: 7.5YR 3/1 (very dark grey). Painted decoration (4 dark brown bands): 2 horizontal rectilinear bands and 2 slightly curvilinear. Traces of the brush? Three different colours of paint. Good preservation. Similar to sherd 543.

Hellenistic?

563, 565, 569-570 AMPHORA ("DRESSEL 1")

(MOU-16) Body. Max. pres. H. 68mm, Max. Pres. W. 95mm, Th. 8-12mm. Fine fabric with mica and a few white-grey inclusions (2-3%), evenly fired, 5YR 6/6 (reddish yellow), exterior slip: 7.5YR 6/6 (reddish yellow). Medium preservation.

Roman

564 OPEN VESSEL (HEMISPHERICAL BOWL?)

(MOU-16) Rim. Max. pres. H. 67mm, Max. Pres. W. 83mm, Th. 5-6mm. Fine fabric with much mica and large pieces of schist (25-30%), unevenly fired, core: 10YR 3/1 (very dark grey), surfaces: 10YR 6/4 (light yellowish brown) and 5YR 5/6 (yellowish red). Traces of thin layer of slip on both surfaces. Cooking vessel? Medium to good preservation.

Late Hellenistic – Early Roman

566 OPEN VESSEL (BOWL?)

(MOU-16) Part of body and base. Max. pres. H. 36mm, Max. Pres. W. 67mm, Th. 7mm (body) 5-6mm (base). Medium to coarse fabric with quartz, black-grey and green schist (15-20%), evenly fired, 7.5YR 5/4 (brown).

Thin dark wash above smoothed surface on exterior. Poor to medium preservation.

Prehistoric?

567 OPEN VESSEL

(MOU-16) Body. Max. pres. H. 18mm, Max. Pres. W. 16mm, Th. 4mm. Fine to medium fabric with many schist pieces (10%), clay: 2.5YR 5/6 (red). Medium preservation.

Hellenistic – Roman?

568 CLOSED VESSEL

(MOU-16) Body. Max. pres. H. 22mm, Max. Pres. W. 31mm, Th. 6mm. Fine fabric with much mica, quartz and white-grey inclusions (15%), evenly fired, clay: 5YR 5/6 (yellowish red), slip: 10YR 4/2 (dark greyish brown). Interior covered with dark-coloured, hard glazed slip; possibly for cooking or storage of liquids. Poorly preserved.

Hellenistic – Roman?

576 BOWL

(MOU-16) Rim. Max. pres. H. 73mm, Max. Pres. W. 93mm, Th. 4-6mm. Fine to medium fabric with quartz and other white, brown and red inclusions (10-15%), unevenly fired, core: 10YR 5/1 (grey), surfaces: 5YR 4/4 (reddish brown). Interior well smoothed and burnished with traces of reddish thin slip. Medium preservation.

Hellenistic?

577 SMALL CLOSED VESSEL?

(MOU-16) Base fragment. Max. pres. H. 17mm, Max. Pres. W. 33mm, Th. 6mm (body) 4-5mm (base). Fine fabric with much mica, schist and white, grey and brown inclusions (20-25%), unevenly fired, interior: 5YR 5/6 (yellowish red), exterior 10YR 4/1 (dark grey). Wheel ridges? Poor preservation.

Hellenistic – Roman?

578, 579, 582, 584-587 CLOSED VESSEL

(MOU-16) Body, handle. Max. pres. H. 55mm, Max. Pres. W. 50mm, Th. 3-4mm (body). Fine to medium

fabric with quartz, mica and other white inclusions (10%), evenly fired, 2.5YR 5/6 (red), thin reddish slip on exterior. Thin walled, wheel-made; firing clouds. Medium preservation.

Hellenistic – Roman?

580 CLOSED VESSEL

(MOU-16) Rim. Max. pres. H. 78mm, Max. Pres. W. 86mm, Th. 10-20mm. Medium to coarse fabric with mainly whitish inclusions (7-10%), unevenly fired, exterior: 2.5YR 5/6 (red), interior: 2.5Y 5/1 (grey). Both surfaces well smoothed. Storage vessel? Medium preservation.

Roman?

581 OPEN VESSEL

(MOU-16) Body. Max. pres. H. 68mm, Max. Pres. W. 79mm. Fine fabric with few inclusions (1%), evenly fired, 5YR 5/6 (yellowish red). Wheel-made; firing clouds on exterior. Poor to medium preservation.

Roman?

583 OPEN VESSEL?

(MOU-16) Base? Max. pres. H. 37mm, Max. Pres. W. 59mm, Th. 5-10mm. Fine to medium fabric with much mica, quartz, schist and grey-white inclusions (15-20%), unevenly fired, surfaces: 5YR 5/6 (yellowish red), core: 10YR 5/2 (greyish brown). Very poor preservation.

Roman?

588 PITHOS

(MOU-16) Base. Max. pres. H. 55mm, Max. Pres. W. 117mm, D. 120-130mm, Th. 9-10mm (body) 19mm (base). Medium to coarse fabric with many white and grey inclusions (20-25%), unevenly fired, interior: 7.5YR 4/2 & 5/1 (brown and grey), red slip: 2.5YR 4/4 (reddish brown). Slap building? Exterior thin layer of reddish slip. Reminiscent of sherd 576. Poor preservation.

Roman or post-Roman?

Stone tools

607 CHIPPED PEBBLE   FIG. 2.1

(KAR-16) L. 47mm, W. 28mm, Th. 18mm, 32g. Marble/limestone. Naturally worn and rounded pebble with natural visible layering. Chipped from two faces.

   Prehistoric

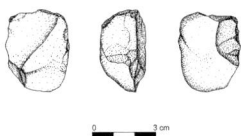

Fig. 2.1. Chipped pebble (KAR-16, Cat.no. 607).

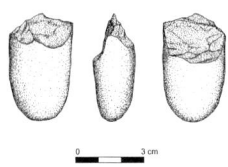

Fig. 2.2. Chipped pebble (ART-9, Cat.no. 608).

608 CHIPPED PEBBLE   FIG. 2.2

(ART-9) L. 35mm, W. 28mm, Th. 21mm, 28.8g. Chipped pebble, quartzite. Naturally worn pebble that has been split. Secondary removals from two faces.

   Prehistoric

610 BIFACIAL HANDSTONE

(MOU-1) L. 53mm, W. 28mm, Th. 18mm, 42.3g. Bifacial handstone, ovate, greenstone. Generally worn to a shine all over. Very thin, but clear and directed striations on one face and on one lateral side (abrasive usewear). Percussive usewear on one end and possibly on other end as well.

   Prehistoric

611 HANDSTONE

(MOU-1) L. 36mm, 30mm, Th. 12mm, 19.8g. Handstone, discoid, marble. Generally smooth all over. No visible usewear.

   Prehistoric

612 HANDSTONE

(MOU-16) L. 88mm, W. 82mm, Th. 31mm, 337g. Handstone, discoid, greenstone. One face is worn and has multi-directional striations. The other side cannot be analysed due to layer of calcite sediment. No visible percussive usewear on ends or lateral sides (of which 40% is covered by calcite).

   Prehistoric

613 COBBLE WITH USEWEAR

(MOU-16) L. 91mm, W. 62mm, Th. 46mm, 402g. Marble/limestone; naturally worn cobble; possible percussive usewear on one end. One face is slightly shiny (abrasive usewear?)

   Prehistoric

Appendix 3

Catalogue of Medieval to modern artefacts

All artefacts have been assigned to one of five Groups (D = Domestic, S = Structural, P = Personal, A = Activities, U = Undefined use), which are further subdivided into 40 functional Categories (Table 1).

Table 2 provides a full overview of the Medieval to Modern artefacts including catalogue number, site designation, artefact type, category, material, number of whole items (W), number of fragments (Frg), and minimum number of items (MNI), description and a tentative date. For instance, a designation of

"D-5" in Table 2 signifies that this is an electrical component belonging in the Domestic Group.

Table 2 provides a quantitative summary of data generated by the analysis of artefacts from the caves on Pelion. The table accompanies the artefact overview in the main reports. The data have been generated by detailed analysis undertaken by the material specialists and further discussion of the artefacts can be found in the specialist reports in Chapter 2.

Table 1. Group, category code and category description used for the description of the survey artefacts (see also Table 2).

Group	Cat. code	Category	Group	Cat. code	Category
Domestic (D)	D-1	Food Prep / Consumption	Activities (A)	A-1	Animal husbandry
	D-2	Food		A-2	Entertainment
	D-3	Food / Food Storage		A-3	Tools
	D-4	Misc. containers		A-4	Firearms
	D-5	Electrical		A-5	Writing
	D-6	Misc. closures		A-6	Furnishings
	D-7	Health or Food		A-7	Military
	D-8	Entertainment		A-8	Automotive
	D-9	Misc. fasteners		A-9	Religious
	D-10	Heating / Lightning		A-10	Agriculture
	D-11	Furnishings	Structural (S)	S-1	Materials
	D-12	Cleaning		S-2	Hardware
Personal (P)	P-1	Social drugs – alcohol		S-3	Electrical
	P-2	Grooming / Health	Undef. use (U)	U-1	Misc. closures
	P-3	Accoutrements		U-2	Misc. containers
	P-4	Toys		U-3	Misc. metal items
	P-5	Footwear		U-4	Misc. fasteners
	P-6	Clothing		U-5	Waste
	P-7	Social drugs – tobacco		U-6	Electrical
	P-8	Health		U-7	Heating / Lightning

Table 2. Catalogue of Medieval to modern artefacts. N.C. = Not collected.

Cat.no.	Site	Type	Cat.	Material	W	Frg	MNI	Description	Date
1	ART-9	Roof tile	S-1	Ceramics	0	1	1		Modern II
2	ART-9	Handle	D-1	Ceramics	0	1	1		Post-Med (19th-20 c.)
4	ART-9	Jar	D-3	Ceramics	0	2	2	1 rim fragment, 1 base fragment, Unglazed Ware, large storage jar (pithos) with wavy incised line on exterior surface	Late Med
5	ART-9	Pot	D-1	Ceramics	0	2	2	1 rim fragment, 1 body fragment, cooking pot, Unglazed Ware	Late Med
6-7	ART-9	Pot	D-1	Ceramics	0	2	2	Base fragments, cooking pot with flat base, Unglazed Ware	Late Med-Post-Med
8-9	ART-9	Vessel	D-1	Ceramics	0	2	2	2 body fragments, with finger-impressed cordons, Unglazed Ware	Late Med-Post-Med
10	ART-9	Jug	D-2	Ceramics	0	1	1	Body fragment, jug with combed decoration, Unglazed Ware	Late Post-Med (probably 19th-20 c.)
11	ART-9	Basin	D-1	Ceramics	0	2	2	Base- and body-fragments of large basin, Unglazed Ware	
12-28	ART-9	Vessel	D-1	Ceramics	0	17	17	Body fragments, Unglazed Ware	Post-Med
29-39	ART-9	Roof tile	S-1	Ceramics	0	11	11	Body fragments, roof tile, Unglazed Ware	Undated
40-44	ART-9	Storage Jar	D-3	Ceramics	0	5	5	Body fragments, large storage jar (pithos), Unglazed Ware	Undated
45-51	ART-10	Vessel	D-1	Ceramics	0	7	7	Glazed Wares	Late Med (13th-15th c.)
52	ART-10	Bowl	D-1	Ceramics	0	1	1	Body fragment, open shape (bowl), 'Zeuxippus Ware Subtype'	Late Med (13th c.)
53	ART-10	Jug	D-2	Ceramics	0	1	1	Body fragment, closed shape (jug), Slip-painted Ware (Green)	Late Med (13th c.)
54-57	ART-10	Vessel	D-1	Ceramics	0	4	4	Body fragments, open shape?, Polychrome Sgraffito Ware	Late Med (late 15th c.)

Cat.no.	Site	Type	Cat.	Material	W	Frg	MNI	Description	Date
58	ART-10	Vessel	D-1	Ceramics	0	1	1	Body fragment, open shape, Monochrome Yellow Glazed Ware	Late Med
59-63	ART-10	Vessel	D-1	Ceramics	0	5	5	Base fragments, all flat base, closed shape, Unglazed Ware	1 fragment Medieval? Micaceous, with wire-marks at bottom, local?; 4 fragments Post-Med (probably 19th-20th c.)
64-66	ART-10	Handle	D-1	Ceramics	0	3	3	Handle fragments, Unglazed Ware	1 fragment (18th)19th-20th c., 2 fragments Late Post-Med (probably same date)
67-69	ART-10	Vessel	D-1	Ceramics	0	3	3	Rim fragments, Unglazed ware	2 fragments Post-Med (16th-20th c.), 1 sherd (18th) 19th-20th c.
70-98	ART-10	Pot / Vessel	-1	Ceramics	0	29	23	Body fragments, Unglazed Ware, 7 fragments Coarse Ware (a cooking pot?), 1 fragment with black slip on exterior surface	Late Post-Med, black slip fragment perhaps 18th c.
99-113	ART-10	Roof tile	S-1	Ceramics	0	15	15	Unglazed Ware	Post-Med
123	POR-2	Lid	U-1	Ceramics	0	1	1	Unglazed Ware, with inscription	Modern II
124	POR-2	Vessel	D-1	Ceramics	0	1	1	Body fragment, Unglazed Ware	Post-Med
125	POR-3	Vessel	D-1	Ceramics	0	1	1	Body fragment, Monochrome Green Glazed Ware	Post-Med
126	POR-3	Vessel	D-1	Ceramics	0	1	1	Unglazed Ware	Post-Med
130-131	POR-4	Basin	D-1	Ceramics	0	2	1	Base fragments, basin or large dish with flat base, Painted Glazed Ware: white slip and yellow lead glaze and green painted decoration on the inside	Late Post-Med (19th-20 c.)

Cat.no.	Site	Type	Cat.	Material	W	Frg	MNI	Description	Date
132	MAK-7	Vessel	D-1	Ceramics	0	1	1	Rim fragment, closed Shape, Unglazed Ware	Late Med – Post-Med?
133-182	ART-1	Vessel	D-1	Ceramics	0	50	1	Base and body fragments, Unglazed Ware	Late Post-Med (19th-20 c.)
183	ART-1	Cup	D-1	Ceramics	0	1	1	Coffee cup, Monochrome White Glazed Ware	Modern
184	KAR-3	Jug	D-2	Ceramics	0	1	1	Spout fragment, jug (probably for water, also known as *ibrik*), Glazed Ware with slip-painted decoration (yellow-brown)	Late Post-Med (19th-20 c.)
185	KAR-4	Jar	D-3	Ceramics	0	1	1	Rim fragment, Unglazed Coarse Ware	Late Med
186	KAR-4	Vessel	D-1	Ceramics	0	1	1	Body? fragment, decorated ware	Post-Med
187-192	KAR-4	Vessel	D-1	Ceramics	0	6	6	Unglazed Ware	Post-Med
193-200	KAR-6	Roof tile	S-1	Ceramics	0	8	8	Unglazed Ware	Post-Med
201	KAR-6	Jar	D-3	Ceramics	0	1	1	Rim fragment, large jar, Monochrome Green Glazed Ware with green glaze on the interior and exterior neck	Late Post-Med (19th-20 c.)
256	MIL-3	Vessel	D-1	Ceramics	0	1	1	Body fragment, Unglazed Ware	Late Post-Med (19th-20 c.)
257	MIL-3	Jug	D-2	Ceramics	0	1	1	Handle fragment, jug (used for water, also known as *kanati*), Glazed Ware: white slip with green lead splashes or blotches on exterior	Late Post-Med (20th c.)
258	MIL-3	Vessel	D-1	Ceramics	0	1	1	Body fragment, Unglazed Ware	Late Post-Med (19th-20 c.)
259	MIL-3	Jug	D-2	Ceramics	0	1	1	Neck-body fragment, jug (used for water, also known as *kanati*), Glazed Ware: white slip with green and yellow lead glaze splashes or blotches on exterior, rouletted decoration on shoulder	Late Post-Med (19th-20 c.)
260	MIL-3	Vessel	D-1	Ceramics	0	1	1	Body or base fragment, Unglazed Ware	Late Post-Med (19th-20 c.)

Cat.no.	Site	Type	Cat.	Material	W	Frg	MNI	Description	Date
261	MIL-3	Pot	D-1	Ceramics	0	1	1	Body fragment, Transparent Glazed Cooking pot (also known as *tsoukali*), from Siphnos	Modern (20th/21th c.)
262	MIL-3	Vessel	D-1	Ceramics	0	1	1	Rim fragment, open shape, Glazed Ware: white slip and yellow lead glaze with green painted stripes on interior	Post-Med (19th-20 c.)
263	MIL-3	Vessel	D-1	Ceramics	0	1	1	Body fragment, Glazed Ware	Post-Med (19th-20 c.)
264	MIL-3	Vessel	D-1	Ceramics	0	1	1	Rim fragment, open shape, Monochrome Green Glazed Ware	Post-Med (19th-20 c.)
265	MIL-3	Pot	D-1	Ceramics	0	1	1	Body fragment, Transparent Glazed Cooking pot (also known as *tsoukali*), from Siphnos	Modern I / II
266	MIL-3	Pot	D-1	Ceramics	0	1	1	Body fragment, Transparent Glazed Cooking pot (also known as *tsoukali*), from Siphnos	Modern I / II
267-271	MIL-3	Vessel	D-1	Ceramics	0	5	5	Body fragments, Monochrome Pale Glazed Ware (on exterior)	Late Post-Med (19th-20 c.)
272	MIL-3	Jug	D-2	Ceramics	0	1	1	Neck fragment, Monochrome Green Glazed Ware (on exterior)	Late Post-Med (19th-20 c.)
273-281	MIL-3	Vessel	D-1	Ceramics	0	11	11	9 body fragments, 1 body with combed decoration; 1 base fragment, Unglazed Ware	Post-Med
281b	MIL-3	Handle	D-1	Ceramics	0	1	1	Handle fragment, Monochrome White Glazed Ware (perhaps Maiolica). See in general, Vroom 2005, 146-9 for decorated Maiolica from Italy and from Greece.	Late Post-Med (19th-20 c.)
281c	MIL-3	Jug	D-2	Ceramics	0	1	1	Rim-neck-handle fragment (upper part), used for water, also known as kanati	Late Post-Med (20th c.)

Cat.no.	Site	Type	Cat.	Material	W	Frg	MNI	Description	Date
282-306	MIL-4	Jug	D-2	Ceramics	0	25	1	Body fragments, jug (used for water, also known as kanati), Glazed Ware: white slip and splashes of green lead glaze on exterior	Late Post-Med (19th-20 c.)
307	MIL-5	Vessel	D-1	Ceramics	0	1	1	Body fragment, open shape, Monochrome Light Green Glazed Ware (on interior)	Late Post-Med (19th-20 c.)
308	MIL-7	Vessel	D-1	Ceramics	0	1	1	Body fragment, Monochrome Glazed Ware	Post-Med?
309	MIL-7	Vessel	D-1	Ceramics	0	1	1	Body fragment, Unglazed Ware	Undated
310-315	MIL-8	Jug	D-2	Ceramics	0	6	6	Body fragments, Unglazed Ware	Post-Med
316	MIL-8	Vessel	D-1	Ceramics	0	1	1	Unglazed Ware	Undated
317-318	MIL-8	Jug	D-2	Ceramics	0	2	2	Body fragments, Monochrome Green Glazed Ware	Late Post-Med (19th-20 c.)
319	AGR-3	Vessel	D-1	Ceramics	0	1	1	Body fragment, Glazed Ware, with slip-painted decoration?	Late Post-Med (19th-20 c.)
320	AGR-3	Vessel	D-1	Ceramics	0	1	1	Body fragment, Unglazed Ware	Late Post-Med (19th-20 c.)
321-323	AGR-3	Vessel	D-1	Ceramics	0	3	3	Rim fragment and 2 body fragments, 1 Monochrome Green Glazed Ware (closed shape) and 1 White Glazed and blue paint on interior (open shape)	Late Post-Med (19th-20 c.)
339	MIL-21	Vessel	D-1	Ceramics	0	1	1	Body fragment, Unglazed Ware	Late Post-Med (19th-20 c.)
340	MIL-21	Vessel	D-1	Ceramics	0	1	1	Body fragment, Glazed Ware, with slip and splashes of green lead glaze on exterior	Late Post-Med (19th-20 c.)
345	KAR-13	Roof tile	S-1	Ceramics	0	1	1		Post-Med
346	KAR-13	Vessel	D-1	Ceramics	0	1	1	Body fragment, Unglazed Ware	Post-Med
347	KAR-13	Vessel	D-1	Ceramics	0	1	1	Body fragment, Monochrome Pale Glazed Ware	Post-Med

Cat.no.	Site	Type	Cat.	Material	W	Frg	MNI	Description	Date
349	KAR-6	Vessel	D-1	Ceramics	0	1	1	Body fragment, Unglazed Ware, with combed exterior?	Post-Med
406	KAR-32	Vessel	D-1	Ceramics	0	1	1	Body fragment, Glazed Ware with slip-painted decoration on inside (almost vanished)	Late Post-Med (18th-20 c.)
407-408	KAR-32	Jug	D-2	Ceramics	0	2	2	Base fragments, Unglazed Ware	Post-Med
409	KAR-22	Vessel	D-1	Ceramics	0	1	1	Body fragment, closed shape, Unglazed Ware with combed? Decoration	Late Roman or Post-Med
414-416	KAR-23	Vessel	D-1	Ceramics	0	3	3	Body fragments, closed shape, Glazed Ware? White slip and transparent lead glaze on exterior. See Andreasen & Pantzou & Papadopoulos 2009, 183, fig. 5.	Late Post-Med (19th-20 c.)
417	KAR-23	Pot	D-1	Ceramics	0	2	2	Rim fragment and body fragment, cooking pot, Coarse Ware with transparent lead glaze on interior	Late Post-Med (19th-20 c.)
458	MOU-5	Jug	D-2	Ceramics	0	1	1	Body fragment, Unglazed Ware, decorated/combed with straight and wavy incised lines	Late Post-Med (19th-20 c.)
459	MOU-5	Jug	D-2	Ceramics	0	1	1	Body fragment, Unglazed Ware, decorated/combed with wavy incised lines	Late Post-Med (19th-20 c.)
533	MOU-7	Jug	D-2	Ceramics	0	1	1	Body fragment, Glazed Ware (in and out), painted in brown stripes on exterior	Modern I
547	MOU-7	Basin	D-1	Ceramics	0	1	1	Rim fragment, large basin (also known as lekane?), Glazed Ware with slip-painted decoration	Late Post-Med (19th-20 c.)
554	MOU-10	Roof tile	S-1	Ceramics	0	1	1	Unglazed Ware	Post-Med
555	MOU-10	Roof tile	S-1	Ceramics	0	1	1	Unglazed Ware	Post-Med

Cat.no.	Site	Type	Cat.	Material	W	Frg	MNI	Description	Date
557	KER-6	Vessel	D-1	Ceramics	0	1	1	Body fragment, open shape, Monochrome Yellow Glazed Ware (on interior)	Late Post-Med (19th-20 c.)
558	KER-6	Vessel	D-1	Ceramics	0	1	1	Body fragment, Unglazed Ware	Post-Med
561	MOU-16	Vessel	D-1	Ceramics	0	1	1	Body fragment, open shape, Monochrome Yellow Glazed Ware	Late Post-Med (19th-20 c.)
572-575	MOU-16	Vessel	D-1	Ceramics	0	4	4	Body fragment, Unglazed Ware (1 decorated)	Post-Med
589	MAK-23	Roof tile	S-1	Ceramics	0	1	1	Unglazed Ware	Post-Med
590	MAK-23	Roof tile	S-1	Ceramics	0	1	1	Unglazed Ware	Post-Med
591	MAK-23	Roof tile	S-1	Ceramics	0	1	1	Unglazed Ware	Post-Med
592	MIL-3	Bottle	P-1	Green glass	0	4	3	Bright green colour; no seems or impurities	Modern II/ Contemporary
593	MAK-12	Beer bottle	P-1	Brown glass	0	11	11	One neck fragment with side mold seam extending from body to lip. The insweep/base of bottle is ribbed, Digits "500" and "18" are embossed at bottom	Modern
594	MAK-12	Soda bottle	D-2	Clear glass	0	10	10	Mark "FIX" is embossed at body	Modern
595	MAK-12	Soda-pop bottle	D-2	Clear glass	0	2	2	Ribbed insweep; Coca-Cola bottle?	Modern
596	ART-1	Beer bottle	P-1	Dark-olive glass	0	5	5	Digits "6" and "9" are embossed at bottom	Modern
597	ART-1	Mirror	P-2	Glass	0	1	1	With back coating	Modern II/ Contemporary
598	KAR-4	Goblet	D-1	Clear glass	0	1	1	Wine glass; no side mold seam	Modern II
599	KAR-4	Goblet	D-1	Clear glass	1	1	2	Drinks glasses; no visible side mold seams, but with termination on rim (handmade?). Some onset of glass sickness.	Modern II

Cat.no.	Site	Type	Cat.	Material	W	Frg	MNI	Description	Date
600	KAR-29	Wine bottle	P-1	Green glass	0	3	3	Very light green colour; neck fragments with collar and with side mold seam extending from body to collar.	Modern II
601	MOU-1	Bottle	P-1	Brown glass	0	1	1	Green-amber colour; neck fragment, partly preserved with collar, no visible side mold seam, "twisted" neck (stops at rim). Impurities and small bubbles in glass showing same direction as the twisting.	Late Post-Med/ Modern I
602	AFE-5	Perfume bottle	P-2	Green glass	1	0	1	"Giovanotta" and "50" are embossed on body and base. Side mold seam ends before rim. Non-homogenous glass w. small bubbles.	Modern I
603	AFE-5	Liqour bottle	P-1	Green glass	1	0	1	Scottish; "V 17" embossed at bottom, with side mold seam.	Modern I
604	AFE-5	Chemical bottle	U-2	Brown glass	1	0	1	"ΧΛΟΡΙΝΗ ΒΙΟΜ ΧΡΟΜΑΤΟΝ Ν ΧΑΤΖΗΑΛΚΟΛΗ" is embossed on body; with screw top; side mold seam continues to rim.	Modern
605	AFE-5	Bowl	D-1	Clear glass	0	1	1		Modern
606	AFE-7	Chemical bottle	U-2	Clear glass	1	0	1	"400" is embossed at base; no visible side mold seam; some onset of glass sickness; throat of bottle narrows to facilitate a cork.	Modern
609	ART-9	Column drum	S-1	Marble	0	1	1	w. 17 visible flutes and support hole (5 cm diam.). One or two incised crosses on upper part of the column. Diam. 28 cm; H:25 cm.	Medieval
614	KAR-22	Horseshoe	A-1	Iron	1	0	1	8 round nail holes and one larger central hole. L:108 mm; front of shoe is heavily worn or broken.	Late Post-Med/ Modern I

Cat.no.	Site	Type	Cat.	Material	W	Frg	MNI	Description	Date
615	KER-7	Horseshoe	A-1	Iron	1	0	1	8 square nail holes, L:97 mm; three nail holes at one side have worn through	Late Post-Med/ Modern I
616	ART-1	Nail	S-2	Iron	1	0	1	Lag bolt w. hexagon head; for use in wood; l:92 mm, head diam. 13 mm.	Modern II
617	MAK-7	Wrought nail	S-2	Iron	1	0	1	L:107 mm, head diam. 19x14 mm, square body, pre-1800s	Late Post-Med/ Modern I
618	MAK-7	Wrought nail	S-2	Iron	1	0	1	L:97 mm, head diam. 14x10 mm, square body, pre-1800s	Late Post-Med/ Modern I
619	ART-9	Bolt	S-2	Iron	1	0	1	Machine bolt w. hexagon head	Modern II
620	MIL-3	Wrought nail	S-2	Iron	1	0	1	L. 92 mm, head diam. 24x20 mm, square body, disintegrating	Late Post Med (pre-1800s)
621	ART-1	Battery	D-5	Composite	1	0	1	Duracell, size A	Modern II
622	MAK-12	Battery	D-5	Composite	1	0	1	Panasonic, size AA, heavy duty/special 1.5 volt	Modern
623	AFE-7	Bucket	U-2	Copper-alloy	0	13	1	Straight-sided vessel with accentuated rim and "foot-rim"; visible side seam; bottom diam. 106 mm; poor preservation.	Modern
624	ART-1	Mesh	A-2	Metal	1	0	1	Wire mesh speaker cover, semi-circular, possibly for transistor radio, with fixtures.	Modern II
625	VOL-4	Screw cap	U-1	Metal	1	0	1		Modern II
626	VOL-4	Nail trimmer	P-2	Metal	1	0	1	Made in Korea; "Dutch" decoration with wind mill and tulips; decoration in yellow, green, orange and gold.	Modern II
627	ART-9	Slag	U-5	Slag	2	0	2	16 and 2 grams.	Undated
628	MAK-23	Coffee boiler	D-1	Metal	1	0	1	White enamel; handle broken off; purble/blue stamp at bottom w. shield and 5 stars.	Modern I/II
629	KAR-4	Ring	U-3	Metal	1	0	1		Modern II

Cat.no.	Site	Type	Cat.	Material	W	Frg	MNI	Description	Date
630	KAR-4	Hair clip	P-3	Metal	1	0	1		Modern II
631	KAR-4	Cap	D-6	Lead	1	0	1		Modern II
632	ART-1	Bottle	D-4	Plastic	0	1	1		Modern II
633	ART-1	Screw cap	D-6	Plastic	1	0	1	From water bottle	Modern II
634	VOL-1	Bowl	D-1	Plastic	0	2	1	Hard, beige plastic with sculpted exterior; embossed with "IKOΠΛΑΣΤ" and "8"; bottom diam. 53 mm.	Modern II
635	MOU-1	Medicine bottle	A-1	Plastic	1	0	1	COVEXIN 8A Inj.Sol. bottle 250ml (125ds). Small ruminant clostridia vaccine.	Contemporary
636	MAK-23	Screw cap	D-6	Plastic	1	0	1	Yellow; with sports bottle closure	Contemporary
637	ART-1	Toy car	P-4	Plastic	1	0	1	Green, no fabrication marks	Modern II
638	ART-1	Casing	D-8	Plastic	1	0	1	Tape cassette cover	Modern II
639	ART-1	Knife handle	D-1	Plastic	1	0	1		Modern II
640	KAR-4	Ring	D-9	Plastic	1	0	1		Modern II
641	KAR-8	Cap	U-6	Plastic	1	0	1	With tool "squeeze marks", possible for eletrical wiring	Contemporary
642	MIL-1	Syringe	A-1	Plastic	1	0	1	Vaccination; white plastic; w. bite marks from goats.	Modern
643	MAK-12	Toothbrush	P-2	Plastic	1	0	1	"Nevadent", heavily worn; alternative use – weapons cleaning?	Modern
644	VOL-3	Comb	P-2	Plastic	0	1	1		Modern II
645	MOU-1	Tubing hose	A-1	Plastic	1	0	1	Clear, flexible plastic; connects to vaccine bottle Cat.no. 635	Contemporary
646	KER-3	Bracket fungus	D-10	Organic	1	0	1	Tinder conk, partly burned	Undated
647	MOU-1	Collar	A-1	Wood	0	1	1	Curved wood	Modern II
648	MOU-1	Collar	A-1	Wood	0	1	1	Curved wood	Modern II

Cat.no.	Site	Type	Cat.	Material	W	Frg	MNI	Description	Date
649	ART-1	Shot	A-4	Composite	4	1	4	Fired birdshot shotgun shells and projectile wad; sizes 8 and 9 (for non-waterfowl); 2 red (8) "Rapid" "Extra" 220 mm. Bottom: "Hellas – MAVR…."; 1 black (9). Marked "ΧΑΤΖΗΣΤΕΦΑΝΟΥ". Bottom: "Hellas 12 – 12"; 1 black (8). Bottom: "OHEDDITE – 12"	Modern II
650	KAR-8	Hammer	A-3	Wood/Iron	1	0	1	Small claw ha mmer w. shaft. Head and claws are damaged; shaft is complete	Contemporary
651	ART-1	Lighter	U-7	Composite	0	1	1		Modern II
652	ART-1	Pen	A-5	Plastic/ Metal	0	4	1	"BIC" pen; two yellow pieces from the pen cylinder have been intentionally cut with a knife.	Modern II
653	KAR-8	Candle	U-7	Stearine	0	2	2	Blue and yellow candles.	Modern I/II
654	MIL-13	Rope	A-6	Organic	0	1	1	With knots, two-stranded.	Modern II
655	KER-1	Insulator	S-3	Porcelain	0	1	1	Power line insulator	Modern I/II
656	KER-7	Icon postcard	A-9	Paper	1	0	1	Laminated; "ΙΕΠΑ ΜΟΝΗ ΦΛΑΜΠΟΥΡΗ"and "Ο Αγιος Συμεων". Back: "Stournaras. Lenorman – Kapaneos 72 Athens (205)".	Modern II
N.C.	AGR-1	Shovel	A-3	Wood/Iron	1	0	1		Modern
N.C.	AGR-3	Soda-pop can	D-2	Aluminum	1	0	1		Contemporary
N.C.	AGR-3	Bottle	D-7	Plastic	1	0	1	Water bottle	Contemporary
N.C.	AGR-3	Snack packaging	D-2	Plastic	1	0	1	Caprice/Papadopoulou	Contemporary
N.C.	AGR-3	Café sachet	D-2	Plastic	1	0	1	Nescafe	Contemporary
N.C.	AGR-3	Condom	P-8	Rubber	1	0	1		Contemporary
N.C.	AGR-8	Rod	U-3	Metal	1	0	1		Modern II/ Contemporary
N.C.	AGR-8	Coil	U-3	Metal	1	0	1	For refrigerator	Modern II/ Contemporary

Cat.no.	Site	Type	Cat.	Material	W	Frg	MNI	Description	Date
N.C.	AGR-8	Shot	A-4	Composite					Modern II/ Contemporary
N.C.	AGR-8	Tool shaft	A-3	Wood	1	0	1		Modern II/ Contemporary
N.C.	AGR-8	Sheet metal	S-1	Iron	1	0	1		Modern II/ Contemporary
N.C.	AGR-8	Board	S-1	Wood	1	0	1		Modern II/ Contemporary
N.C.	AGR-8	Lid	U-1	Iron	1	0	1		Modern II/ Contemporary
N.C.	AGR-8	Oil drum	U-2	Iron	1	0	1		Modern II/ Contemporary
N.C.	AGR-8	Fertilizer bag	A-10	Plastic	2	0	2		Modern II/ Contemporary
N.C.	AGR-9	Foil	U-3	Aluminum	1	0	1		Modern II/ Contemporary
N.C.	ART-1	Wire	U-3	Iron	1	0	1	Packing band	Modern II
N.C.	ART-1	Sack	U-2	Plastic	1	0	1		Modern II
N.C.	ART-1	Hand towel	P-6	Fabric	1	0	1		Modern II
N.C.	ART-1	Trousers	P-6	Fabric	1	0	1		Modern II
N.C.	ART-1	Blouse	P-6	Fabric	1	0	1		Modern II
N.C.	ART-1	Polystyrene	S-1	Polystyrene	0	1	1	Large piece of expanded polystyrene	Modern II
N.C.	ART-1	Sheet metal	S-1	Iron	1	0	1		Modern II
N.C.	ART-1	Roof tiles	S-1	Elenit					Modern II
N.C.	ART-1	Sofa	D-11	Wicker	0	1	1		Modern II
N.C.	ART-1	Table	D-11	Wicker	0	1	1		Modern II
N.C.	ART-6	Sheet metal	S-1	Iron	1	0	1	Corrugated iron	Modern II
N.C.	ART-6	Post	S-1	Wood	3	0	3		Modern II
N.C.	KAR-10	Wire	U-3	Iron	1	0	1	Chicken wire	Modern II
N.C.	KAR-10	Wire	U-3	Iron	1	0	1		Modern II
N.C.	KAR-10	Staff	A-1	Wood	1	0	1		Modern II
N.C.	KAR-11	Carpet	D-11	Fabric	1	0	1		Modern II
N.C.	KAR-11	Pole	S-1	Iron	1	0	1		Modern II
N.C.	KAR-11	Rope	A-6	Plastic	1	0	1		Modern II
N.C.	KAR-14	Roof tiles	S-1	Ceramic					Undated
N.C.	KAR-27	Battery	D-5	Composite	1	0	1		Modern II

Cat.no.	Site	Type	Cat.	Material	W	Frg	MNI	Description	Date
N.C.	KAR-27	Fodder bag	A-1	Plastic	1	0	1		Modern II
N.C.	KAR-29	Soda-pop bottle	D-2	Clear glass	1	0	0	Pepsi bottle	Modern II
N.C.	KAR-29	Conserves can	D-3	Aluminum	1	0	1		Modern II
N.C.	KAR-29	Rope	A-6	Plastic	0	1	1		Modern II
N.C.	KAR-4	Work coat	P-6	Fabric	1	0	1	Blue	Modern II
N.C.	KAR-4	Scarf	P-6	Fabric	1	0	1		Modern II
N.C.	KAR-4	Hand bag	U-2	Plastic	1	0	1	Black, "Lucky Duck"	Modern II
N.C.	KAR-5	Sheet metal	S-1	Iron	1	0	1		Modern II
N.C.	KAR-5	Post	S-1	Wood	0	1	1	Burnt fragment	Modern II
N.C.	KAR-6	Wire	U-3	Iron	1	0	1		Modern II/ Contemporary
N.C.	KAR-6	Box	D-4	Plastic	1	0	1		Modern II/ Contemporary
N.C.	KAR-6	Cup	D-2	Plastic	1	0	1	For yoghurt	Modern II/ Contemporary
N.C.	KAR-6	Plastic sheet	S-1	Plastic	1	0	1		Modern II/ Contemporary
N.C.	KAR-7	Foil	U-3	Aluminum	1	0	1		Modern II/ Contemporary
N.C.	KAR-7	Plastic sheet	S-1	Plastic	1	0	1		Modern II/ Contemporary
N.C.	KER-5	Icons	A-9	Composite	5	0	5		Modern II
N.C.	KER-5	Incense burner	A-9	Metal	1	0	1		Modern II
N.C.	KER-5	Cross	A-9	Metal	1	0	1		Modern II
N.C.	KER-5	Candles	A-9	Stearine	2	0	2		Modern II
N.C.	KER-7	Matchbox	U-7	Cardboard	1	0	1	Contains matches	Contemporary
N.C.	KER-7	Lighter	U-7	Composite	1	0	1		Contemporary
N.C.	KER-8	Lighter	U-7	Composite	1	0	1		Modern II/ Contemporary
N.C.	KER-8	Bowl?	U-2	Metal	1	0	1		Modern I (1930-40s)
N.C.	KER-8	Belt	P-6	Leather	1	0	1		Modern I (1930-40s)
?	KER-8	Helmet	A-7	Metal	1	0	1	Italian M33 combat helmet (x 272)	Modern I (1930-40s)

Cat.no.	Site	Type	Cat.	Material	W	Frg	MNI	Description	Date
N.C.	KER-9	Large can	D-4	Plastic	1	0	1		Modern
N.C.	KER-9	Cigarette bud	P-1	Tobacco	0	1	1		Modern
?	MAK-12	Rifle bullet	A-4	Lead	1	0	1		Late Post-Med/ Modern I
N.C.	MAK-2	Conserves can	D-3	Aluminum	1	0	1		Modern II/ Contemporary
N.C.	MAK-23	Pistachio Shells	D-2	Nut Shell	0	1	1		Contemporary
N.C.	MAK-3	Pipe	S-1	Iron	2	0	2		Modern II
N.C.	MAK-3	Water trough	A-1	Wood	1	0	1		Modern II
N.C.	MAK-5	Walnut Shell	D-2	Nut Shell	0	1	1		Modern II
N.C.	MAK-5	Picture frame	D-11	Wood	1	0	1		Modern II
N.C.	MAK-7	Pistachio Shells	D-2	Nut Shell	1	0	1		Modern
N.C.	MAK-7	Shot	A-4	Composite	1	0	1		Modern
N.C.	MIL-1	Sack	U-2	Plastic	1	0	1		Modern II
N.C.	MIL-1	Shirt	P-6	Fabric	1	0	1	Red	Modern II
N.C.	MIL-1	Jacket	P-6	Fabric	1	0	1	Navy uniform jacket w. buttons	Modern II, Junta period, 1967-74
N.C.	MIL-1	Rope	A-6	Organic	0	1	1		Modern II
N.C.	MIL-1	Board	S-1	Wood	1	0	1		Modern II
N.C.	MIL-1	Post	S-1	Wood	2	0	2		Modern II
N.C.	MIL-1	Spark plug	A-8	Composite	1	0	1		Modern II
N.C.	MIL-1	Fertilizer bag	A-10	Plastic	1	0	1	Ceratex 5	Modern II
N.C.	MIL-13	Cigarette bud	P-1	Tobacco					Modern II
N.C.	MIL-16	Olive	D-2	Fruit Stone	0	1	1		Modern II
N.C.	MIL-19	Shoe	P-5	Leather	1	0	1		Modern II
N.C.	MIL-19	Thermo flask	D-1	Metal	1	0	1		Modern II
N.C.	MIL-20	Conserves can	D-3	Aluminum	1	0	1	Fish conserves	Modern II
N.C.	MIL-21	Newspaper	A-2	Paper	0	1	1		Modern II

Cat.no.	Site	Type	Cat.	Material	W	Frg	MNI	Description	Date
N.C.	MIL-3	Bed frame	D-11	Wood	0	1	1		Modern I
N.C.	MIL-9	Bottle	U-2	Clear glass	0	1	1		Modern II/ Contemporary
N.C.	MIL-9	Battery	D-5	Composite	1	0	1		Modern II/ Contemporary
N.C.	MIL-9	Foil	U-3	Aluminum	1	0	1		Modern II/ Contemporary
N.C.	MIL-9	Cup	D-4	Plastic	1	0	1		Modern II/ Contemporary
N.C.	MIL-9	Matchbox	U-7	Cardboard	1	0	1		Modern II/ Contemporary
N.C.	MIL-9	Café sachet	D-2	Plastic	1	0	1	Nescafe	Modern II/ Contemporary
N.C.	MIL-9	Coffee milk cup	D-2	Plastic	1	0	1		Modern II/ Contemporary
N.C.	MOU-1	Wire	U-3	Iron	1	0	1		Modern II
N.C.	MOU-1	Staff	A-1	Wood	1	0	1		Modern II
N.C.	MOU-10	Icons	A-9	Composite	41	0	41		Contemporary
N.C.	MOU-10	Incense burner	A-9	Metal	7	0	7		Contemporary
N.C.	MOU-10	Candle	A-9	Stearin	6	0	6		Contemporary
N.C.	MOU-10	Lighter	U-7	Composite	1	0	1		Contemporary
N.C.	MOU-10	Dustpan	D-12	Plastic	1	0	1		Contemporary
N.C.	MOU-10	Bracelet	P-3	Composite	1	0	1		Contemporary
N.C.	MOU-10	Flower pot	D-11	Ceramics	2	0	2		Contemporary
N.C.	MOU-10	Candle tripod	A-9	Wood	1	0	1		Contemporary
N.C.	MOU-10	Oil bottle	D-10	Plastic	1	0	1		Contemporary
N.C.	MOU-10	Glass	D-1	Clear glass	1	0	1		Contemporary
N.C.	MOU-17	Cigarette bud	P-1	Tobacco			10+		Contemporary

Cat.no.	Site	Type	Cat.	Material	W	Frg	MNI	Description	Date
N.C.	MOU-24	Beam	S-1	Wood	1	0	1		Undated
?	MOU-24	Rifle shell	A-4	Metal	1	0	1	Unfired (X 261)	Modern I
N.C.	MOU-28	Conserves can	D-3	Aluminum	1	0	1		Modern II
N.C.	MOU-3	Beam	S-1	Wood	1	0	1		Undated
N.C.	MOU-7	Post	S-1	Wood			10+		Modern II
N.C.	MOU-7	Pallet	S-1	Wood	2	0	2		Contemporary
N.C.	MOU-9	Bottle	D-4	Plastic	1	0	1	Water bottle 1.5 litres	Contemporary
N.C.	MOU-9	Candle	U-7	Stearine	7	0	7	White; "cache" of candles	Contemporary
N.C.	VOL-1	Mat/Carpet	D-11	Fabric	1	0	1		Modern II
N.C.	VOL-2	Undef.	D-4	Clear glass	0	1	1		Contemporary
N.C.	VOL-2	Wire	U-3	Iron	1	0	1		Contemporary
N.C.	VOL-2	Nozzle	A-3	Metal	1	0	1	For water hose	Contemporary
N.C.	VOL-2	Bottle	D-4	Plastic	1	0	1		Contemporary
N.C.	VOL-2	Screw cap	D-6	Plastic	2	0	2	From water bottles	Contemporary
N.C.	VOL-2	Small box	D-4	Plastic	1	0	1		Contemporary
N.C.	VOL-2	Ring	U-4	Plastic	1	0	1		Contemporary
N.C.	VOL-2	Band	U-4	Plastic	1	0	1	Packing band	Contemporary
N.C.	VOL-2	Almond Shells	D-2	Nut Shell	0	1	1		Contemporary
N.C.	VOL-2	Shoe sole	P-5	Leather	0	1	1		Contemporary
N.C.	VOL-2	Duvet	D-11	Fabric	1	0	1		Contemporary
N.C.	VOL-2	Shoe	P-5	Leather	1	0	1		Contemporary
N.C.	VOL-2	Sock	P-6	Fabric	1	0	1		Contemporary
N.C.	VOL-2	Carpet	D-11	Fabric	1	0	1		Contemporary
N.C.	VOL-2	Matchsticks	U-7	Wood			10+		Contemporary
N.C.	VOL-2	Lighter	U-7	Composite	1	0	1		Contemporary
N.C.	VOL-2	Newspaper	A-2	Paper	0	1	1		Contemporary
N.C.	VOL-2	Razor	P-2	Plastic/Metal	1	0	1		Contemporary
N.C.	VOL-2	Pen	A-5	Plastic/Metal	1	0	1		Contemporary
N.C.	VOL-2	Juice carton	D-2	Paper	1	0	1		Contemporary

Cat.no.	Site	Type	Cat.	Material	W	Frg	MNI	Description	Date
N.C.	VOL-2	Transistor radio	A-2	Plastic/ Metal	0	1	1		Contemporary
N.C.	VOL-2	Plank	S-1	Wood	1	0	1	With nails	Contemporary
N.C.	VOL-2	Window sill	S-1	Marble	0	1	1	Small fragment	Contemporary
N.C.	ZAG-1	Grill	D-1	Metal	1	0	1	For barbeque	Contemporary
N.C.	ZAG-4	Beer bottle	P-1	Brown glass	1	0	1	Amstel, 500 ml.	Contemporary
N.C.	ZAG-5	Conserves can	D-3	Aluminum	1	0	1		Contemporary

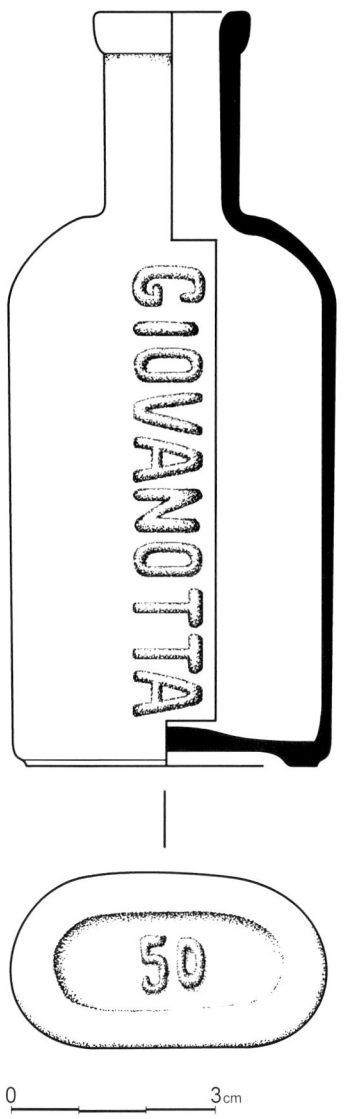

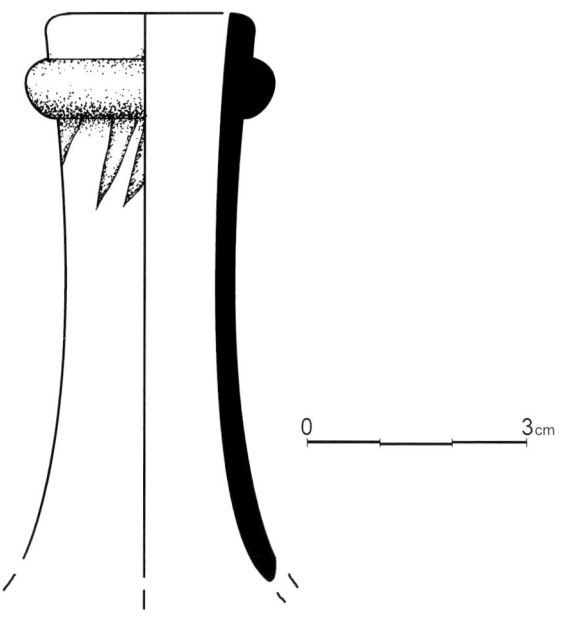

Fig. 3.2. Hand-blown wine bottle collar with "twisted" collar (MOU-1, Cat.no. 601).

Fig. 3.1. Light green 50 ml perfume bottle marked "Giovanotta" (AFE-5, Cat.no. 602).

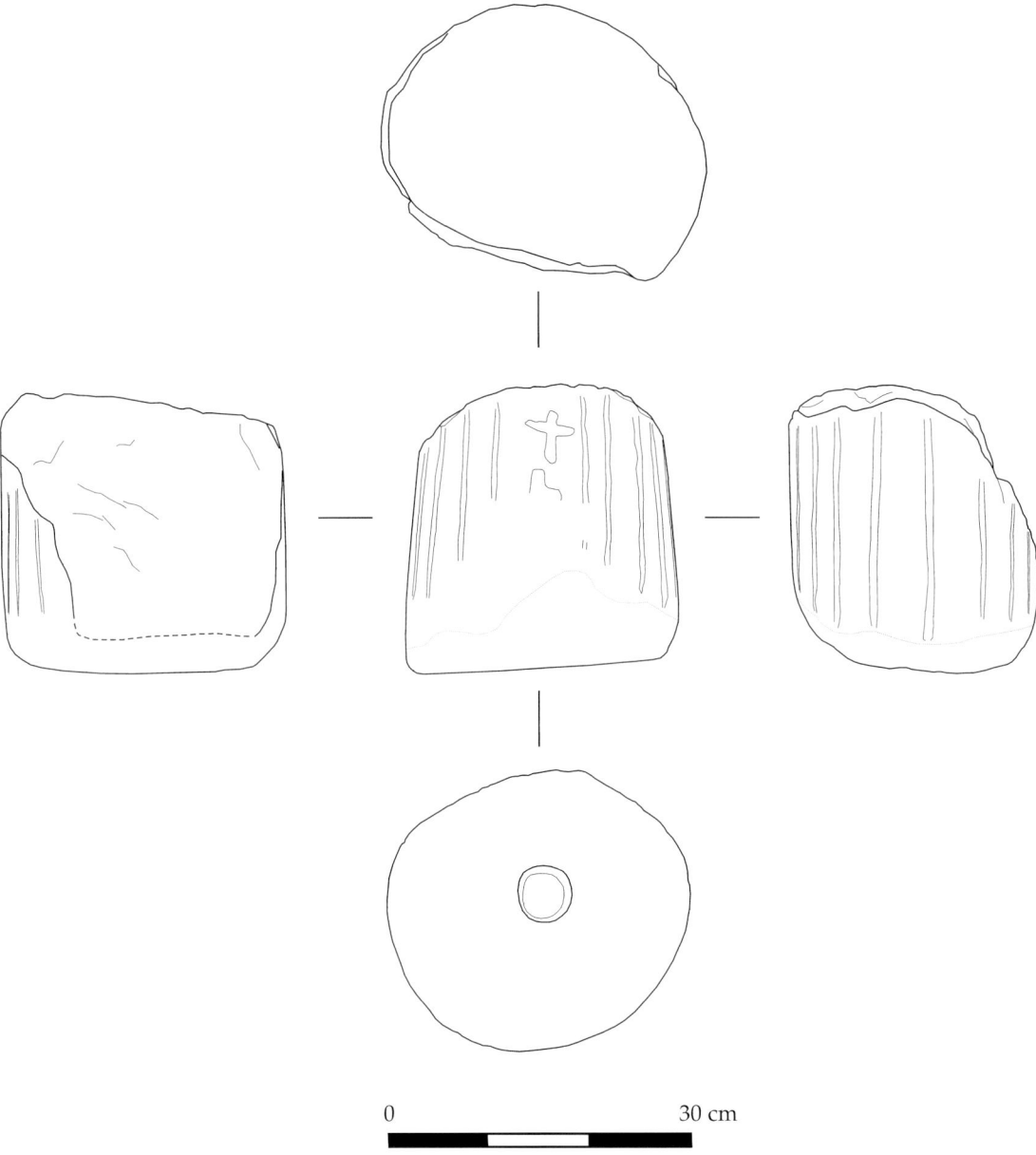

0   30 cm

Fig. 3.3. Fluted marble column drum with two crosses. Byzantine? (ART-9, Cat.no. 609).

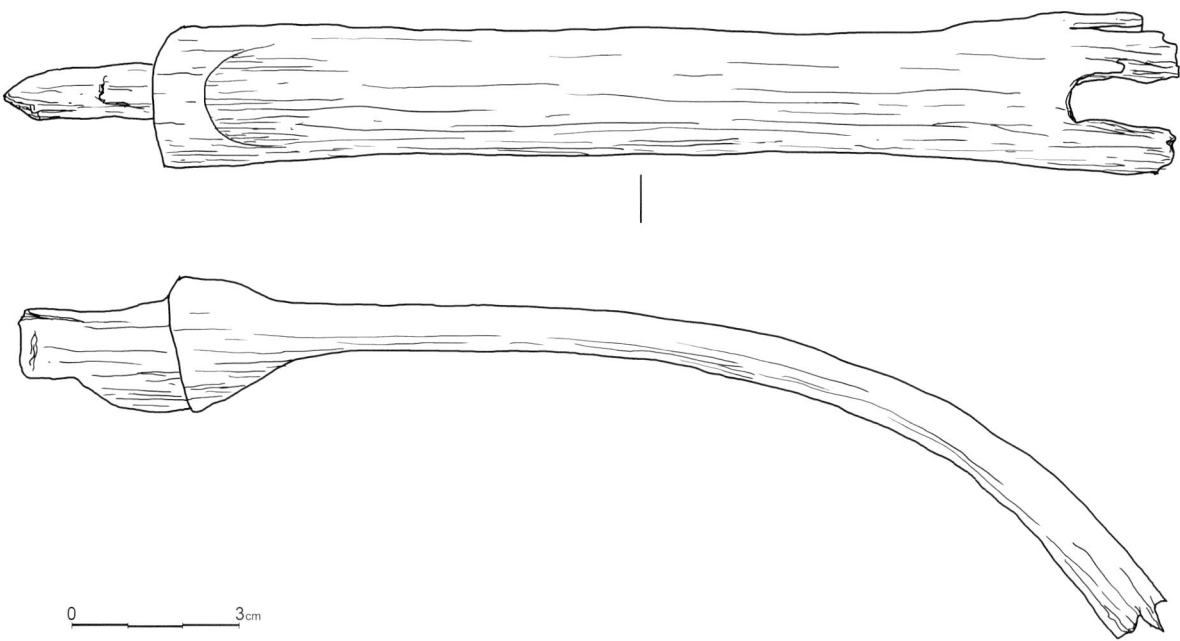

Fig. 3.4. Goat collar of curved wood (MOU-1, Cat.no. 647).

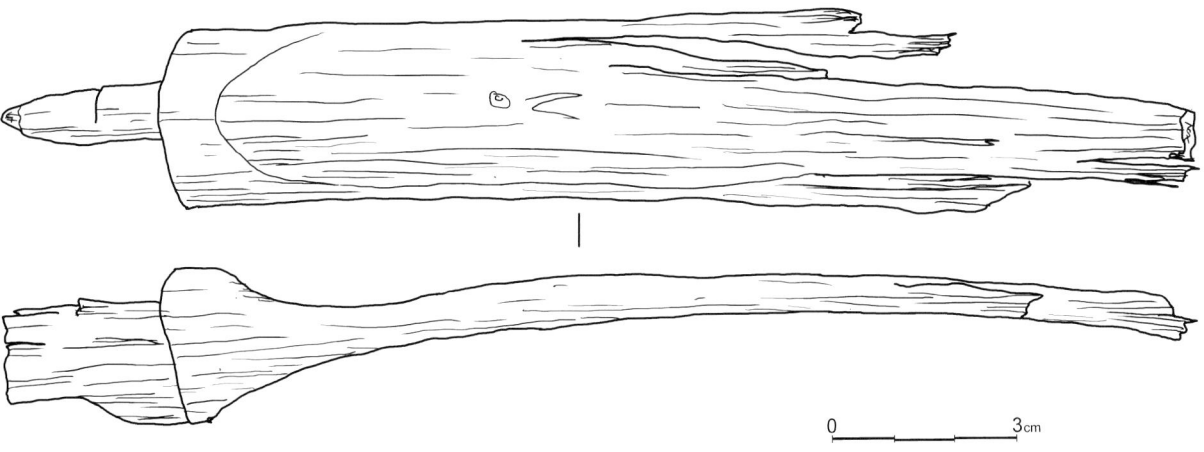

Fig. 3.5. Goat collar of curved wood (MOU-1, Cat.no. 648).

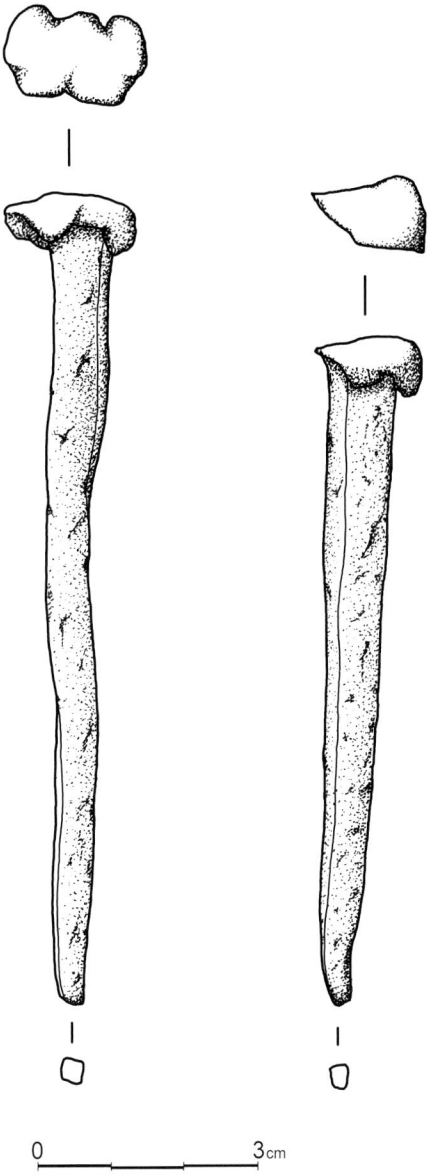

Fig. 3.6. Wrought iron nails (MAK-7, Cat.no. 617 & 618).

List of Figures

Fig. 2.2.6. Telephone poles, local tree stems and ply-wood as building elements in a covered animal pen near Milies (MIL-1).

Fig. 2.2.7. Pig enclosure with wooden posts and chain link fence (KAR-1).

Fig. 2.2.8. Pig enclosure with metal gates and metal doors in front of shallow rockshelter (KAR-9).

Fig. 2.2.9. Drystone wall and lath fence with metal sheet covering (KER-9).

Fig. 2.2.10. Entrance of hand-dug tunnel at Lake Karla with mortar-reinforced walls (KAR-13). Age unknown.

Fig. 2.2.11. Entrance section of MOU-7 with remains of drystone walls and lath fence (V=vertical post; U/L=upper/lower horizontal beam).

Fig. 2.2.12. Remains of drystone wall in front of Mil-4.

Fig. 2.2.13. Collapsed storage facility in Kanalia village (KAR-31).

Fig. 2.2.14. Remains of cement render above small cave (KAR-2).

Fig. 2.2.15. Wood-framed window in drystone wall in front of rockshelter (MIL-2).

Fig. 2.2.16. *Mandri* below Makrinitsa built against two shallow rockshelters. The interior of the *mandri* is equipped with electrical lightning, dividing walls and feeding troughs.

CHAPTER 2.3

Fig. 2.3.1. Skull of red fox in situ at *Tsounaga* (inner chamber).

Fig. 2.3.2. Red fox skull (MOU-1) and mandible and hedgehog mandible (KAR-8) as examples of wild and intrusive animals.

Fig. 2.3.3. Chicken bones collected during the survey.

Figs. 2.3.4. – 2.3.6. Chopping butchery marks for marrow extraction. Bone samples from MOU-12 and MIL-3.

Fig. 2.3.7. Filleting butchery marks on sheep/goat bone from KAR-8 or MOU-1.

CHAPTER 2.4

Fig. 2.4.1. Limpets (*Patella caerulea*?) from KAR-23.

Fig. 2.4.2. Topshells (*Monodontinae*) from MOU-7.

CHAPTER 2.5

Fig. 2.5.1. Human remains from KAR-8: (a) right radius, left ulna, left radius, left clavicle, and femoral fragment (from left to right); (b) ribs (superior aspect); (c) inferior aspect of cervical and thoracic vertebrae (upper row), dorsal aspect of left third metacarpal and manual proximal phalanx (lower row); (d) mandible (occlusal aspect).

Fig. 2.5.2. Unfused femoral head from MOU-1.

Fig. 2.5.3. Left first rib and right fifth metatarsal (from left to right) recovered from VOL-3.

Fig. 2.5.4. Left and right maxillae and zygomatics recovered from MOU-16.

Fig. 2.5.5. Visceral surface of the fractured left rib (PCP6) showing bony callus (posterior is to the left) (KAR-8).

Fig. 2.5.6. Magnified view of the interproximal groove on the distal surface of the mandibular right first molar (KAR-8).

Fig. 2.5.7. Maxillary sinus of the right maxilla (PCP23) showing lobules of bone indicative of chronic maxillary sinusitis (MOU-16).

CHAPTER 2.6

Fig. 2.6.1. Fragment of coarse ware vessel from KAR-8 (Cat. No. 127)

Fig. 2.6.2. Fragment of coarse ware vessel from KAR-8 (Cat. No. 203)

Fig. 2.6.3. Rim from handmade, burnished vessel (Cat No. 255)

Fig. 2.6.4. Fragment from bowl with high polished light-coloured slip (Cat No. 378)

Fig. 2.6.5. Handle from closed vessel (Cat No. 379)

Fig. 2.6.6. Fragment from closed vessel with part of handle (Cat No. 405)

Fig. 2.6.7. Upper part of small stirrup jar (Cat No. 207, 248, 249 and 254)

Fig. 2.6.8. Fragment of plain ware open vessel. Late Helladic dipper? (Cat No. 252)

Fig. 2.6.9. Possibly Late Helladic decorated vessel (Cat No. 212)

List of Tables